INTEGRATING RELIABILITY INTO MICROELECTRONICS MANUFACTURING

DESIGN AND MEASUREMENT IN ELECTRONIC ENGINEERING

Series Editors

D. V. Morgan,
Schools of Electrical, Electronic
and Systems Engineering,
University of Wales
College of Cardiff
Cardiff, UK

H. L. Grubin,
Scientific Research Associates Inc,
Glastonbury, Connecticut, USA

THYRISTOR DESIGN AND REALIZATION
P. D. Taylor

ELECTRONICS OF MEASURING SYSTEMS
Tran Tien Lang

DESIGN AND REALIZATION OF BIPOLAR TRANSISTORS
Peter Ashburn

ELECTRICAL CHARACTERIZATION OF GaAs MATERIALS AND DEVICES
David C. Look

COMPUTERISED INSTRUMENTATION
Tran Tien Lang

PRINCIPLES AND TECHNOLOGY OF MODFETs VOLUMES 1 AND 2
H. Morkoc, H. Unlu, G. Ji

REALIABILITY OF GALLIUM ARSENIDE MMICs
Edited by Aris Christou

INTEGRATING RELIABILITY INTO MICROELECTRONICS MANUFACTURING
Aris Christou

INTEGRATING RELIABILITY INTO MICROELECTRONICS MANUFACTURING

Aris Christou
CALCE Electronic Packaging Research Center
University of Maryland, USA

JOHN WILEY & SONS
Chichester • New York • Brisbane • Toronto • Singapore

Other Wiley Editorial Offices

John Wiley & Sons, Inc., 605 Third Avenue,
New York, NY 10158-0012, USA

Jacaranda Wiley Ltd, 33 Park Road, Milton,
Queensland 4064, Australia

John Wiley & Sons (Canada) Ltd, 22 Worcester Road,
Rexdale, Ontario M9W 1L1, Canada

John Wiley & Sons (SEA) Pte Ltd, 37 Jalan Pemimpin #05-04,
Block B, Union Industrial Building, Singapore 2057

British Library Cataloguing in Publication Data

A catalogue record for this book is available from the British Library

ISBN 0 471 94407 6

Typeset in 10/12pt Times by Laser Words, Madras
Printed and bound in Great Britain by Bookcraft (Bath) Ltd.

Dedicated to

the students and the staff at the University of Maryland
CALCE Electronic Packaging Research Center
and to those who
will follow in their footsteps

my wife Carol, daughter Sharon and son Chris
who assisted me in preparing the manuscript

CONTENTS

PREFACE

The manufacture of microcircuits begins with the silicon or gallium arsenide wafer and, after several processing operations, results in a fully packaged electronic component. For semiconductor manufacturing, the processing operations allow the industry to design-in reliability through the proper selection of materials, processing parameters, and technologies. The student of manufacturing must therefore be taught the critical relationships between manufacturing, reliability, and quality. These relationships are even more critical in fabrication steps which involve elevated temperatures, e.g., thin film deposition, oxidation and diffusion/implantation. The rapid progress achieved in microelectronics is the result of a planar technology using resists and lithography, which has made it possible to process thousands of transistors in a single circuit.

Solid state devices are today based on the field effect transistor and the bipolar transistor. At present, the FET (field effect transistor) and n–p–n, p–n–p transistors are used in computers, calculators, communications, entertainment equipment, and process control systems. Since 1970, the growth in the number of components per chip has increased steadily by 10^6 and is projected to reach one billion per chip by the year 2010. At the same time, the requirement for zero defects has placed the emphasis on reliability in the manufacturing stage. Hence, the objective of this book is to examine the combined effects of manufacturing and reliability, and to establish a methodology for meeting the technology challenges of VLSI (very large scale integration) listed below.

Technology challenges in microelectronics

High lithography resolution with no resist defects.
Achievement of scaled tolerances.
Improved etching processes with no radiation damage.
Reliable thin gates.
Shallow implants with no enhanced diffusion.
Low temperature processes.
Reliable interconnect technology.
Removal of process-induced radiation damage.
Enhanced package reliability.

In the first three chapters of the book, the background relationships between manufacturing reliability and quality are established. Chapters 1 and 3 deal with the relationship between yield and reliability, while Chapter 2 progresses to the requirements for dual use electronics as the priority for the 1990s. Chapter 4 introduces manufacturing and the necessary building blocks of microcircuits. Chapters 5–9 review fabrication technology as affecting microcircuit quality and reliability. Chapters 10 and 11 cover the manufacturing technology of packages, including plastic packages. The remaining chapters introduce new technologies which will become important in the 1990s: microelectromechanical systems, robotics, and microwave integrated circuits.

The major intent of this book is to explain and examine the principles, processes, and materials of reliability manufacturing. It is aimed at students and technologists who will be applying reliability principles to manufacturing, as well as researchers who must select materials and processes in order to achieve maximum yield at the lowest manufacturing cost in order to achieve the quality necessitated by the competitive market.

1

MICROELECTRONICS INDUSTRY AND MANUFACTURING TRENDS IN RELIABILITY

1.1 INTRODUCTION

The advent of the information age in 1950 also heralded the age of microelectronics, a field which is prominent in today's economy. In this section, we indicate the importance of the microelectronics industry in contemporary life. Its size worldwide is 500 billion dollars and clearly employs the largest percentage of technically trained people. The employment opportunities in microelectronics also encompass materials science, physics and chemical engineering. Figure 1.1 shows that we are now in the third period of our industrial development, mainly the age of information or microelectronics. Accompanying this phase has been an expected decrease in heavy industry as the primary driving force.

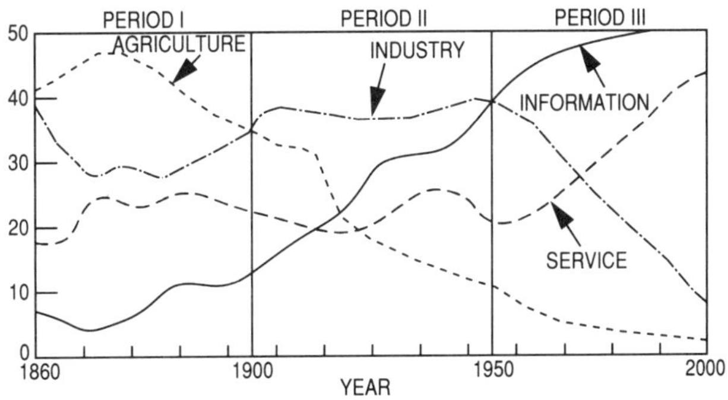

Figure 1.1 Changing composition of the workforce in the United States.

From Figure 1.1 we may also conclude that one of the key reasons why we study the fabrication of microelectronics is to achieve new breakthroughs in technology. Namely,

- Economic breakthroughs resulting from associated industries in the fabrication of microelectronics.
- The progression toward smaller size and higher reliability.
- Reliability and yield are interrelated, as will be shown in subsequent chapters. The importance of reliability is implied by the complexity of our circuits. Manufacturing success, or yield, is the basic indication that a given circuit design for a product is correct.

The advent of microelectronics has resulted in an increase in the number of transistors per chip, with a decrease in critical sizes of the transistors. Therefore, as we approach the year 2000, we can expect the fabrication of 100 million transistors on one silicon chip. In order to achieve this, transistor dimensions must shrink. The oxide thickness has decreased to below 100 Å, while the channel length is now below one micron. These numbers show the direction of microelectronics in the present research environment. Figures 1.2 and 1.3a show the present trends in channel and oxide dimensions. Connected to the required decreases in key dimensions are the necessary increases in manufacturing yield and its optimization. All of the trends indicate that zero manufacturing defects will be attained by the year 2000.

Of prime importance in the attainment of small feature size is the clean room. We are now required to process ICs in class 10 and class 1 laboratories. A clean room environment of class 10 indicates that only 10 particles per cubic foot are present, thanks to air circulating and air filtration equipment. In order to achieve the CMOS circuits of the future, the tolerable particle size must also be less than one micron, as shown in Figure 1.3b. Therefore, manufacturing yield will only be

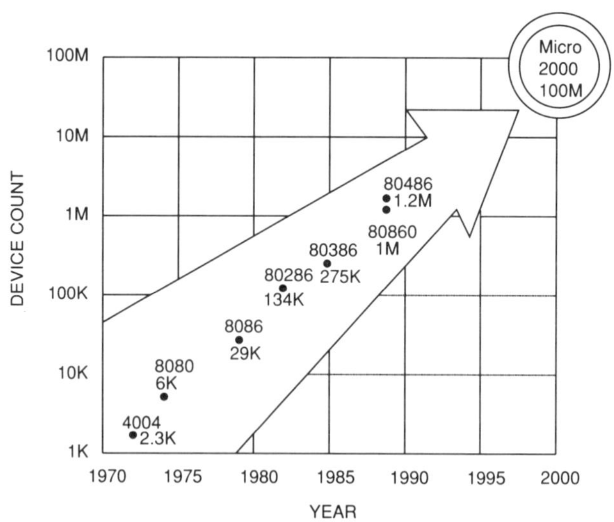

Figure 1.2 Microcomputer transistor count per chip versus time.

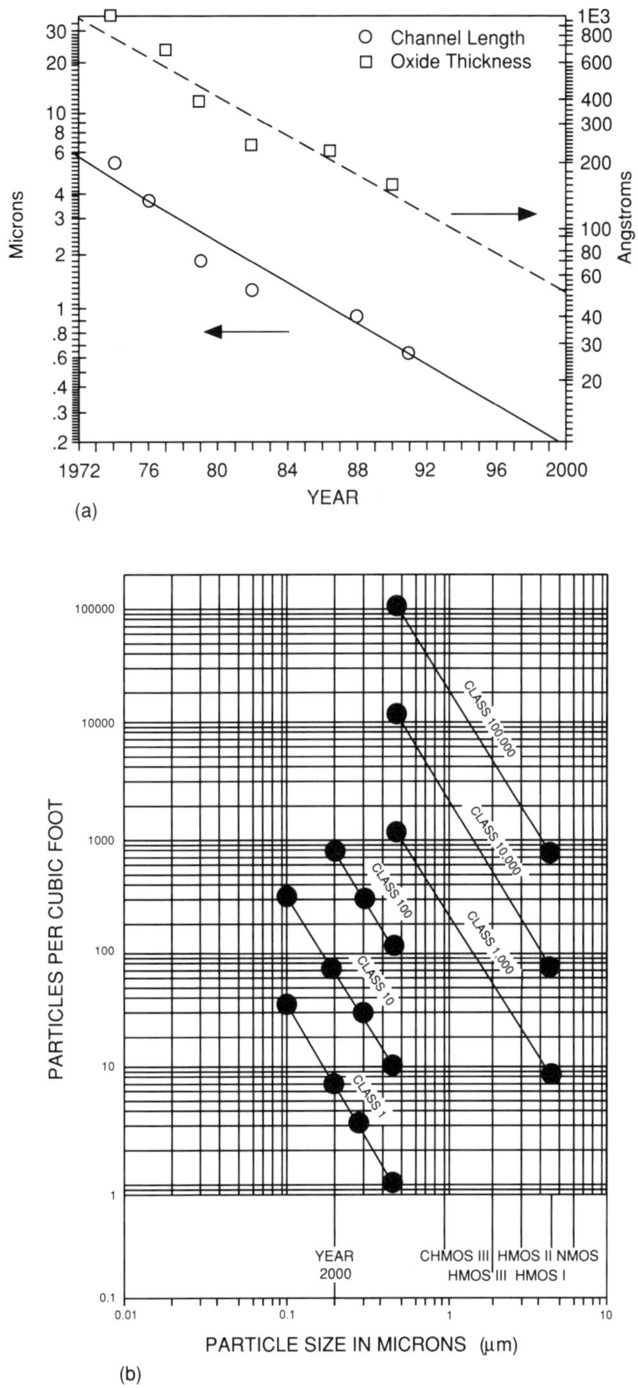

Figure 1.3 (a) VLSI channel length and gate oxide thickness trends. (b) Clean room requirements.

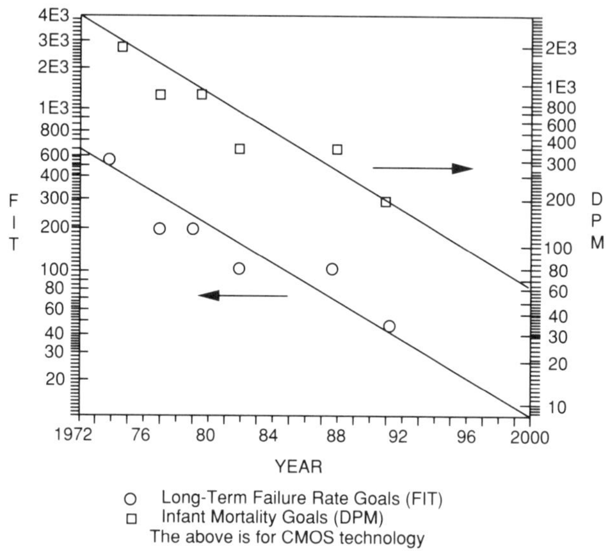

Figure 1.4 Infant mortality and long-term failure rate goals at intel versus time.

achievable in a highly optimized clean room fabrication facility, which must be designed to operate continuously. A submicron CMOS circuit with one million transistors must be manufactured in a class 10 environment in order to achieve a successful product.

Initially, the performance, reliability and manufacturing yield of integrated circuits was plagued by defects and processing mistakes, which resulted in a low long-term failure rate but in a very high initial failure rate. This can be seen in Figure 1.4, where the failure rate is defined as failures per 10^9 hours of device operation. One DPM is one failure for the first 100 hours of operation. Infant mortality problems are caused mainly by manufacturing defects which have not been screened or detected, and accounted for the high initial failure rate.

The defects introduced by processing and manufacturing result both in infant mortality and long-term reliability problems. Figure 1.4 shows the importance of relating manufacturing yield and defects to reliability. Reliability is then the measure of the lifetime of an integrated circuit. This parameter encompasses all defects, including those induced by manufacturing processes. The goal of all manufacturing processes is to minimize the long-term failure rate while eliminating infant mortalities. This goal has been achieved, as shown by Intel's data, which project zero failures by the year 2000.

1.2 BUILT-IN RELIABILITY APPROACH

The achievement of zero failures and the complexity of the integrated circuit implies that reliability and high manufacturing yield must be designed into the product. Hence, manufacturing must not introduce any changes in material properties. Reliability structures must be designed into the manufactured product in order to minimize reliability testing. Testing can identify reliability problems, which may then be eliminated. The new approach, however, is to achieve reliability through elimination of all product weakness at the design phase of the product. Although the initial cost is high, a reliable product will result in low overall costs.

However, increasing the manufacturing reliability increases cost, as shown in Figure 1.5. Reliability improvement may be attained through testing or by improving the manufacturing technology and by reducing defects through process defect elimination. There is a crossover point where product costs increase due to the large cost of testing. It is only beyond the crossover point where built-in reliability approaches are required. Elimination of reliability problems from integrated circuits implies that all failure mechanisms are identified and understood and have been eliminated in subsequent design iterations.

The focus of the manufacturing industry is therefore on the failure mechanism rather than on numbers related to "time to failure". Each product element is then separated into its fishbone diagram in order to determine all the elements affecting its reliability. Figure 1.6 shows such a fishbone diagram approach for microelectronics.

The manufacturing trend of the 1990s is toward understanding a product's failure mechanisms, so as to eliminate them through process improvements. Figure 1.6 shows the difference between the two approaches. The old approach is based on testing and then labelling the finished product as reliable in terms of its mean time to failure. Ideally such a diagram would be created for each failure mechanism, such as electromigration and oxide breakdown. In the case of electromigration, practically all possible processing variables will affect it, as shown in Figure 1.7.

If we wish to understand electromigration, the fishbone diagram contains all of the process variations, manufacturing and material, and then determines the causes of failure. These causes are then eliminated through manufacturing improvement, in order to attain an optimized microcircuit. For a given metallization system, which has been made specific to a product by the machine parameters, the susceptibility to electromigration failure would be minimized at the design stage.

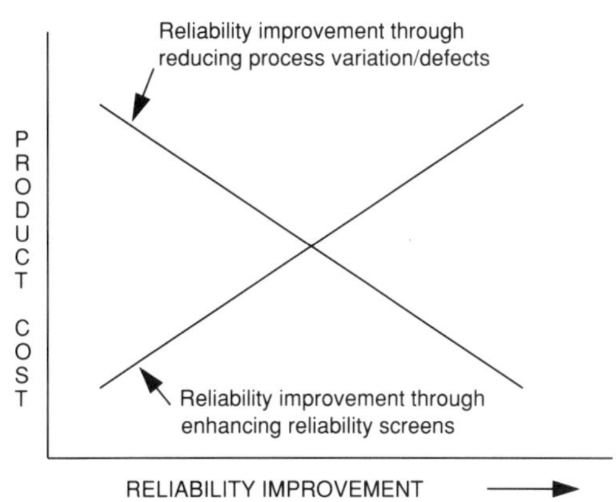

Figure 1.5 Relationship between chip cost and improved chip reliability based on philosophy of reliability screening versus reducing process variation.

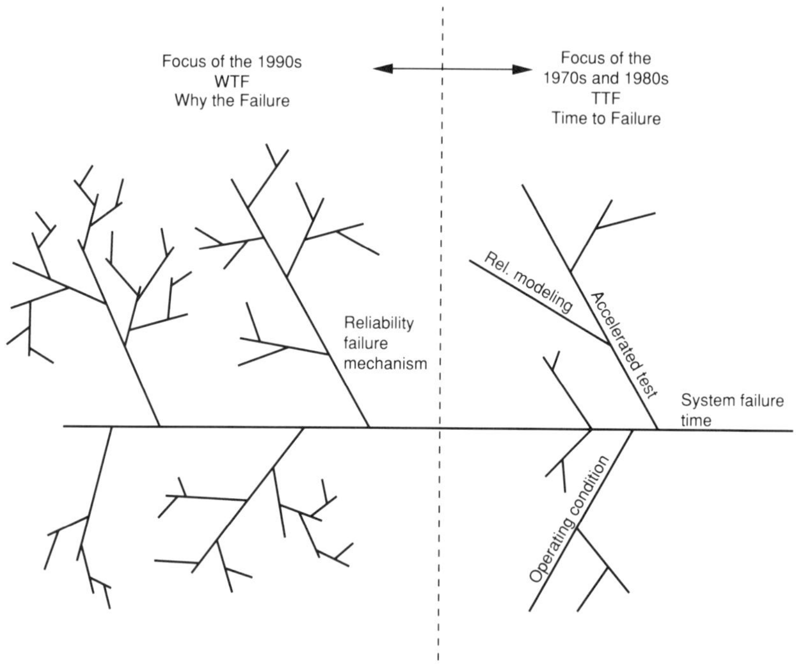

Figure 1.6 Fishbone diagram showing the reliability engineering focus in the past and in the future.

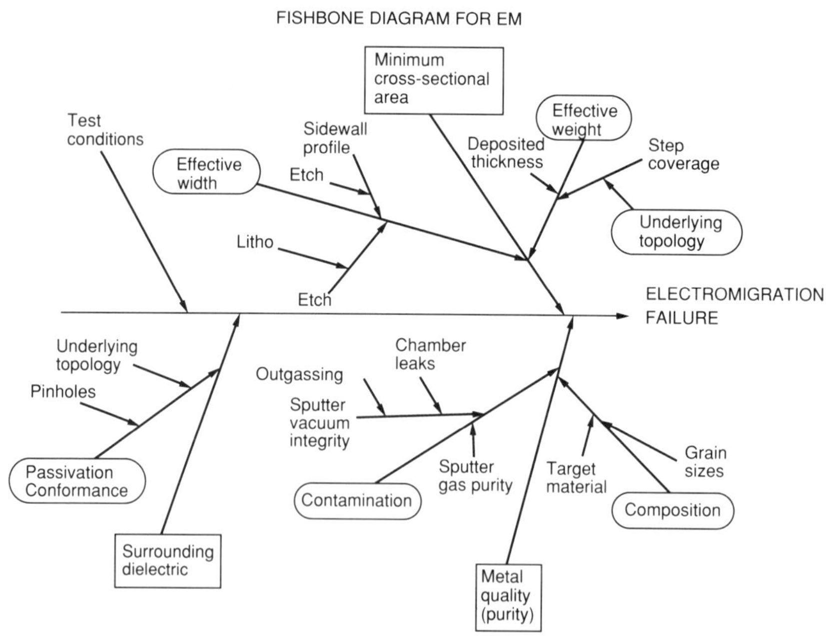

Figure 1.7 Fishbone diagram showing input variables which affect electromigration time to fail.

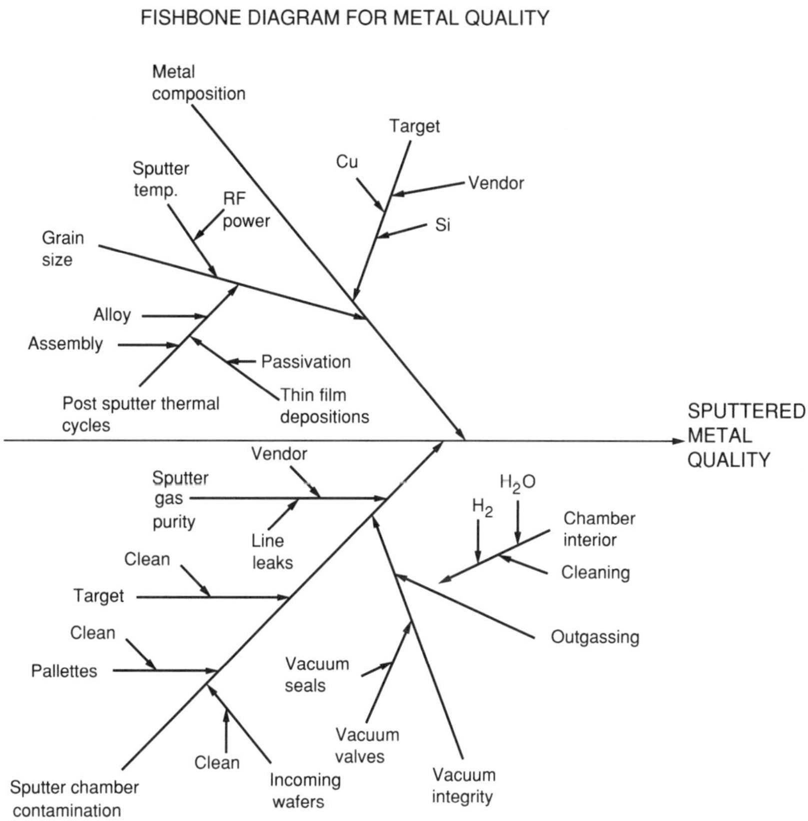

Figure 1.8 Expanded fishbone diagram showing input variables which affect metal quality.

Electromigration has traditionally been the primary failure mechanism at the circuit level. However, metallizations have failed due to high current densities and voids present in the metallic films as manufactured. It is important to note that such defects may be identified at the manufacturing design phase.

A similar fishbone diagram for sputtered metal can also be constructed. The process variables are grain size, passivation, impurities and thin film deposition conditions. All of the variables lead to a sputter metal film quality for gates or ohmic contacts. Figure 1.8 shows an expanded diagram for sputtered metal.

1.3 MANUFACTURING YIELD AND RELIABILITY

Manufacturing yield and reliability are interrelated. The manufacturing yield depends on the number of defects found during the manufacturing process, which in turn determine the reliability. The field of reliability has evolved from a stepchild to a driving force in manufacturing and marketing. The demands for ever greater advancements in device density, performance and reliability, and for new products always on the horizon are also changing the microworld of the semiconductor industry. Earlier, attempts to assure high IC reliability involved the use of functional testing, life testing and accelerated stress tests of the entire circuit. The approach of product testing was becoming too expensive, too time consuming, and less able to identify effectively the causes for parametric and functional failure. A new system of studying reliability was suggested by Schroen, utilizing test structures sensitive to specific failure mechanisms. By stressing these structures, more accurate information about the reliability of the circuit would be obtained in a shorter time than possible by the use of traditional methods. Schroen also saw this as a means of reducing the dependence on costly, time consuming burn in testing. These new techniques led to the development of the wafer level structures and test procedures for oxide and metallization reliability.

In the wafer level reliability technique, both processing and design affect reliability. The best approach to improving reliability and thus yield of any manufacturing process is to be proactive rather than reactive. Being proactive means that new questions will have priority. These questions represent a shift in attention from output to input. They have to do with uncovering the root causes of a given reliability problem and finding ways either to avoid or to reduce and control the problem. The fundamental question is, what are the input parameters to a given process which have to be either eliminated, modified or controlled to improve the reliability of the product? In this context, controlled means that the variability of the parameter is minimized. An input parameter would be, for example, a concentration of some impurity in a processing gas that would adversely affect reliability. An input parameter that might have to be modified would be a metal alloy that is not adequately resistant to failure due to electromigration. An input parameter that will have to be controlled so that all variations are minimized would be, the processing temperature. An example of being proactive is: anticipation of reliability problems derived from particles causing shorts between adjacent metal lines and the reduction of breakdown voltages of oxide films. The traditional approach would ask questions that are directed to the output. What tests shall be used to detect the existence of such short circuits and oxide defects? How should they be used? How can they be improved? Two types of test structures might be selected to assess the densities of such random, particle caused defects: large, interdigitated, comb serpentine, metal line test structures and capacitor type test structures having a range of areas. There might be questions about the effectiveness of such structures.

More recently, under pressure to qualify small quantities of highly reliable circuits needed by the U.S. government at affordable cost, another evolutionary step has been taken. It is the Qualified Manufacturers Listing (QML), with the purpose of bringing in new technologies for earlier marketing, proving circuits with improved reliability and quality, and doing so at reduced costs. It evolved over the last five years through close working relationships between government and industry, as a highly systematized procedure to qualify not just complex application-specific integrated circuits (ASIC), but standard products as well. The QML approach is one that places a heavy emphasis on documentation. It also depends on monitoring and controlling the processing steps, and on tests to monitor critical failure mechanisms. It does so in an environment that explicitly demands a heightened sensitivity and commitment to reliability and quality throughout the fabrication process. In doing so, it incorporates a greater flexibility in controlled process changes introduced by the manufacturer for enhanced reliability. It also provides the option for phasing out screen tests if this will not degrade reliability. It has been demonstrated that through traditional approaches there would not be enough test time or test parts to resolve failure rates as low as 10 FITs (FIT = failure rate, 1 FIT = 1 failure per 10^9 hours), predicted for the end of the century. If one could no longer measure with confidence the reliability of a product, then the rules of the reliability game would be changed.

1.4 RELIABILITY, YIELD AND QUALITY APPLIED TO MICROELECTRONICS

The process of the manufacture of integrated circuits and how its yield can affect the quality of the outcoming product will be the primary focus of this book. The number of defects produced during the manufacturing process can be effectively controlled by introducing test points at crucial times, rather than throughout the assembly line. This improves not only the quality of the outcoming product, integrated circuit in this case, but the yield of the manufacturing process is also enhanced significantly, thus improving the quality of the overall system. It was noted that the introduction of test points was effective only in critically identified areas. Random distribution of test points in the process was also observed not to yield the desired results of high quality and low defects. A new approach to reliability will be required. The traditional approaches are characterized by attention to the development and use of methods to measure effects, and the use of measurement results to predict product reliability. Reliability is appraised based on the reliability driven rules that are implemented. Crook's description of the new approach to reliability centered on the wafer-processing aspects of the fabrication of ICs. But the approach is much more than that. It encompasses all aspects of the fabrication process and applies particularly to any process where it is impractical to measure directly the reliability of the product. As such, and in the context of this chapter, the approach is applicable.

The basic ideas of building-in reliability are not new, and evidence of their use is scattered in the literature. An initial, organized use of the approach in industry, and a recognition of its benefits has recently been reported. The proactive approach leads to a line of questions that is directed to the cause. What are the sources for the particles? How can they be eliminated or controlled? How are the particles created? How are they transported? How and when are they being introduced onto the wafer? And how, where, and when can the presence of these particles be best monitored? There has been a preoccupation with tests to detect the effects of reliability problems, and with the prediction of product life using numbers from these tests. This mindset tends to regard success as being able to predict failure rather than to prevent it. In the preoccupation with testing, we have submerged the need to identify and eliminate the causes of these problems. Efforts to make the connection between an effect and the cause have too often been made at the level of just finding which knob to turn and by how much. To visualize how to begin to change the reactive mindset when there are few role models to follow is difficult. Many seem to know the words but have no idea how to translate them into practice. There are no marching orders to follow. Walls exist between the various segments of the manufacturing line and the suppliers. These walls impede any effective integration of reliability considerations. Operating in relative isolation are the circuit designers, the process developers, the process and assembly engineers, the materials and parts suppliers, the processing and test equipment manufacturers, and the reliability engineers. The latter see an especially high barrier because of their traditional policing role, which promotes adversarial relationships. Particles can be introduced by people, but there are other and more serious sources. Machine wear and material deterioration can be significant. Hence fundamental questions regarding machine design and materials science come into play, in order to find ways to minimize these as significant sources of particles. Gas lines, valves, orifices and interior surfaces can form, collect, trap or release particles depending on gas flow, mechanical vibrations and other conditions. Hence, there are fundamental questions about the physics and chemistry of microscopic particles regarding their creation, transport and interactions within the processing environment. Answers can lead to ways to avoid or reduce the problem by appropriate control of gas flow and redesigns of equipment. Another source for particles may be processes for depositing, etching, and removing the various films involved. Hence a detailed knowledge of these processes may be needed to fully control the particle problem. There are questions about how, when and where the presence of particles is to be measured within the processing equipment. The possibility of random bursts of particles will require continuous measurements. Incoming gases and liquids are a potential source, hence the supplier must be involved in the reliability effort. This example also serves to illustrate other aspects of the approach, such as a generic solution to the problem of particles and their degradation effects on product reliability. It goes beyond simply the problem of particles causing short circuits and oxides with reduced breakdown voltages. And it leads to results that are useful to allied industries where particles are a concern.

There are other input parameters that affect intrinsic reliability, such as particle contamination. In such cases, detailed understanding of both the involved processes and the failure mechanisms will likely be required for the successful application of the approach. Suppose that there is concern about the failure of metallization interconnects due to electromigration. One part of the response might be to optimize the process for sputter depositing the metallization. Answers to questions about what metal film characteristics are desirable require an understanding of the electromigration failure mechanism. It might be concluded that it may be desirable to control the mean grain size and the grain size distribution. The task would then be first to identify the input parameters that affect these characteristics, second to determine how to monitor them, and third to minimize variations of these parameters. To do this would require a detailed understanding of the deposition process. Note that at an interim stage, there would be a need to monitor the grain character of the metallization. But, once the critical input parameters have been identified and controlled, there would no longer be a need for such tests. The proactive approach opens up a direction of inquiry that leads explicitly to many technical and scientific challenges related to understanding of the deposition process.

Considering VLSI reliability, projections indicate that by the turn of the century, microcomputer chips will have 100 million transistors, and failure rates of less than 10 FIT. As critical input parameters are identified, measured and controlled in manufacturing, and as a methodical process of discovering causes for potential reliability problems is implemented, alternative means for applying reliability will be used. Product reliability will be appraised on the basis of documented control of critical input parameters and of reliability driven rules implemented in the design and manufacture of the product. This approach will first supplement, then gradually replace the traditional approaches for assuring product reliability as it becomes less and less feasible to obtain explicit determinations. A natural result of these developments is that the customer will join with industry and its allied suppliers as an active partner in this change.

The building-in reliability approach involves concepts that, while not new in themselves, are diametrically opposed to our traditional ways of thinking about reliability. The idea of being proactive runs counter to the traditional role of the reliability engineer whose task has been typically to ensure that others have done their job. Thus it also runs counter to the way many of us have been trained to think. The traditional approaches to reliability have led to an ingrained mindset of reactive testing. There is a disinclination to reduce the barriers and to work as a team unless there are incentives to do so. Such incentives have to come from top management. Long-term accountability needs to be built in. Process designers and process developers have to remain available and responsible for the developed circuit in production. This can be done by having the process design and development engineers in close proximity. Allied industries, such as the processing and test equipment manufacturers, need to be brought under the reliability tent by entering into collaborative projects that involve experiments to identify critical input parameters and to devise effective ways to monitor them.

1.5 SUMMARY

We have examined the reasons why reliability is important in failure phenomena such as metallization. We have also looked at how the yield of the manufacturing process can affect the quality of the outcoming product. The number of defects produced during the manufacturing process can be effectively controlled by introducing test points at crucial times, rather than throughout the assembly line. This not only improves the quality of the outcoming product, but also enhances significantly the yield of the manufacturing process, thus increasing the quality of the overall system. Test points are effective, as noted, only at the critically identified areas and their random distribution in the process was observed not to yield the desired results of high quality and minimal defects density.

REFERENCES

[1] Hong, C. and Crook, D., "Breakdown energy of metal (BEM) — a new technique for monitoring metallization reliability at wafer level", *IRPS*, 1985, 108.

[2] Root, J. and Turner, T., "Wafer level electromigration tests for production monitoring", *IRPS*, 1985, 100.

[3] Crook, D.L., "Evolution of VLSI reliability engineering", *Proc. IRPS*, 1990, 2.

[4] Schroen, Walter H., "Process testing for reliability control", *Proc. IRPS*, 1978, 81.

[5] Kenzo Sasaoka, "Reliability, the key to success", *Proc. ISRM*, 1990, 43.

2

DUAL USE MICROELECTRONICS: A RELIABILITY APPROACH

2.1 INTRODUCTION

Advanced technologies are key to the next generation of captive electronic systems, as well as commercial products. These technologies include infrared focal plane arrays, microwave devices and circuits, and microelectronics. In the past, the Defense Department's investment in the industrial base has not strategically defined research and development, nor has it established a commercial product base. This investment, however, has tied up almost 20% of the manufacturing gross national product. As a result, this has been detrimental to the international competitiveness of American industries, especially the electronics industry. In this chapter the present status of the microelectronics industry is assessed in terms of its economic constraints which include the role of government funding and the existence of product standards documentation.

The Defense Department (DoD), through its interdependent relationships with manufacturers, has created industries lacking in international competitiveness and other industries which depend on DoD contracts for survival. Consequently, the direction for the future of the microelectronics industry must be the identification of dual commercial-defense requirements, followed by an assessment of these requirements against the nation's industrial base. Strategies must be developed which will allow DoD to meet its requirements with the available commercial resources, and at the same time enhance the nation's industrial base. This chapter will present the strategy of dual use technologies for the electronics industries, developed to increase the nation's competitiveness in electronics. The present misconceptions and operating procedures for advanced technologies will be discussed, followed by proposed changes for the 1990s.

Initially, dual use industries must come to grips with commercial factors, if they are to function competitively in the international marketplace for advanced electronics products. These factors, which include standardization, product development approaches, and training, can only be stimulated by the Defense Department. The emphasis of a leadership role in the area of research and development, therefore, shifts from DoD to industries and universities, through the incorporation of the dual use approach. Unit cost reductions, and improvement in quality and reliability, must now become the driving forces behind DoD's investment strategy. The cornerstone of this effort in microelectronics and communications is the dual use development of new technology, as well as the development of weapon systems concurrent with established commercial manufacturing capabilities.

The dual use technology strategy has four thrusts: (1) forging a relationship with industry which establishes the commercialization (non-military use) of products as the top priority; (2) improving the acquisition, quality and reliability base; (3) establishing military strategic plans in-line with industrial directions; (4) ensuring the availability of the industrial base for dual use products.

The greatest challenge lies in the development of innovative commercial solutions to the nation's industrial base shortfalls in microelectronics, electronics, communications and advanced materials. The improvement in industrial competitiveness must be accomplished with a decrease in the defense budget. The key is therefore in stimulation of the dual use private sector initiatives, in lieu of government funded or government researched solutions (for example: its in-house research laboratories). This also includes the creation of an attractive environment which will enable private industry ingenuity to flourish without DoD interference. The Defense Department must therefore regularly seek industrial and university advice in the area of industrial base deficiencies in electronics, microelectronics and communications. Specifically, three separate approaches for microelectronics and communications will be discussed in this chapter:

(a) Increase in the total use of commercial manufacturing processes and products, in place of military specified processes and products.
(b) Emphasis on the development of new products based on industrial, commercial, applied research.
(c) Utilization of life cycle costs for new products and processes as the key acquisition evaluation criteria.

The advantages gained by the use of advanced electronics in smart weapons, communications systems, computers and controls has been seen as a major reason for the success of the U.S. military in recent encounters. However, the availability of military parts and the large use of foreign electronic parts was also realized at this time.

A specific example of the current problem is the status of packaging and interconnects. Military electronics are typically isolated from mainstream packaging and interconnection developments due to unique requirements and specifications. In fact, the DoD procurement policies generally discourage insertion of commercial or dual-use packaging and interconnection technologies. One result is that military electronics research and development has not gained many of the advantages reaped from commercial industry and has not adequately focused on packaging and interconnection technologies. Large investments in development of advanced packaging and interconnection strictly for military applications are now becoming difficult to justify. The use of alternate electronics (commercial as opposed to mil-spec) is being seen as a means to offer advantages associated with speed, functional integration, availability, weight and smaller footprint, as well as cost. It is apparent that commercial electronic components and assemblies, especially in the packaging area, are essential in meeting the objectives for future military systems.

For the reader to understand the fundamental issues and challenges faced by the DoD in the area of electronics, it is necessary to discuss DoD requirements and policies, including military standards and handbooks, and the potential use of alternate electronics components and assemblies. The present discussion focuses on the DoD, although similar issues and challenges confront the Department of Commerce.

2.2 APPROACHES AND MISCONCEPTIONS

In 1987 and 1989 the Defense Science Board addressed the concerns of the semiconductor dependency problem and stated that technological leadership is irrevocably tied to manufacturing capacity and leadership. The revenues generated by successful manufacturing are essential to achieving the levels of research and development required for technological leadership. Without technological leadership, the DoD cannot expect industry to produce affordable high quality, reliable electronics and communications systems. Contrary to this belief, DoD's present initiatives for radar, communications, and electronic warfare, based on the MIMIC (microwave monolithic circuit) and VHSIC (very high scale integrated circuit) programs, are in the development of such systems which are affordable and available, but only for narrow military applications. There has been a strong emphasis on affordability without the concurrent or previous development of commercial products and a commercial product base and requirements. Hence the traditional market approach has been reversed.

The approach taken by both VHSIC and MIMIC has been serial, with development of materials, device technology, design, processing, testing and packaging in the initial phases, followed by an "affordable manufacture" phase, and then by system applications or an insertion phase. The pilot production lines are based on the small throughput requirements for DoD, and not on the presently available commercial practices of computer integrated manufacturing. The standard three-phase approach taken by DoD typically excludes any possibility of dual technology development. Strict military specifications and acquisition policies also preclude product or manufacturing innovations.

The standard three-phase DoD approach is as follows:

Phase 0 — Studies

- Identify specific problems to be overcome
- Identify transition windows of opportunity
- Develop technology roadmap

Phase 1 — Development
- Process technology development
- Design methodology
- Demonstrate affordability
- Testing requirements
- System brassboard demonstration
- Demonstrate producibility

Phase 2 — Advanced Development
- Extension of technology development
- Advanced demonstrations
- Affordability demonstrations

Phase 3 — Future Development
- Concurrent research with phases 1 and 2
- Investigation of additional capabilities and tools.

A typical DoD program has up to 50 companies participating in phase 0, reducing to 2–4 industrial teams in phase 2. Such an approach, which leaves up to 40 companies stranded after phase 0, is wasteful in many respects, since the industrial investment for each project is often two to three times the DoD contract award. The companies which do not continue into the subsequent phases do not use the results of phase 0 investigations for commercial products.

In terms of communications systems, the technological approaches may be discussed with respect to device technology and frequency range. The microwave region, for systems operating at less than 30 GHz, is where DoD has concentrated its effort in seven companies: Hughes, GE, ITT, Martin Marietta, Raytheon, Texas Instruments (TI) and TRW. The millimeter wave region, for systems operating at greater than 30 GHz, is where system development has been carried out at ITT, Martin Marietta and TRW. Infrared (IR focal plane arrays) development investigations are typically led by Hughes Aircraft, while electro-optical system development has been carried out by TI, Honeywell, and ITT. The systems, which are totally DoD dedicated, are based on conventional MESFETS (metal semiconductor field effect transistors) or HBTs (heterojunction bipolar transistors). The digital systems are based on CMOS integrated circuits. In the next sections, we will investigate how the present approach leads to a DoD captive product base which is wasteful and does not enhance the nation's industrial commercial product competitiveness.

2.2.1 Product demonstrators

Communications systems, electronic warfare, radar and smart weapons requirements are based on wide-band chip sets which lead to component demonstrators and ultimately to commercial products. The demonstrators then become technology decision points. Communications systems chip sets are millimeter wave chips and the technology demonstrators are missile front ends, Ka-band or Q-band amplifiers. At no point are manufacturing requirements or commercial system (non-military) requirements inserted. Product development is based on systems such as jammers, decoys, airborne radar, or satellite (i.e. Milstar). Therefore, at no point in the initial product development stage are commercialization requirements factored into the design of these systems.

All technology insertion efforts at the initial development phase are also based on the low volume military requirements. For instance, passive radar requirements call for amplifier pairs and a production volume of less then one hundred systems per month for the next five years. This requirement may be decreased to only 10 systems per month or not more than 20 IC chips required per month. Clearly, if this is to become cost effective, a commercial industrial base must be developed, where the amplifiers are manufactured for a commercial system with a large share of the analog amplifier market. One example may be the direct broadcast satellite system where millions of x-band amplifiers are required per year.

2.2.2 Manufacturing technologies and innovation

The entire development of an advanced electronics technology product base has been limited by the lack of computer integrated manufacturing techniques and advanced electronics manufacturing equipment. Computer integrated manufacturing methodology is presently available in universities and may be transferred to industry. However, the decline of manufacturing equipment industries has seriously impacted the development of microelectronics and communications systems.

Also limiting the effective development of dual technologies in the area of microelectronics and communications is the absence of normal market incentives. This is clearly evident by the three-phase approach DoD has taken for most systems. With such an approach, unit cost reductions, quality improvements and shortened delivery times do not stimulate demand for additional units or systems. Nor do they provide a certain company a greater market share. Unit cost reductions do not result in increased profit for each company and more often, the reverse has been true. The VHSIC and MIMIC programs have shown that increased investment in plant, new equipment and training will not yield higher profits, new procurement contracts, new products or an improved market share. The absence of such normal market incentives for VHSIC and MIMIC related products has held back the commercial competitiveness for US goods in the area of digital electronics and communications. The absence of normal market incentives for product development is compensated by a massive DoD effort in Research and Development which becomes an additional barrier to industrial innovation, due to the tight controls and technical reviews carried out by Government Science officials.

Additional procurement policies and regulations further inhibit the achievement of dual technologies and manufacturing excellence. Communications system procurement has placed emphasis on affordability, which forces shortcuts and reduces incentives for quality and reliability. The evaluation criteria for most systems focus on performance at the expense of producibility. The procurement process forces contractors to assume the risk for funding new test equipment and, therefore, also the absence of risk sharing and risk taking for innovation. The emphasis has been on low procurement cost or affordability, to the complete exclusion of low life cycle costs (emphasizing quality, reliability, availability, maintainability). In addition, the entire process restricts profits to levels which virtually guarantee that each contractor will have limited ability to invest in future commercial products or productivity programs.

We will now investigate typical wafer manufacturing developments for military electronics, and then assess their impact on industrial competitiveness. Wafer manufacturing developments are military application intensive, solely fine-tuned for products such as smart weapons, and EW and narrow band communications systems. These are funded by DoD as follows:

A. Automated gate recess,
B. Optical lithography for 0.5 micron gates,
C. Electron beam lithography for 0.25 micron gates,
D. Automated metal deposition and liftoff,
E. Gate position control,
F. Backside processing,
G. Test vehicle processing for line qualification.

Even after incorporation of the above manufacturing improvements, the yield is still at the 25–30% level, clearly too low for dual technology product development. The available chip sets developed for demonstration are all at the maximum 30–40% yield level. The present chip sets for EW, radar and communication systems are divided among wide band (i.e., 6 chips) Ku-band (i.e., 4 chips) and millimeter wave (i.e., 3–4 chips).

In the above approaches to manufacturing enhancements, DoD has heavily relied on process and manufacturing line production specifications, such as the 0.5 micron or 0.25 micron gate size requirement, even for x-band products. Therefore, the resulting effect is that components must be manufactured by processes which are not state-of-the-art for the best product at the best price. The manufacturer typically has little voice in component design and manufacturing process selection. This lack of consistency and uniformity in technology and process specifications also exists among the various government agencies (such as MESFETs for Navy, HBTs and HEMTs for Air Force), and also has adverse impact on contractor operations and costs [1–3].

2.2.3 Communications system life cycle costing and R/D selection

Although most communications system acquisition data support requirements can provide information for an effective life cycle cost analysis, DoD personnel rarely use such data for life cycle cost procurement. DoD personnel always focus on acquisition costs and ignore the concept of life cycle costing. The risk involved in life cycle projections may be limited by providing adequate means for monitoring and evaluating actual life cycle costs. Projections may be effectively made through existing life cycle costing models, again available at universities [2, 3].

Other than the inadequacy of present life cycle cost projections, most of the existing communications systems also suffer from inadequate product management techniques (statistical process control, etc.). DoD is dedicated to testing in quality and reliability, rather than applying process controls to building-in quality and reliability. Some changes are occurring, however, as is evident by qualified manufacturing lines for MIMIC products [1]. Likewise, strategic planning for Research and Development has become totally ineffective for industrial companies, since R and D productivity is controlled by government budget decisions and not by the corporate culture. Non-defense companies therefore tend to be more efficient and innovative because of the existence of corresponding incentives. Incentives for defense industrial innovation are almost non-existent since the only reward is perhaps a new DoD contract and not a new multi-million dollar commercial product.

2.2.4 Military specifications and practices

A serious obstacle to the development of dual use advanced technologies is the policy of military specifications and qualification procedures which in many cases has been adopted by the commercial sector. During the 1960s, a policy with an associated set of documents was developed by the military and government, whereby suppliers which produce parts accordingly are said to have military compliant parts. The specific terminology is JAN (Joint Army–Navy) compliant parts. The requirements for microcircuits were and remain defined in the military document MIL-M-38510 and the test methods and procedures are defined in the military standard MIL-STD-883. Rome Laboratories (an Air Force Laboratory) out of Rome, New York is the preparing facility for the documents and has the final technical interpretation of the specifications and standards. Rome Labs, however, are separate from the industrial innovation centers and therefore cannot appreciate the complexity of each product. The Defense Electronics Supply Center (DESC) located in Dayton, Ohio is the agent and the qualifying activity, which also certifies and audits manufacturers, and publishes more reports. Once certified by DESC, a manufacturer can qualify parts to the performance and mechanical specifications (documented in a M38510/0000 "slash sheet") on a part-by-part basis. Standardization is achieved when enough manufacturers produce a part to the same specifications and test it via military standard MIL-STD-883. Technical issues are resolved through coordinated activities with the Joint Electronics Devices Engineering Council (JEDEX 13.2) and another group called G-12.

Initially, the government's purpose in developing specifications was to help in the standardization of microcircuits used in DoD programs and to ensure that a highly reliable part was delivered from the manufacturer. As will be seen, the system caused significant problems in effective implementation, and in up-dating the documents, and with the numerous exceptions which arose as technologies were modified and new technologies were developed. As a result, the number of problems and documents continued to grow, and the quagmire of regulations and specifications has developed into an obstacle to the military electronics industry.

To investigate further, MIL-M-38510 titled "General Specifications for Microcircuits" defines the general requirements for monolithic, multichip, and hybrid microcircuits, and the quality and reliability requirements that must be met in the acquisition of military microcircuits. The manufacturing lines where these devices are built are certified to another military standard, MIL-STD-976, titled "Certification Requirements for Microcircuits". The manufacturing processes and the changes to these processes are under strict control, as defined in MIL-M38510, and require the manufacturer's customers to be notified.

Because qualification to MIL-M-38510 is a long (six months to several years) and expensive process, most of the parts available to the specification were/are old technology (obsolete). In order for subcontractors, on the other hand, to comply with contractual requirements, Standard Military Drawings (SMD) are used. However, with the SMD program, another layer of bureaucracy has been formed whereby subcontractors or microcircuit manufacturers develop drawings under the direction of DESC. Because DESC is the controlling government agency, the drawings are commonly referred to as DESC drawings. The DESC parts are manufactured in accordance with MIL-STD-883. The electrical specifications are controlled by the government as in the MIL-M-38510 devices, although the extensive JAN qualification is not necessary. Changes to manufacturing processes or functions are in accordance with another military standard, MIL-STD-480, which requires the part manufacturers to notify its customers only if they change the fit, form or function of the device.

Another class of drawings, the Specification Control Drawings (SCD) is made when the manufacturer's "off-the-shelf" part is compliant to MIL-STD-883, and their standard electrical characteristics, published in their data books, are sufficient for the application. Similar in nature to a SMD, the device electrical characteristics are tabulated, but they are under vendor control as opposed to DESC. The parts are marked with the manufacturer's standard catalog part number. As with the SMD parts, change notification is in accordance with MIL-STD-480. Vendor SCDs are made when a SMD is not available for the part of choice. In all of the MIL-STD systems, the bureaucratic impediment for dual use advanced technology products is great. It is obvious that DoD must streamline or eliminate these procedures.

The Mil Equivalent approach is an informal system evolved from the JAN system, whereby manufacturers advertize parts as equivalent to JAN parts but without the qualification by DESC. Because most electronic parts are produced off-shore, the manufacturer can self-certify the facilities, which in turn reduces the part cost. In addition, off-shore facilities generally produce a highly consistent level of quality due to high volume runs with more automated equipment. Unfortunately, DoD policy penalizes the use of such parts, under the assumption that there is a risk associated with the lack of a check on the testing or whether the part actually meets the manufacturer's data sheets.

The military has also been concerned with reliability. Mission success and the fear of an accident are some of the primary drivers for reliability at the systems level. In terms of reliability, MIL-STD-785, "Reliability Program for Systems and Equipment, Development and Production" is the fundamental specification in the expanse of reliability specifications. The standard specifies both general requirements and specific tasks for managing reliability programs. Each task described therein is usually addressed by more specific military standards and associated military handbooks. For instance, the MIL-STD-785 addresses three types of tasks:

- *Reliability accounting tasks* provide information for the acquisition, operation, and support management of the system/equipment.
- *Reliability engineering tasks* focus on the prevention, detection, and correction of reliability design deficiencies, weak parts, and workmanship defects.

- *Reliability management tasks* deal with the management responsibilities of the reliability program.

Once again, the documents associated with reliability are out-dated, difficult to use, and generally inconsistent with present commercial practices for advanced electronics products. For example, one military handbook associated with the prediction of reliability, is MIL-HDBK-217. "Reliability Prediction of Electronic Equipment," is used to specify reliability and to predict failure rates of all electronic parts used by the military and government. It also provides a basis for design decisions, such as choice between alternative concepts or of part quality levels, derating to be applied, use of proven versus state-of-the-art techniques, etc. Only steady state temperature (not rate of temperature change, temperature gradient, vibration, etc.) is explicitly accounted for with the Arrhenius equation. The reliability prediction is based on the collection of field data without knowledge of the actual "root cause" of the failure, and is as much as 20 years out of date with respect to some electronic parts (i.e. connectors, [4]). In other cases, knowledge from an obsolete CMOS technology has been transitioned in order to predict failure rates for GaAs integrated circuits. In an attempt to help the electronics designer, the models suggest that hermetic parts be used, and that parts be extensively screened, run at cooler temperatures, and derated. Unfortunately, there is a vast amount of evidence that these so-called reliability enhancement measures will add extensive costs, often with little reliability benefit. In fact, in many cases, reliability can be decreased by following this out-dated methodology for modern electronics. Discussions of failure prediction methodologies and scientific alternatives to the MIL-HDBK-217 failure prediction methodology can be found in [4, 5].

In the following paragraphs, additional reliability related standards will be discussed. The obvious limitations to technology development are so great and costly that most industries spend as much money meeting the standards as they do in research and development. MIL-STD-781, "Reliability Testing for Engineering Development, Qualification and Production" specifies the general requirements and specific tasks for reliability testing during development, qualification, and production of systems and equipment. Tasks specified in this standard include descriptions for Reliability Development/Growth, Reliability Qualification Testing, Production Reliability Acceptance Test, and Environmental Stress Screening. Associated with this standard is a handbook, MIL-HDBK-781, "Reliability Test Methods, Plans, and Environments for Engineering Development, Qualification, and Production", which explains the techniques used in reliability testing and provides a menu of test methods, test plans and environmental test profiles to tailor the testing procedures as per specific requirements of the reliability program. MIL-STD-454, "Standard General Requirements for Electronic Equipment" is the technical baseline document for the design and construction of electronic equipment for the Department of Defense. MIL-STD-454 covers specific requirements used to provide general guidelines for the design and construction of various types of electronic equipment. Each requirement includes a procedure, a process, or the selection and application of parts and materials.

There are two basic approaches to the application of environmental stress screening. In one, the government explicitly specifies the screens and screening parameters to be used at various assembly levels. Failure-free periods are sometimes attached to these screens as an acceptance requirement. MIL-STD-2164 takes this approach. The second is to have the contractor develop and propose an environmental stress screening program which is tailored to that product and is subject to the specific approval of the procuring activity. DoD-HDBK-344 (USAF) takes the second approach. In MIL-STD-883, MIL-STD-810, MIL-STD-2164, and DoD-HDBK-344, the failure mechanisms for screening and testing are not explicitly identified. Therefore, failures caused by the tests recommended in these standards could conceivably be induced by mechanisms other than those occurring under normal operating conditions. Although research has been conducted on statistical inference methods for life testing, the physical requirements of a test, and especially an accelerated test, have not focused on the root cause of failure.

The problem of specifications and standards is a critical one. Not only is it cost ineffective, but it establishes an adverse environment for technological innovation and has prevented the rapid transition of technological breakthroughs to new products.

2.3 INTEGRATION OF DEFENSE AND COMMERCIAL PRODUCTS: THE PROPOSED SOLUTION

2.3.1 Microelectronics

Silicon and gallium arsenide based integrated circuits (ICs), as strategic components of contemporary communication systems, along with advanced electronics products, are having a growing impact on the economy of the United States. Through their use in areas far beyond communications technologies, as diverse as those of home appliances or automobiles, consumer electronics and precision mechanics, they bring decisive added value and competitiveness to products. Incorporated into manufacturing systems, robots or data communication networks, the communications systems ICs allow renewed flexibility and precision in the realization of complex tasks, thus influencing the effectiveness of production facilities and services. In addition, a dual use technology IC industry must assume a leadership role to stimulate a generation of new demands for advanced integrated circuits. It must also create synergic relationships with users to which the continuous increase in the level of integration will provide the scope for a new competitive edge. This makes the timely availability of key ICs and their diffusion into the market place key factors for the economic development of the nation. It is within this framework that the Defense Department must meet its needs.

Three main classes of ICs represent the bulk of industrial consumption for commercial products:

- Volume produced ICs, such as memory chips used in a variety of consumer and computer products. DoD must employ these as the vehicle for acquiring advanced skills applicable to other IC products and to drive technology (see for example [4]).
- Application Specific ICs (ASICs), which are necessary to obtain fully optimized and competitive low-cost semiconductor solutions for electronic systems

producers. These ASICs, either in silicon or GaAs, must meet all of DoD's communications systems requirements.

- Programmable standard hardware (microprocessors), which must support a wide range of applications, both commercial and military, and offer flexibility through programmability.
- Advanced packaging concepts for ICs, such as multi-chip modules and optical interconnects.

These classes of products are tightly linked together. State-of-the-art silicon and gallium arsenide solutions are necessary for both commercial and military communications systems. Such solutions may be extended to cover information technology (IT industry). DoD must capitalize on industrial strengths and help to develop required system engineering skills and capabilities. In order to secure these objectives, however, the entire spectrum of problems and products would have to be addressed, with the commercial sector taking the lead. This includes Memories to provide volume production and to generate the most advanced technologies, manufacturing skills and capabilities, computer assisted design (CAD) and testing (CAT) tools and systems. The latter would facilitate price competitive design and production of communications systems and IT required ASICs of ever growing complexity, including single or multi-chip dedicated architectures, MMICs, micro-processors and micro-controllers, and multi-chip modules. In order to complement the above, dual use specialized equipment and materials, as well as specific training actions, will have to be promoted.

2.3.2 Availability of commercial components

Over 90% of all integrated circuits are commercially packaged in plastic [4, 5, 6], making the overhead costs per production unit lower by virtue of volume production. Almost all microelectronic device functions are commonly developed first in plastic packages. It is estimated by suppliers that at any given time, approximately 30% more part functions are available in plastic than the military required ceramic or metal [4]. The non-plastic packages are developed only if there is sufficient market interest or if it is necessitated by technology. Therefore, plastic packaged integrated circuits are available sooner than the military required ceramic or metal packages. Such a commercial approach for components must also be adopted by DoD.

Table 2.1 shows some of the tradeoffs among technology, package and complexity, indicating the existence of a high degree of diversity in products. In addition, the failure rate of plastic packages has decreased from about 100 failures per million device hours in 1978 to about 0.05 failures per million device hours in 1990. Encapsulation of integrated circuit chips with fast, automated transfer molding, makes plastic packages cost effective. Ceramic packages, on the other hand, are sealed by slower, more difficult glass sealing or brazing techniques, with more expensive ceramic material. A reduction in package costs means a substantial saving in overall equipment costs. Furthermore, the increased use and acceptance of surface-mounted packages has increased the attractiveness of plastic. Ceramic

Table 2.1 Plastic ICs and technologies.

Type of component	Technology	Package	Complexity (transistors)
DA converter	BiCMOS	SO16	80
DA converter	CMOS	PLCC 28	82
SRAM 256K	CMOS	SO 28	1536 000
DRAM 1MB	CMOS	SOJ 26	1024 000
OPT-EPROM	CMOS	PLCC 32	512 000
ASIC	CMOS	PQFP 144	120 000
LOGIC SSI	BiCMOS	SO 16	50

surface-mounted packages are more expensive than ceramic insertion-mounted packages, while there is little cost difference between plastic surface-mounted and plastic insertion-mounted packages.

In 1991, plastic packages with 10 to 50 failures per billion hours were routinely produced at a small fraction of the cost of ceramic packages — a few cents versus a few dollars [7]. Furthermore, some common commercial plastic parts are superior to those made in ceramic (Figure 2.1) [4, 8, 9]. In environments such as those characterized by high levels of mechanical shock and vibration, plastic packages are decisively superior to ceramic or metal packages [5, 10]. Signetics Corporation, Sunnyvale, CA, reports that the failure rate of ceramic packages is twice that of plastic packages in continuous-operation tests in a dry environment. The reason is that moisture is driven out of a plastic package but is trapped inside a hermetic package [7]. A plastic package also has advantages of light weight and small size

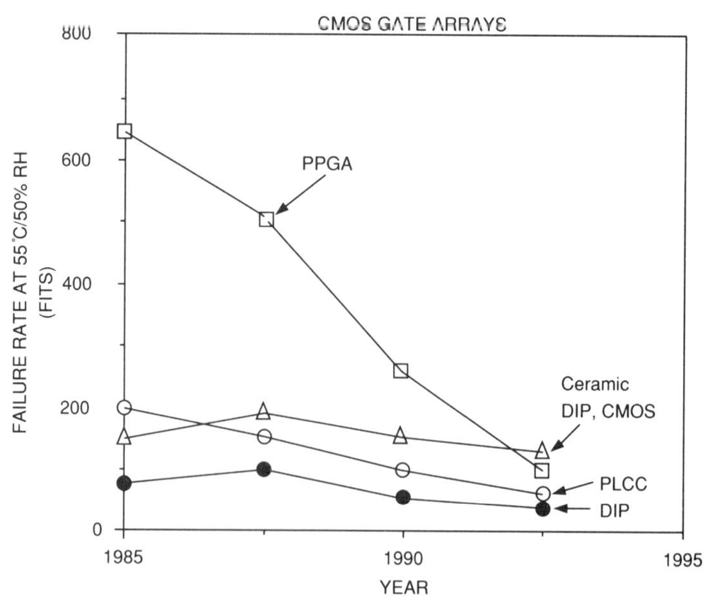

Figure 2.1 Plastic package reliability improvement and comparison with a ceramic DIP.

compared with a ceramic or metal counterpart and is also cost effective. The relative costs of various packages normalized to the cost of plastic dual in-line packages (DIP) have been summarized previously [10]. Dicken [11] investigated a typical LSI circuit, in which the die area is approximately 50,000 square mils, and which is assembled in a 64-pin, $5 package. It is also noteworthy to analyze the reliability data between the chip and the plastic package. As shown in Figure 2.2, plastic package reliability is at the same level as chip reliability for bipolar ICs.

In the first phase of the plastic Component Plan supported by the French Ministry of Defense [6], the cost reduction was found to be 48% for a VHF/FM manpack transceiver implemented with plastic packages instead of ceramic ones, for each of the 12 SMT PCBs of the equipment. Studies by ELDEC Corporation, Bothell, WA show that the cost reduction potential for a proximity switch electronics unit is 12% of the purchased component cost if plastic parts are used [10].

In a study to compare plastic and hermetic packages [10], circuit card assemblies consisting of both plastic and hermetic packages were temperature cycled between −55 and +85 °C through 1000 cycles. No differences between the plastic and hermetic devices were observed in terms of failure rates or shifts in any of 26 measured parametric values. Conservative estimates of minimum product lifetimes indicate acceptable performance with respect to failures caused by temperature cycling environments, in the 1–25 year range for both plastic and hermetic packages.

The Electronic Systems Group (ESG) at Westinghouse Electronic Corporation conducted research on the use of commercial plastic-encapsulated microcircuits in an uncontrolled environment [12]. The study concluded that properly designed plastic parts could be used in environments with uncontrolled humidity and temperature cycling.

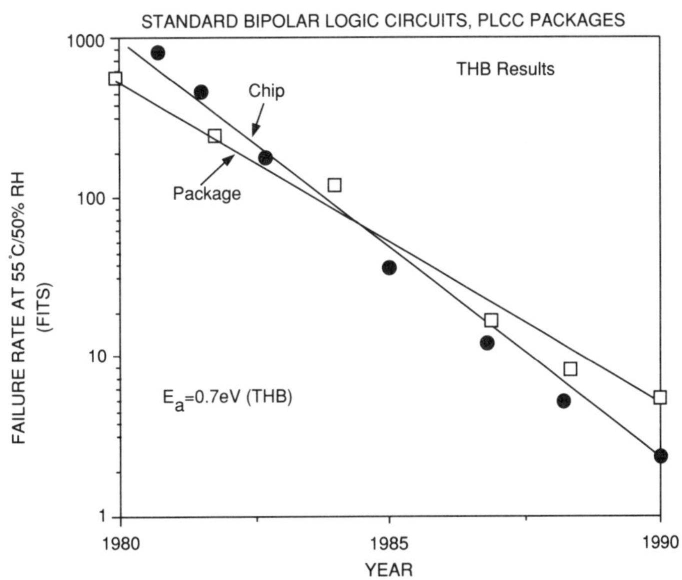

Figure 2.2 Plastic package and chip reliability for bipolar logic circuits.

Hughes [13] evaluated the reliability of a LSI device in a plastic leaded chip carrier and concluded that the continued improvement in the reliability of plastic encapsulated devices has provided an economic alternative to ceramic encapsulated parts. The exceptions are where the operating temperature is greater than +125 °C or where devices are continually exposed to both high temperature and humidity. The improvements in plastic encapsulated devices have been achieved primarily by increased purity of the plastics, and the use of high quality passivation layers, which minimize the main reliability hazard of moisture induced failures. To achieve a consistently high level of reliability, it is important to maintain quality control of supplies.

In a study [14] on burn-in for GaAs ICs, it was determined that burn-in is no longer an inexpensive and totally effective screen for today's high performance, high I/O count devices. In fact, costs have been estimated at $8–30 per unit to perform burn-in after the tooling [14]. The specific conclusions from this study is that correct materials and process choices with effective process control of the assembly operation can eliminate infant mortality and the need for burn-in.

Boeing Aerospace Co. in Seattle, Washington, evaluated the design, qualification, screening and application requirements for plastic encapsulated solid-state devices for space applications [15]. Data was collected on 1035 plastic encapsulated devices and 75 hermetically sealed control group devices that were purchased from each of five different manufacturers in the categories of low power Schottky TTL (bipolar) digital circuits, CMOS digital circuits, operational amplifier linear circuits, and NPN transistors. The devices were divided into different groups and screened with three sets of procedures. All devices are then subjected to an accelerated life test to validate the screening procedures. The conclusions were:

- Burn-in is not necessary as a reliability assurance technique for plastic encapsulated LSTTL (Large Scale Transistor–Transistor Logic) devices.
- Double electrical testing seems to have no effect in improvement of subsequent life test reliability.
- The most important single screen that should be invoked for plastic encapsulated LSTTL devices is 100% electrical measurement at two temperatures (25 and 100 °C).
- LSTTL devices are suitable for use in plastic encapsulated form.

Boeing's screening cost analysis found plastic parts subjected to burn-in to be more expensive than JAN Class B hermetically sealed parts in all cases. The cost difference was mainly due to the small quantity handling and setup costs for screening single lots of parts, as opposed to the large quantity production runs made by manufacturers of JAN Class B parts. If the manufacturer performed the screening of plastic encapsulated parts, there would be a 2:1 cost advantage in favor of plastic encapsulated MSI bipolar TTL parts. The cost advantage of plastic parts disappears if screening is to be performed by the user for small quantities of parts.

Hughes [13] and others [16] evaluated the effects of 100% burn-in and temperature cycling on the corrosion resistance of plastic encapsulated TTL integrated circuits. He showed that failure distributions in THB were similar for burned-in and non-burned-in humidity-sensitive devices. Distributions were also similar for two temperature cycling ranges. When normal (not humidity sensitive) devices were THB tested after an excessive temperature cycling of 110 cycles, there was a very small, but statistically significant effect. Burn-in produced similar effects. He concluded that commonly used burn-in and temperature cycling were not effective screens for moisture related corrosion of aluminum metallization.

Flood [17], while analyzing the failure modes of plastic encapsulated devices, examined the stress tests used to detect failures, as well as the design and process improvements to eliminate these failures. He concluded that plastic encapsulated devices have displayed good reliability growth as a result of systematic design and process improvements. The failure mechanisms and the stress tests to accelerate these failure mechanisms are well understood. The technology for plastic packaging is still evolutionary and will produce even higher reliability as new materials and techniques are applied.

Delco Electronics Division of GMC [17] performed a comprehensive study to determine the suitability of state-of-the-art plastic packaged integrated circuits for military applications, and to develop methods and techniques for assessing and assuring the reliability of these packages. They found that plastic encapsulated integrated circuits are capable of performance approaching that of their hermetic equivalents under most test conditions, that the use of a barrier coating over the chip significantly increases the resistance of plastic devices to thermal shock and pressure cooker stresses, and that the level of internal power dissipation has a definite effect on the failure rate during biased humidity conditions. They also found that some screens (burn-in, thermal shock) used in their study do not significantly improve the device reliability. There are more cases where the reliability of a screened group is poorer than that of the unscreened group.

The derating factor, mathematically defined, is the ratio of rated level of a "stress" parameter for a part to its actual operating level. The derating factor is a margin of safety or "margin of ignorance," determined by the criticality of any possible failure and by the amount of uncertainty inherent in the reliability model and all its inputs. Ideally, this margin should be kept to the minimum required, so as to maintain the cost-effectiveness of the design. This puts the responsibility on the reliability engineer to identify as unambiguously as possible the rated strength, the relevant operating stresses and the reliabilities.

Current military and government derating guidelines, such as RADC Derating Guidelines and the Navy NAVMAT, are not only out of date, but also inaccurate and misleading. In general, the derating methods are based on the reliability prediction methods from MIL-HDBK-217. As such, they do not address the mechanisms which cause failure.

Screening carries substantial penalties in capital, operating expense, and cycle time, and its benefits diminish as a product approaches its reliability entitlement. Furthermore, products near entitlement will not benefit from stress screening, and may be harmed by any additional damage imposed by this process [18]. Any commitment to screening must include the necessary funding and staff to determine root cause and corrective actions for all failed units. The type of screening should be input from the design, manufacturing, and quality teams. A 100% screen may be necessary during the early stages of production, although a plan which includes the earliest possible reduction in the sample size through corrective action is strongly recommended. If almost all the products fail in a properly designed screen test, it shows that the design is incorrect. If a large number of products fail, a revision of the manufacturing processes is required. If the number of failures in a screen test is negligible, the processes should be in control and any observed faults may be beyond the control of the design and production process. At the time the process matures and screening rejects decrease, the decision to screen is economical, as it may be appropriate to replace a 100% screen for statistical process control purposes. High product reliability can only be assured through the use of robust product designs, capable processes which are known to be in control, and qualified components and materials from supplier processes which are also capable and in control.

2.4 ELECTRONIC ASSEMBLIES

Major contributors to the cost of an electronic assembly include components, the printed wiring board (PWB), manufacturing processes, and thermal management. In this section, costs of an electronic assembly associated with the PWB in terms of geometric factors, materials, thermal management, assembly, and mounting techniques, are discussed in order to indicate how dual use technologies may be possible at the assembly level. The cost of a PWB is largely driven by the manufacturing costs associated with layer count, trace width, trace spacing, materials, via density, and via diameter [19]. As the line width decreases, the yield decreases and cost increases. If the layer count goes beyond eight layers for a given trace width, the cost of a PWB increases significantly. In such a case, trade-offs between large and fine lines and spaces must be considered.

Component mounting technology can have a significant impact on costs. For example, PWB costs for a surface mount board, as compared to an insertion mount board are lower if the component count is the same. This reduced cost results from lower packaging density on the surface mount board, leaving more space for routing, and enabling a reduction in the number of substrate layers required to achieve the same functionality. Generally, a reduction of layer count from eight to six results in PCB cost savings of about 25%. However, the real estate gained from the reduction of via diameters leads to an increased via aspect ratio.

A problem can arise when components are mounted on some of the standard electronic modules (SEM). When dimensional requirements for adhesive layers, ceramic substrate, circuit and solder are subtracted from the module thickness, only 72 mils are left for the package height. This is significantly less than JEDEC height allowances of 100 mils and 120 mils for even LCCs. In fact, no LCCs on the Standard Parts List meet the 72 mils height requirement. This case is worse for leaded devices which are often commercially available. Approaches that can be taken to solve this problem include the following:

(1) Increase the pitch of the SEM to permit the use of standard commercial packages. The advantages are ensured availability, continued reliability of the components and minimal cost. While components will increase in thickness as they become wider in the future due to required thicker wafer, increasing the pitch provides a long-term solution. However, it can pose a logistics problem,

in the short term, due to the number of systems already existing with standard pitch.

(2) Retain the same module pitch or stack height, but redesign the module so that more space is available for package height. This means reduced height for component to board interconnections, relaxation of the processing tolerances, and a reduction of the number of substrate layers. Reduced open space will also limit the accessibility of components during repair. This is costly and not recommended.

(3) Use non-standard components on a drawing to drawing basis. This will adversely affect availability of components and at the same time increase costs. This is not recommended.

Thermal management considerations are a significant part of each product design and cost and trade-offs must be conducted throughout the design process. The effects of temperature on PWB assemblies [20] can be categorized as follows:

- Effects due to changes in electrical characteristics that affect device (component) parameters and associated compatibility with overall function operation.
- Effects that produce mechanical stress in components, boards and interconnects, due to temperature extremes, temperature changes or thermal cycling.
- Effects of temperature for the first category were also addressed in this chapter. The second category involves understanding the effects of temperature on the material properties and the architecture. Temperature is generally considered to be a key parameter in the design of electronic equipment and its effects on reliability are widely documented. While some studies suggest that temperature is the most critical stress influence on microelectronic device failures, the actual failure mechanisms have generally not been quantified in terms of whether a steady-state temperature, temperature change, rate of temperature change, or spatial temperature gradient are responsible for failures.

Temperature reduction has an associated cost because of added hardware. The misapplication of the reliability concepts may cause failure avoidance efforts like temperature reduction, without yielding anticipated overall results, or expensive system implementations whose costs and complexities exceed the anticipated benefits in reliability. While certain types of failures can be accelerated by temperature effects, the solution does not necessarily lie in the direction of temperature reduction. More insight into failure mechanisms can yield solutions in terms of design and processes. Merely attacking the temperature issue, without understanding the root cause of the problem, can lead to acceptance of expensive and complex temperature control measures.

The evolution of cooling systems over the years, in terms of complexity, power consumption and weight, reflects the perception that steady state temperature dramatically affects electronic equipment reliability. Such systems, which have increased in complexity over the years, increase the total dollar costs of installed electronic equipment by large amounts, add several hundred pounds of weight, and increase the electrical power consumption. The result may be decreased reliability as a consequence of added complexity. In addition to penalties related to weight, volume and cost, penalties in terms of power supply for driving the coolant, resulting noise, delicate maintenance, and dependence on the active cooling system, can cause potential reliability problems in electronic systems.

The wide acceptability of steady-state temperature dependence of electronic equipment reliability is a consequence of existing failure prediction methodology and the Arrhenius failure acceleration model. The implied precision of the Arrhenius acceleration concepts, along with the exhaustive numerical values in the failure prediction methodology tabulations, as though representing a comprehensive database, lead to an aura of authority and accuracy that encourages a large number of trade studies common in the industry. In a mid-October 1989 meeting at Rome Laboratories, application of the MIL-HDBK-217 failure prediction methodology as a reverse engineering process to increase reliability was strongly discouraged. The validity of the assumption of constant failure rate, exponential acceleration of failures due to temperature, and benefits from linear derating as a function of temperature, is thus questionable [4].

A possible approach to improvement in the thermal design of cooling systems has been suggested by passive cooling in commercial avionics [3]. The test results have shown that there are great advantages in employing passive cooling, including no dependence on a mechanical system. This limits the interactions that could potentially result in loss of functionality, reduced power consumption for cooling machinery, reduced weight and volume, by virtue of elimination of the cooling system ducts, and fans. In addition, no complicated maintenance is needed for the active cooling system. A series of tests carried out at Boeing Commercial Airplanes [15] has proven that passive cooling incurs no identifiable penalties.

Assembly cost includes component placement, soldering, cleaning, repair, and testing. The larger the number of components, the more expensive the assembly cost. Further, for different geometry and types of components, different auto-insertion equipment is employed, leading to increased setup time. When surface mount technology is used, the same pick-and-place equipment can handle all types of surface mount components. Therefore, assembly costs for surface mounting are considerably lower than for insertion mounting, the cost saving 30% or better in placement alone. Reduction in material and labor costs associated with automated assembly provides additional savings.

Reducing the number of parts by integrating structural and electronic functions into the substrate can achieve space and cost reductions while improving product reliability. Commercial electronic assemblies offer cost, size, and weight advantages by combining structural, mechanical, electronic and circuit functions into the design and architecture of substrates for low power circuits. However, confidence in the argued reliability [21] of such innovative substrate design remains to be gained through field tests.

2.5 PRIORITIES FOR THE 1990s IN ELECTRONICS

The communications IC and IT industries, together with the DoD, must help mobilize human, financial and technological resources. In light of recent positive experiences in joint ventures or consortia, U.S. IC companies now show a determination to progress further on the road of integration of joint efforts and to subscribe to an unprecedented commitment in cooperative R and D and to the subsequent necessary production investments. In light of the above priority themes, the plan for the 1990s in communications systems should emphasize High Density (Advanced CMOS, ASIC and SOI 0.3 micron MOS) and High Speed (GaAs integrated circuits) together with the assembly procedures. In peripherals, the emphasis must be in magnetic storage and input/output devices (an I/O device combining displays and sensors). Components for dual use technologies must make use of all cost reduction practices, including plastic packages.

The rationale behind the suggested approach lies in the progression towards an ever larger level of integration; the evolution of IC technology is characterized mainly by a continuous reduction in device geometries. This evolution is expected to continue at least until the mid-nineties, when 0.3 μm silicon technologies will be used. These advanced technologies are essential for:

- Large standard systems
- A number of application specific circuits such as real-time signal processing and artificial intelligence. There is no alternative for U.S. industry but to continue to develop these advanced processing technologies for commercial products. The resources needed for the development of advanced memories are, however, so large, that success is only possible through an efficient collaboration between commercial and military product interests.

With the reduction in geometry size, CMOS processes are becoming more complex and more expensive. The SOI technology can provide an alternative to conventional bulk silicon and could be cost effective around 0.5–2.5 microns. Indeed, SOI offers advantages such as simpler processing (removal of complicated isolation, no wells), reduced processing time, higher performance, higher packing density, ease of design. Substantial progress has been made on the materials side, with SIMOX now looking the most promising.

In conjunction with the high density CMOS circuits, a commercially driven processing technology of III–V materials for high-speed systems must be developed. The commercial development of monolithic microwave circuits, either as individual circuits or as mixed digital and analog circuits on the same die, is urgent for U.S. industry. Commercial systems applications are emerging, in which real-time processing will be a bottleneck if fast progress in the III–V technologies is not made. Such applications can be found in mobile communications, small direct radio links, transportation systems and satellite systems. The DoD requirements must follow the commercial product technology, in order to truly allow the normal market driven development of a low-cost high frequency technology. In the following paragraphs, the dual use technology requirements for high density and high speed communication ICs and systems will be discussed.

In order to fully implement a dual use technology for high density communications systems, a commercially driven development of submicron CMOS (0.3 μm) technologies, optimized for very high density ASIC applications, must take place. High speed, high packing density are crucial requirements. The technology development will take into account the integration of various blocks (microprocessor, RAM, ROM) on large size chips. These technical and manufacturing objectives can be met with the development of logic technologies derived as directly as possible from memory process, and aimed at either the production of low cost, high volume standard products, or standard products customized with other process modules (as bipolar transistor or EPROM cells). For any dual environment, it is necessary to initiate the investigation of the suitability of the developed technology modules to ASIC applications, so as to prepare the close interaction between various CMOS products. Necessary also, is the rapid development of related manufacturing engineering methods to ensure timely and cost effective production. Progress in manufacturing engineering also requires a dedicated large-scale effort. Again, cooperation in as many technical areas as possible is required without giving up individual company strategy. Since the resources (investment and manpower) needed for the independent development of, for example, a new memory generation or a new communications systems receiver are forbiddingly large, cooperation on all possible levels exploiting the large degree of synergy in sub-technologies, not only reduces costs and development risks, but would also accelerate the overall program. Manufacturing equipment development, automation, clean room technologies, including waste management and safety issues, are areas which lend themselves to a high degree of cooperation and burden sharing between the military and the commercial sector.

The demonstrators for high density technologies must be in commercial products. This approach represents a clear break for DoD, from the present practice where demonstrators are related to military systems and commercialization means that a military product may be purchased from a given company. Therefore, state-of-the-art will dictate the requirements for competitiveness and for the demonstrators. The latter should be chosen first to demonstrate the capabilities of the process and second to be useful in areas which is, or will be, of prime importance to the nation's economy. The area of video codes using bandwidth reduction for video conferencing, TV and HDTV applications is a possible example. These are systems which involve a large amount of signal processing and thus can serve as very good examples of the benefits which can be achieved with dual use VLSI technology. Such an approach represents a break from typical DoD demonstrators, which were the outcome of the VHSIC programs. The VHSIC demonstrators were entirely for military systems with very few benefits to commercial large market products. In addition, coordination with other dual use technology thrusts must take place, including projects on manufacturing equipment, CAD, simulation, and SOI 0.3–0.5 micron CMOS. Already, circuits with complexity up to 64K bit have been demonstrated.

The second commercial objective previously discussed is in the development of dual use technology for high speed systems based on GaAs ICs. Consequently, the overall objective would be to integrate analog, as well as digital functions of GaAs ICs for real-time signal processing commercial applications. As the targets are systems and sub-systems rather than loose components, the availability of technologies in the U.S. is essential for synergy with the U.S. communications industry. In order to achieve this, the following activities are necessary:

- Develop processes to a high maturity level, demonstrated by high performance integrated circuits. It should be suitable to manufacture analog devices with bandwidths above 20 GHz and dividers for 30 GHz.

- Develop mixed function analog and digital ICs of as high a frequency range as possible; for example, a 5 Gbit/s sampling rate with 6–8 bit resolution.
- Demonstrate manufacturing environment yield and reproducibility.

A real-time signal processing demonstrator for dual use technology is the key element in a number of products for communications systems and for information technology. The best features of compound semiconductor technology will be used in different building blocks of the system, e.g., HEMTs front end, HBT A/D converter, etc. As frequencies increase, the role of each technology may change. Ultimately, a low cost single signal processing chip for high volume commercial telecommunications systems would be developed. The design, technology and chip sets from this commercially driven approach would be used by the military to meet its EW, radar and smart weapons system requirements. The involvement of systems and equipment manufacturers in the definition and assessment of the circuits, and future applications of these technologies must be strongly encouraged.

In addition to the development of dual use technology for high density and high speed systems, special attention must be paid to wafer fabrication equipment (WFE). Although the WFE market size is small ($2.7 billion) compared to the semiconductor market ($44.5 billion), this sector is fundamental in the chain that feeds into the huge electronic systems market ($525 billion). It is an enabling industry that permits the production of state-of-the-art systems, hence its strategic character. It is universally recognized that modern electronic systems, component and device manufacturing requires the timely availability of advanced high quality equipment. Dependency on foreign suppliers for equipment cannot guarantee this availability. In recognition of this need, the substantial involvement of equipment manufacturers in the definition, development and validation of dual use technology projects is essential. Automation is of high priority in view of the requirements of modern manufacturing science. Automation, of course, goes beyond the strict equipment field, and future dual use technology developments should have a strong link between process science and automation. The specific types of dual use technology developments in the area of equipment are listed below:

- multiprocessing
- lithography
- automation
- metrology

Before entering into a more detailed description of the enabling dual use technologies for communications systems, some general remarks applicable to all of them are called for:

(a) *Quality.* Both manufacturers and users of semiconductor systems and production equipment consider quality of the highest importance. Quality is characterized by:
  - Short Mean Time Between Failures (MTBF).
  - High producibility and uniformity
  - Low particle count and elimination of contamination.

  Achieving the above three objectives, however, constitutes a major task.

(b) *Integrability/flexibility.* Different production machines will be coupled to one another so that several production steps follow each other without breaking the controlled environment. This will enhance the uniformity of each process.

(c) *Standardization.* Since more and more equipment will be integrated at the user site, it is of great importance that standardization receive more attention.

The dual use technology for multiprocessing will allow industry to realize various processes within load locked, fully automated, cassette to cassette processing equipment. The benefits are reduced cost, higher throughput, ease of process standardization and easier transferability of each process from one industry to another. For instance, possible process combinations include: in situ etching of native oxide followed by doped poly-silicon deposition, and eventually a metal layer, multichamber metallization and analysis between sputtered and CVD metal layers. Transport between processes should take place under a vacuum or a controlled atmosphere. Improved process and equipment monitoring and control techniques should be envisaged.

In the area of lithography, the following technological steps should be combined:

- lithography
- resist processing
- etching
- metrology and inspection

Currently, the above processes are used as separate process steps with only limited interfacing. For future state-of-the-art systems, more integrated concepts are imperative. The upgrading of automation is presently limited by the communication protocol, and is almost non-existent in defense manufacturing. In addition, the current defense industry has very little process equipment under CAM system control. Process equipment disposes a "standard" communication protocol. If all of the process equipment were connected directly to the CAM computer, the latter would be overloaded. A hierarchical architecture with data sharing between host and cell controllers can solve this crucial problem. Finally, we must address the problems associated with quality and process control. Since dual use technologies for the future require much more precise process control, there is a very strong demand for in-situ and on-line measurement systems. There are end point detection systems available, but manufacturers are not implementing wellknown off-line measurement techniques for in-process control. Many of the parameters for control can be measured with optical or electrical sensors. In many cases, adjustment of the measurement system probably has to be done automatically, i.e., precise mechanical systems will have to be employed. Data acquisition and evaluation must be performed by sophisticated software.

2.5.1 Peripheral technologies

The market for communications systems will become the world's second largest by 1995 (information systems is first), with the peripherals market comprising a significant portion of it. Dual use technology, therefore, represents an opportunity for leveraging research and development investments in this area. Some actual market requirements can be identified as follows:

- In mass storage, more and more capacity is needed for a smaller and smaller form factor;
- In displays and in all devices, more intelligence is required.

The trend and, therefore, the objective of dual use technology in the field of optical or magnetic mass storage is to increase the recording density using advanced magnetic thin films. Additionally, dual use technology is required for the development of new concepts in input and display. The merging and collection of several different functions into the same input equipment provides new opportunities to improve the marketability of communications and information systems. In order to achieve this, the strong involvement of manufacturers with capability and experience in flat panel displays and large area sensors is necessary.

2.5.2 Commercial specifications and processes

Procurement of communications systems for utilization in the frequency range from L-band to W-band is typical of the inefficiencies of the present approach, since quality and reliability are tested-in according to the well established MIL STDS. The transition to a dual use technology in communications must also be accompanied by a 100% use of commercial specifications so that the same resources may produce both military and commercial products. This also implies that the specialized security and testing requirements for the manufacture of military products must be eliminated. The potential benefits include reduced costs and improvement in quality and reliability.

It is generally believed that the separation in the industrial base between defense and commercial products is obsolete. However, very few firms produce military and commercial products in the same plant. The industries which serve both markets (i.e., TI, Raytheon, GE, Hughes, etc.) maintain a rigid separation between the two lines of business. This separation is reflected in accounting procedures, management techniques, industrial training and marketing techniques. The barriers to integration are well known: the requirements for process and product specifications, excessive testing requirements which are often obsolete before the completion of the tests, and cost accounting standards. In VLSI and analog related products and process technologies, commercial practice has surpassed defense practice, resulting in a higher per unit cost for less advanced products for the Defense Department. A typical example is the present commercial market for an 8–12 GHz Direct Satellite Broadcast Amplifier, which sells for approximately 20 dollars, versus a 9–10 GHz band, which sells for 500 dollars. The basic technology is the same; however, the testing and packaging requirements imposed by the DoD has forced industry to create an entirely separate and dedicated product line with a low throughput and low manufacturing yield for essentially the same product which exists in the commercial sector. The following table compares the costs of commercially equivalent systems with the analogous military products.

	Commercial	Military	Cost/Unit (dollars)
L-band cellular receiver	x		20
L-band radar module		x	2250
DBS amplifier	x		20–22
Band 9/10 decoy		x	500–550
W-band amplifier	x		125
GPS amplifier		x	1750–2200

In addition to the wide acceptance of commercial specifications and procedures in a dual use technology, the rapid transfer of products from research and development to the market is essential in order to meet the pressures of international competition. This necessarily requires the integration of concurrent process, design and manufacture activities into the research and development cycle and into the acquisition life cycle of each process. Therefore, the acquisition of a product would also contain activities for product development, process science research and manufacturing technology. Concurrent product development thus would eliminate the lengthy procurement activities which at the present time take place at each stage of the serial life of future products.

The approach, finally, for dual use technology to be adopted by DoD must contribute to an improvement in the acquisition processes and must provide selective incentives based on competitive cost reductions or technology advances. The present system of scientific monitoring and government project management presents unnecessary barriers to management motivation for the achievement of manufacturing excellence and industrial competitiveness.

2.6 WHERE IS THE COMPETITION?

While the U.S. has enjoyed dominance in the electronics technologies over the past four decades with the development of the transistor, integrated circuit and microprocessor, it is now lagging the Japanese, with present trends indicating the emergence of a new group of leading electronics powers. The Department of Commerce has already warned that the U.S. is in danger of being surpassed in many of the emerging technologies, which may not only be crucial to the industrial well-being of the nation, but to its military security as well.

Defense research has long dominated Federal R&D. In 1960, of $13.2 billion spent on U.S. R&D, $8.7 billion came from the Federal government and $4.5 billion came from industry, with 80% of Federal spending being defense related. In 1992, the DoD may have absorbed over 60% of Federal R&D spending. The result is that "the Department of Defense has become the nation's Ministry of Technology by default," according to Robert M. White, head of the National Academy of Engineering (NAE), Washington, D.C.

While the U.S. spends more on R&D than Japan, Germany, France and the United Kingdom combined, these and other countries focus their R&D efforts on helping their citizens gain advantage in the civilian marketplace. In Japan, emphasis is on transferring ideas to the marketplace, with negligible defense-oriented R&D.

Efforts by Japanese companies are coordinated by a variety of organizations led by the Ministry of International Trade and Industry. Similarly, the 12 countries of the European Community (EC) emphasize commercial competitiveness by addressing issues ranging from basic research to prototype products.

2.7 CURRENT HOPES AND SUMMARY

Avionics reliability engineers have understood that the use of commercial electronics can offer competitive advantages associated with performance and density, availability, weight, as well as dollar cost. In commercial avionics, Boeing Commercial Airplanes, Honeywell, Eldec and Collins Commercial Avionics have been actively investigating the use of commercial electronics. Each of these companies has been using some commercial products over the past seven years and is presently conducting investigations and tests to qualify and/or set guidelines for the use of commercial plastic packages. In particular, Boeing Commercial Airplanes has developed a preliminary document titled "Electronic Component Selection Requirements" for their suppliers. The reliability of alternative packages is being extensively studied, and most of the failure mechanisms and associated stress tests to accelerate these mechanisms are well documented [10, 13, 17, 22, 23]. One general conclusion is that, as a result of design, material, and process improvements, commercial integrated devices have shown very good reliability growth.

Although affordability has become a key issue in the development of military systems because of budget cuts, the military has typically been slow to enact change. For example, the U.S. Army, Fort Monmouth has written position papers since 1974 supporting the use of commercial packages in Army electronics. However, there has been little implementation of commercial products in army electronics because of the authority given to Rome Laboratories (and their view that commercial parts have not yet been proven viable). Similarly, Wright Patterson Air Force base does sponsor an effort, called the Avionics Integrity Program (AVIP), whose mission is to focus on understanding and proactively addressing the fundamental causes of failure, rather than using handbooks and standards to mandate reliability. Unfortunately, the fact that the AVIP program does not offer a documented alternative with supporting standards and handbooks has been the downfall of the program. The AVIP principles have been supported by many of the leading avionics firms in the U.S., but there have been few success stories.

In order to help in the establishment of closer ties between the military and commercial electronics industries, various committees and working groups have been formed. While many of these have indirectly focused on ways to obtain military funding in a time of military budget cuts, the Electronics Packaging/Interconnect Task Force [24] has provided some constructive and objective recommendations. These include:

- the modification of present military standards and specifications for packaging and interconnection to allow non-traditional materials and fabrication methods;
- the development of a reliability science to qualify innovative packaging and interconnection methods, especially protective coatings for integrated circuit

chips to replace hermetic packages, and new interconnect substrates to replace printed wiring boards, along with the associated new materials and process technologies;

- the support of qualification testing at the module or sub-assembly level without regard to the packaging and interconnection methods used within the module or sub-assembly; and
- the establishment of industry-led commercial/military cooperative efforts for the incorporation of advanced packaging and interconnection technology into military electronics.

With increased defense cutbacks, the advantages of cost, size, weight, and availability of commercial electronics are beginning to motivate the military to modify their views about dualuse electronics. For example, qualified sources in military equipment report that Rome Laboratory is revising MIL-I-38535, the specification for integrated circuit manufacturing, to allow the use of commercial plastic encapsulated devices. As another example, the U.S. Army Electronics Laboratory Command has complexed a document, "Microcircuit Application Guidebook," suggesting that equipment designers do not automatically specify hermetically sealed ceramic or metal packages, but consider commercial plastic packages.

Perhaps the most progressive effort is associated with a new and unique standard developed by the Navy, titled MIL-STD-2036, which supersedes MIL-E-1600H. This standard provides guidance and requirements for the use of commercial off-the-shelf and ruggedized equipment, in addition to equipment manufactured to military specifications. This applies to all electronic equipment, including mission critical systems, and for shipboard applications, including submarines. There is, however, a statement in the standard which requires that supplied equipment meet service and functional requirements, a point which often implies that all other military standards be met, and indirectly that military products be used.

Finally, there has been increased activity by consortia of companies in the electronics area. One such consortium which was awarded a National Science Foundation grant as an Industry–University Cooperative Research Center, is the University of Maryland CALCE Electronics Packaging Research Center. This Center has 28 members who support the mission to address the reliability, screening and qualification of alternate packages. Most of the members are suppliers to the military or Government.

The Defense Department, through its unique interdependency relationships with key industries, has inhibited the nation's commercial competitiveness in advanced technologies both at the chip and component level. The solution and direction for the future has been outlined in this chapter, and specifically calls for an increase in the total use of commercial manufacturing processes and products in place of military specified ones, and for emphasis on the development of new products based on commercial initiatives. The cornerstone of this effort in electronics, microelectronics and communications is the dual use development of new technology, and the development of manufacturing processes which will ensure product reliability.

REFERENCES

[1] "Defense Critical Technologies Plan", 1989, Committee on Armed Services of the Senate and House of Representatives, PL 100-456.

[2] Report on the Workshops on Microelectronics and Peripherals, DG XIII EEC XIII/147/89-EN.

[3] Jones, P.T. and Teece, D.J., "Research agenda on competitiveness", Berkeley Business School, Berkeley, CA 1987, pp 45-51.

[4] Condra, L.W., "Using plastic-packaged ICs in avionics and military products", *Technology Review/ELDEC*, **5**, February, 1990, 1-5.

[5] Taylor, C.H., "Just how reliable are plastic encapsulated semiconductors for military applications and how can the maximum reliability be obtained?", *Microelectronics and Reliability*, **15**, 1976, 131-134.

[6] Lasseur, J., "Electronic equipment in very hostile environments", *ESREF Proceedings*, 1991, 73-84.

[7] Watson, G.F., "Plastic-packaged ICs in military equipment", *IEEE Spectrum*, February 1991, 62-63.

[8] Hakim, E.B., "Case history: procurement of quality plastic encapsulated semiconductors", *Solid State Technology*, 1977, 71-73.

[9] Setliff, J.E., "A review of the commercial microcircuit qualification and reliability methodology", *Advanced Microelectronics Technology Qualification, Reliability and Logistics Workshop*, August 13-15, 1991, Seattle, WA.

[10] Condra, L.W. and Pecht, M., "Commercial microcircuit options in military avionics systems demand reliability", *Defense Electronics*, December, 1991b, 43-47.

[11] Dicken, H., *Physics of Semiconductor Failures*, 3rd ed, DM Data Inc., Scottsdale, AZ, 1988.

[12] ESG, "Report on the 2 use of plastic encapsulated microcircuits in commercial designs", *Report # RE-9100*, Westinghouse Electric Corporation, Baltimore, MD, 1991.

[13] Hughes, J., "A practical assessment of current plastic encapsulated microelectronic devices", *Quality and Reliability Engineering International*, **5**, 1989, 125-129.

[14] Fayette, D.F., "MIMIC QML status", *Advanced Microelectronics Technology Qualification, Reliability and Logistics Workshop*, pp 245-261, August 13-15, 1991, Seattle, WA.

[15] Leonard, C.T., *Passive Cooling for Avionics Can Improve Airplane Efficiency and Reliability*, NAECON, 1987.

[16] Olsson, C., "Reliability of plastic encapsulated logic circuits", *Quality and Reliability Engineering International*, **5**, 1989, 53-72.

[17] Flood, J.L., "Reliability of plastic encapsulated integrated circuits", *IRPS*, 1972, 95-99.

[18] Leonard, C.T. and Pecht, M., "How failure prediction methodology affects electronic equipment design", *Quality and Reliability Engineering International*, **6**(4), October, 1990, 243-250.

[19] Prasad, R.P., *Surface Mount Technology: Principles and Practice*, Van Nostrand Reinhold, New York, 1989.

[20] Dasgupta, A., Barker, D. and Pecht, M., "Reliability prediction of electronic packages", *Journal of Institute of Environmental Sciences*, **33**(30), May, 1990, 36-45.

[21] Kellogg, M.N., *Electronics Manufacturing Engineering*, **6**(2), 1991.

[22] Maynard, J. and Mroczkowski, R., "Connector reliability predictions", *Trans. on Reliability*, 1991, in press.

[23] Rawleigh, S., private communications with Westinghouse Electric Company, October 10, 1991.

[24] NSIA, "Electronic packaging/interconnect: the next crisis for cost-effective military electronics", *Electronics Packaging/Interconnect Task Force*, National Security Industrial Association, October, 1990.

3

MANUFACTURING YIELD AND RELIABILITY

3.1 MANUFACTURING YIELD AND RELIABILITY

In the first two chapters the relationship between manufacturing and reliability was established. These two subjects are interrelated since the process yield is affected by defects introduced during manufacturing, which then affect reliability. Yield and reliability for VLSI circuits has been quantified. For a majority of VLSI circuits, both processing and design affect reliability, as will be shown in subsequent chapters which will introduce the basic engineering concepts.

In a properly fabricated wafer, we expect 100% yield within a given specification, which includes both the circuit design and fabrication selection. Once a fabrication sequence has been selected, then the processing and circuit sensitivities must be considered. Processing sensitivities are related mainly to photolithographic errors introduced during manufacturing. Circuit sensitivities indicate the fact that some circuits are more difficult to manufacture than others and hence appear to have a lower manufacturing yield.

3.1.1 Mechanisms of yield loss

The reduction of manufacturing success is quantified by the empirical determination of yield. The three mechanisms which reduce yield are given below.

(1) Processing effects
   (a) Photolithography defects — line defects, photoresist products on surfaces,
   (b) Variations in active layers — uneven etching and rough surfaces,
   (c) Variations in dimensions — introduced by etching and lithography errors.

(2) Circuit sensitivities
   (a) Threshold voltage — variations and sensitivities in applied voltage,
   (b) Channel length variations — these sensitivities result from lithography effects.

(3) Point defects
   (a) Dust particles (200 μm) — which affect metal line deposition,
   (b) Defects in the wafers — incorporated during growth,
   (c) Surface inhomogeneities — resulting from dislocations and stacking faults.

IC yield is then a function of all of the above effects:

$$Y = Y_0 Y_1(D_0, A, k_i) \tag{3.1}$$

where D_0 = density of point defects/unit area,
A = area of chip,
Y = good chips per wafer/number of chip sites per wafer (total wafer yield),
$(1 - Y_0)$ = Fraction of chip sites that yield bad chips due to processing related effects,
$(1 - Y_1)$ = Fraction of related chip sites which yield bad chips due to other effects.

Since yield is a statistical parameter and implies a probability function, the yield functions would be multiplied in order to attain the total yield of equation (3.1). Y is then the total wafer yield, which is a measure of good chips per wafer normalized by the number of chip sites per wafer. The function of Y is shown to be a function of the chip area.

If we ignore lithography defects, then point defects are the major defects present. We also consider the case that the point defects are randomly distributed, and then discuss the case for various non-uniform distributions of defects. The random distribution of defects implies that the processing steps have not introduced a new set of defects which would change the distribution function. This type of calculation would result in the determination of the minimum wafer yield, only affected by material defects.

3.1.2 Uniform density of point defects

In the ideal case where processing does not degrade yield, then the loss of good chips is from randomly distributed point defects in the starting wafer, metallization and dielectrics. If n defects are distributed randomly among N chips, then the probability P_k that a given chip contains k defects is

$$P_k = \frac{n!}{k!(n-k)!}\frac{1}{N^n}(N-1)^{n-k} \tag{3.2}$$

If $n/N = m$, then we have the Poisson distribution function given by P_k:

$$P_k \cong e^{-m}\frac{m^k}{k!} \tag{3.3}$$

The yield for the wafer is defined as the probability that the chip contains no defects, or $k = 0$:

$$Y_1 = P_0 = e^{-m}$$

The probability that a chip contains one defect is then obtained from equation (3.3).

$$P_1 = m\mathrm{e}^{-m}$$

If A is the chip area, the total area of the chips = NA, then the density of defects is given as D_0, where,

$$\frac{n}{NA} = D_0$$

This case applies only to material defects, without the inclusion of manufacturing defects, randomly distributed over the chip area and over the entire wafer. The average number of defects/chip as a function of chip area is then given as D_0A.

The yield $Y_1 = P_0 = \mathrm{e}^{-D_0A}$ is now a function of chip area and defect density. A priori we know that for large chips, the yield should decrease. This model tends to predict a very low yield, since the actual yield is considerably greater than the calculated yield. We now consider a non-uniform distribution of defects over a wafer, which is a more realistic situation and takes into account processing defects.

3.1.3 Simple non-uniform distributions of defects (D_0)

If D_0 is non-uniform across a wafer, as one would expect from the inclusion of manufacturing defects, then a defect distribution function is required, as shown in equation (3.4):

$$Y = \int_0^\infty \mathrm{e}^{-DA} f(D)\,\mathrm{d}D \tag{3.4}$$

where:

$f(D)$ = normalized distribution function of defect density

$$\int_0^\infty f(D)\,\mathrm{d}D = 1$$

The yield for the three types of distributions most likely to occur is listed below (Figure 3.1):

(a) Delta function

$$Y_1 = \mathrm{e}^{-D_0A}$$

(b) Triangular

$$Y_2 = \frac{1 - \mathrm{e}^{-2D_0A}}{D_0A}$$

(c) Rectangular

$$Y_3 = \frac{1 - \mathrm{e}^{-2D_0A}}{2D_0A}$$

for which we have in the limit of large D_0A,

$$\lim_{D_0A \geq 1} Y_1 = \mathrm{e}^{-D_0A}$$

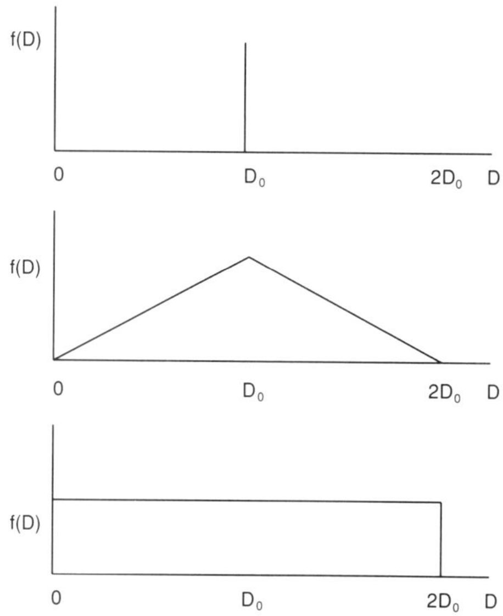

Figure 3.1 Distribution function used to calculate wafer and chip yield.

$$\lim_{D_0 A \geq 1} Y_2 = \frac{1}{(D_0 A)^2}$$

$$\lim_{D_0 A \geq 1} Y_3 = \frac{1}{2 D_0 A}$$

These distributions, however, are not physically realistic, with the exception of the rectangular distribution. A more realistic choice to predict yield is the gamma distribution function, with parameters which can be set to each physical case. The gamma distribution function (equation (3.5)) allows the selection of each parameter to be fitted to a specific manufacturing situation.

$$f(D) = \frac{1}{\Gamma(\alpha)\beta^{\alpha}} D^{\alpha-1} \mathrm{e}^{-D/\beta} \tag{3.5}$$

$$D, \alpha, \beta > 0$$

where α, β are the two distribution parameters, $\Gamma(\alpha)$ = gamma function. $D_0 = \alpha\beta =$ average density of defects, and the variance of defects is given by $\mathrm{var}(D) = \alpha\beta^2$. From these parameters every significant detail about the distribution of defects may be calculated. The probability of a chip having k defects is P_k, as before, and where $\int_0^{\infty} f(D)\,\mathrm{d}D = 1$. We define $Y_4 = P_0$ for zero defects as the yield of a wafer having a gamma distribution of defects. Each of the pertinent parameters is shown below in terms of α and β:

$$\frac{\sqrt{\mathrm{var}(D)}}{D_0} = \frac{1}{\sqrt{\alpha}} \quad \text{and} \quad S = \frac{1}{\alpha}$$

$$P_k = \int_0^{\infty} \mathrm{e}^{-m} \frac{m^k}{k!} f(D)\,\mathrm{d}D$$

$$= \frac{\Gamma(k+\alpha)}{k!\Gamma(\alpha)} \frac{(A\beta)^k}{(A\beta+1)^{k+\alpha}}$$

$$Y_4 = P_0 = \frac{1}{(A\beta+1)^\alpha} = \frac{1}{(1+SD_0A)^{1/s}} \tag{3.6}$$

where

$$S = \frac{\text{var}(D)}{D_0} = \frac{1}{2}$$

The yield Y_4 is a function of defect density and chip area. In the limiting case, gamma becomes the delta function, and the yield becomes the Poisson yield, which is similar to the delta function distribution.

For $S \rightarrow 0$ $f(D) = S(D - D_0)$
For $S \simeq 0$, $\beta = SD_0$, then Y_4 is the Poisson yield

$$Y_4 = e^{-D_0A}, \quad S \rightarrow 0$$

The gamma yield function gives the best approximation to wafer yield, due to the inclusion of the distribution function parameters, which are adjustable. Since many types of defects affect the yield parameters, D_0 and S may vary considerably. Very often a chip will have the critical component and additional circuitry, which may be redundant both from a performance and circuit design point of view. The above formalism may now be extended to the case of both circuit redundancies and multiple defects.

3.1.4 Yield of chips with redundant circuitry and multiple defects

The basic yield equation may be changed to include redundant devices:

$$\text{Yield: } Y_1 = P_0 + nP_1 \tag{3.7}$$

P_0 = probability of a chip containing no defects.
P_1 = probability of chip containing one defect.
n = probability that a chip may be repaired with a redundant column of MOS devices, or MESFETs in the case of GaAs.

Therefore, one must redistribute common components in order to group them redundantly for yield enhancement.

If the wafer contains a multiple number of defects, then each defect is assumed to affect yield independently. Typically yield is affected by a number of defects as shown below.

(a) gate oxide defects = defects are only in the gate region
(b) oxidation induced stacking faults = source and drain regions
(c) substrate dislocations

Each of these defects can be described by its own distribution function.

Recall from the gamma function

$$P_0 = Y_4 = \frac{1}{(1 + SD_0A)^{1/s}}$$

Then the yield from each type of defect is:

$$Y_n = \frac{1}{(1 + S_n A_n D_{n0})^{1/S_n}}$$

The parameters s_n, A_n and D_{n0} must be determined individually. The total yield is then determined by taking the product:

$$Y = \prod_{n=1}^{N} \quad Y_n = \prod_{n=1}^{N} (1 + S_n A_n D_{n0})^{-1/S_n} \tag{3.8}$$

The above equation also assumes that each term is approximately one, or $S_n A_n D_{n0} \ll 1$, as is required if the microcircuit is a commercial, mature product.

For a mature product, then, each term is close to one, or $S_n A_n D_{n0} \ll 1$ and the natural logorithm of the yield function becomes:

$$\ln Y = \sum_{n=1}^{N} \left[-\frac{1}{S_n} \ln(1 + S_n A_n D_{n0}) \right]$$

Since,

$$S_n A_n D_{n0} \ll 0$$

then,

$$\ln(1 + S_n A_n D_{n0}) \approx S_n A_n D_{n0}$$

$$\ln Y \simeq \sum_{n=1}^{N} [\quad A_n D_{n0}]$$

$$Y = \exp\left(-\sum_{n=1}^{N} A_n D_{n0} \right)$$

or

$$Y = \mathrm{e}^{-AD}$$

where A and D is the total area and defect density, respectively.

This approximation indicates that manufacturing success for a mature product should be close to 100%. Yield is again an exponential function, with D a function of the type of circuit, circuit geometry and the fractional area of the chip covered. High density integrated circuits such as VLSI and VHSIC circuits cover a greater percentage of the chip area, increasing the probability that a material or processing defect would be present at an active site. If we have M multiple chips, then yield is the exponential function of M-multiple chips:

$$Y_M = \mathrm{e}^{-MAD}$$

The relationship between reliability, quality and manufacturability in microelectronics is always cost. The manufacturing of ICs must result in a reliability level in excess of 10^7 hours at room temperature for the product to be a commercial success. For the above reasons we must examine basic reliability concepts, in order to understand the relationship between yield, cost and reliability.

3.2 FAILURE DISTRIBUTIONS/RELIABILITY FUNCTIONS IN MICROELECTRONICS

Reliability is something a manufacturing process attempts to attain at each stage, in order to extrapolate the results at each level to a full product reliability prediction. For an IC we first define a successful operation, and then a failure as a deviation from its electrical operation, resulting in the product not meeting its specifications. The concepts of electronic reliability and availability must also be explained, since their definitions are clearly different. Reliability is defined as the probability that the IC will operate within its specified limits for a given length of time. Availability on the other hand is the probability that an item will operate when needed, rather than a specified continuous operation.

The basic concepts may be expressed in terms of mathematical functions and statistical distribution functions. Reliability is always defined for a given type of device and then extrapolated to include all device types of the same technology.

3.2.1 Cumulative distribution function (cdf)

The cumulative failure distribution (cdf) is the basic reliability function used in microelectronics. It is defined as $F(t)$, the probability that the device will fail at or before time t. The function cdf has the following boundary conditions:

$$F(t) = 0, \qquad t < 0$$

Therefore, the distribution of failures will vary according to a function which must have the following limits:

$$F(t) \to 1, \qquad t \to \infty$$

$$0 \leq F(t) \leq F(t'), \qquad 0 \leq t \leq t'$$

This then allows us to define the reliability function as the complement of $F(t)$, or

$R(t)$ = probability that the device will survive to time t, or:

$$R(t) = 1 - F(t)$$

This function plus the cumulative cdf must be equal to one. At any specific time, we also require the rate of change of the cumulative density function, or the failure rate. This introduces a new function called the probability density function, which allows us to derive the time dependence of the failure rate. The probability density function (pdf) is the derivative of $F(t)$, expressed as:

$$f(t) = \frac{\mathrm{d}}{\mathrm{d}t} F(t)$$

$$F(t) = \int_0^t f(x)\,\mathrm{d}x$$

$$f(t) = -\frac{\mathrm{d}}{\mathrm{d}t}R(t)$$

It is noted that the pdf is just the negative of the reliability function. In addition to the rate of change of the reliability function, most IC manufacturers must determine the instantaneous failure rate of their product. In most circuit applications, we are also concerned with the instantaneous failure rate, known as the hazard rate. This can be derived by considering the difference between good devices at time t and failed devices at time $t + \Delta t$. The fraction of devices good at time t and which fail at time $t + \Delta t$ is:

$$F(t + \Delta t) - F(t) = 1 - R(t\Delta t) - (1 - R(t)) = R(t) - R(t + \Delta t)$$

The average failure rate during the time interval Δt is:

$$\mathrm{FR} = \frac{1}{\Delta t}\frac{R(t) - R(t + \Delta t)}{R(t)}$$

as $\Delta t \rightarrow 0$, then, we obtain $\lambda(t)$, the instantaneous failure rate, as follows:

$$\lambda(t) = -\frac{1}{R(t)}\frac{\mathrm{d}R}{\mathrm{d}t}$$

The instantaneous failure rate is critical in determining at which point a given circuit is in its projected life, and in allowing us to calculate the mean time to failure. The reliability function in terms of the failure rate is:

$$\lambda(t) = -\frac{1}{R(t)}\frac{\mathrm{d}R(t)}{\mathrm{d}t} = \frac{f(t)}{R(t)} = \frac{f(t)}{1 - F(t)} = -\frac{1}{\mathrm{d}t}\ln R(t)$$

The reliability function becomes:

$$R(t) = \exp\left[-\int_0^t \lambda(x)\,\mathrm{d}x\right]$$

The engineering number found in most of the literature is the mean time to failure, which can be defined by the probability density function. This function must be weighted by the time parameter, or integrated over the projected mission life of the microelectronics product. The mean time to failure is then:

$$\mathrm{MTTF} = \int_0^\infty t f(t)\,\mathrm{d}t$$

which is the device's average age at its failure point, for a device population whose density function is $f(t)$.

The MTTF then defines the life of a device of the same type, independent of its failure mechanism. All possible failure mechanisms are included in this calculation.

The microelectronic device has three periods in its total life before failure, as shown in Figure 3.2. The failure rate variation versus time indicates a region of

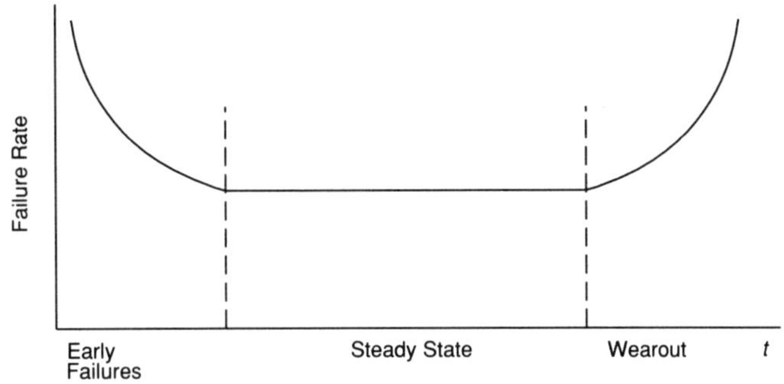

Figure 3.2 Failure rate for a microelectronic device showing the three regions of failure.

decreasing failure rate where all of the early failures are identified and removed from the distribution. The constant or steady-state region, then, defines the useful life of the product before the failure rate starts to increase. The final region is the wearout region where λ increases exponentially. The most important region of the product's life is when the failure rate is zero or constant, its useful life period.

A constant failure rate is characterized by an exponential distribution function, as shown through the following analysis:

$$\lambda(t) = \lambda_0 = \text{constant}$$

$$R(T) = \mathrm{e}^{-\lambda_0 t}$$

$$F(t) = 1 - \mathrm{e}^{-\lambda_0 t}$$

$$f(t) = \lambda \mathrm{e}^{-\lambda_0 t}$$

$$\text{MTTF} = \int_0^\infty t\lambda_0 \mathrm{e}^{-\lambda_0 t}\,\mathrm{d}t = \frac{1}{\lambda_0}$$

As derived above, the MTTF is constant and equal to $1/\lambda_0$. Weibull distributions are often used for integrated circuits, specifically when failure rate varies as a power of the age of the device. The most general formalism is to assume that the failure rate is proportional to time to a power $(\beta - 1)$ since by setting $\beta = 1$ a constant may be obtained. In general:

$$\lambda(t) = \frac{\beta}{\alpha} t^{\beta-1} \qquad \alpha, \beta = \text{constants}$$

Two regions are also defined by the value of β which may describe both early failures and the wearout region.

(a) $\beta < 1$, $\lambda(t)$ decreases with time, describes early failures.
(b) $\beta > 1$, $\lambda(t)$ increases with time, describes wearout period.

$$R(t) = \exp\left\{-\frac{1}{\alpha}t^\beta\right\}$$

$$F(t) = 1 - \exp\left\{-\frac{1}{\alpha}t^\beta\right\}$$

where Γ is the well known gamma function. The parameter β therefore defines the early failures when its value is less than one. The probability density function then may be expressed as:

$$F(t) = \frac{\beta}{\alpha}t^{\beta-1}\exp\left\{-\frac{1}{\alpha}t^{\beta}\right\}$$

$$\text{MTTF} = \alpha^{1/\beta}\Gamma\left(1+\frac{1}{\beta}\right)$$

The Weibull distributions are often used to describe failures during the wearout period of a solid state device, since excellent agreement with the data reported in literature has been obtained.

A number of plotting or graphical techniques exist which allow one to quickly determine the critical reliability of a microelectronic device. Often, we wish to predict devicefailure rates when the number of device failures is small, for instance in the case of a new circuit or in a prototyping environment. The technique is to plot the logarithm of average failure rate (AFR) versus the log of time (see Figure 3.3). The average failure rate (AFR) is given as:

$$\ln \text{AFR} = -S\ln t + \ln K$$

$$\text{AFR} = \frac{F(t)}{t}$$

From Figure 3.3, the constants K and S are determined, and the calculation of $\lambda(t)$ may be carried out as shown below:

$$F(t) = Kt^{1-s}$$

$$f(t) = K(1-s)t^{-s}$$

$$\lambda(t) = \frac{f(t)}{1-F(t)} = \frac{K(1-s)t^{-s}}{1-Kt^{(1-s)}}$$

If only a small fraction of devices failed, then $F(t) \ll 1$ and $\lambda(t)$ may be approximated as:

$$\lambda(t) \simeq K(1-s)t^{-s}$$

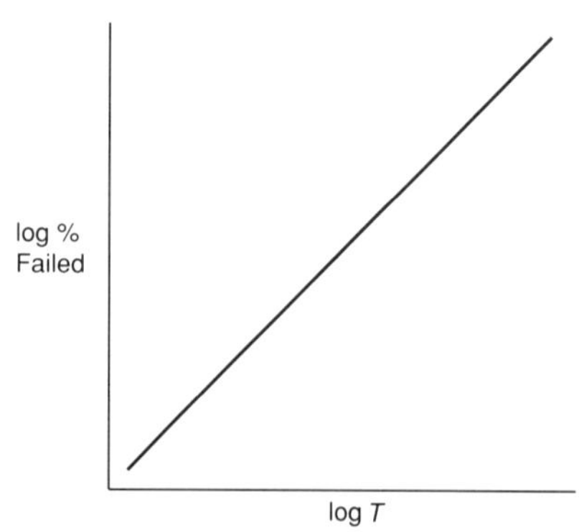

Figure 3.3 Graphical treatment of the average failure rate.

or

$$\lambda(t) = (1 - s)\text{AFR}$$

which is the failure rate one uses if a small lot size of devices are used in the reliability analysis. This type of approach is taken only in the case of prototypes, a new product or when the product design has not been finalized. For a mature device circuit, one utilizes log-normal statistics.

The log-normal distribution function has successfully described failure statistics of semiconductor devices over long periods of time. This function can represent any three periods of the circuit's life and is given as:

$$f(t) = \frac{1}{\sigma t\sqrt{2\pi}} \exp\left[-\frac{1}{2}\left(\frac{\ln t - \mu}{\sigma}\right)^2\right]$$

Median time to failure is given by t_{50} defined as:

$$t_{50} = \mathrm{e}^{\mu}$$

then the probability density function may be written as:

$$f(t) = \frac{1}{\sigma t\sqrt{2\pi}} \exp\left[-\frac{1}{2}\left(\frac{1}{\sigma}\ln\frac{t}{t_{5}0}\right)^2\right]$$

The average time to failure is expressed as:

$$t = \exp\left(\mu + \frac{\sigma^2}{2}\right)$$

and

$$\sigma \cong \ln\frac{t_{50}}{t_{1b}}, \qquad [t_{1b} = t_{15.866\%}]$$

The cumulative failure density function $F(t)$ is given as:

$$F(t) = \frac{1}{\sigma\sqrt{2\pi}} \int_0^t \frac{\mathrm{d}x}{x} \exp\left[-\frac{1}{2}\left(\frac{\ln x - \mu}{\sigma}\right)^2\right]$$

where the failure rate is

$$\lambda(t) = \frac{f(t)}{1 - F(t)}$$

Failure rates for integrated circuits are given in terms of FITs where 1 FIT is $1/10^9$ hours of testing. In general, a single slope MTF curve indicates the existence cf a single failure mechanism. This, however, usually exists only with devices which are well within the wearout region.

The failure rate for a device is complicated by the existence of more than one failure mechanism. If we assume that:

(a) Each mechanism is governed by a log-normal distribution
(b) σ, μ are different for each failure distribution, then the total failure rate is:

$$\lambda_T(t) = \frac{\varepsilon_s f_s(t) + (1 - \varepsilon_s) f_M(t)}{1 - \varepsilon_s F_s(t) - (1 - \varepsilon_s) F_m(t)}$$

where ε_s = fraction of early failures, and $f_s(t)$, $F_s(t)$, f_M, F_M are the distribution parameters of the two populations.

Such failure distributions are called bi-modal as shown in Figure 3.4, and indicate the presence of early failures or two distinct types of failure mechanisms. Many such failure distributions were observed by AT&T Bell Laboratories in optical equipment and in high speed switches.

Accelerated life tests are conducted through the effective selection of an acceleration parameter in order to determine the device's failure rate or mean time to failure. Three different acceleration factors will be discussed, which are all typical of microelectronic devices. In the case of temperature acceleration, the reaction rate as a function of temperature is assumed to be exponential and characterized by an activation energy E_a:

$$R = R_0 \exp(-E_a/kT)$$

The ratio of times to failure due to increasing temperature is the acceleration factor given by:

$$\text{AF} = \frac{t_1}{t_2} = \exp\left[\frac{E_a}{k}\left(\frac{1}{T_1} - \frac{1}{T_2}\right)\right]$$

where t_1 and t_2 are the times to failure at temperatures T_1 and T_2 respectively.

This is based on an exponential reaction kinetics law for each temperature. The function of AF allows us to determine the activation energy.

Many failure mechanisms may be accelerated only by voltage and current and not by temperature. This may apply to the following cases

(a) dielectric breakdown,
(b) charge injection, hot electron effects,
(c) corrosion.

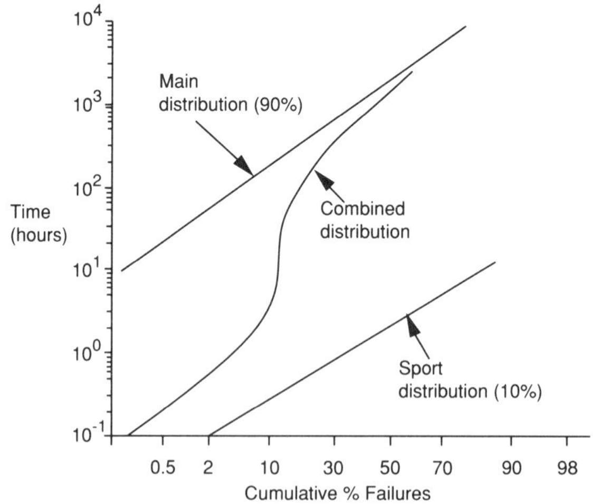

Figure 3.4 Failure distribution curves for two distributions of device failures.

These effects usually indicate that the failure rate R may be expressed as

$$R(T, V) = R_0(T)V^{\gamma(T)}$$

Although the acceleration factor for voltage has been used, the temperature appears as a secondary effect in the above cases. For current acceleration (i.e. electromigration), the failure rate is expressed as:

$$R(T, J) = R_0(T)J^{\gamma(T)}, \quad \gamma(T) \simeq 1\text{–}4$$

Again, in this type of model the reaction rate also includes current effects, and even temperature effects.

Water vapor quickly permeates plastic packaging, then penetrates the passivation, resulting in corrosion of the interconnects. This is a common failure mechanism which is a function of voltage, temperature and relative humidity and is given as

$$R(T, RH) = R_0(RH)^{\gamma(V,T)}$$

where $\gamma = 1$ to 2.

In the case of current acceleration, the MTF equation which describes electromigration is given by

$$\text{MTF} = j^{-n}\text{e}^{Q/kT}$$

where Q is the activation energy and j is the current density. For $Q = 0.4$ eV and $j = 10^5$ A/cm^2 then MTF = 206.7 hours, which is the room temperature MTF for an Aluminum metal stripe with an activation energy of 0.4 eV.

An example of a yield calculation will now be reviewed for a specific type of circuit. Yield of MMICs (monolithic microwave integrated circuits) is made up of process yield and the total line yield.

The total process yield is the product of two components, DC yield and RF test yield

$$Y_{\text{Process}} = Y_{\text{DC}} \times Y_{\text{RF}}$$

The DC yield now must take into account the different components on an integrated circuit, such as transistors and capacitors. The total DC yield is the product:

$$Y_{\text{DC}} = Y_{\text{ACT}} \times Y_{\text{PASS}}$$

where, Y_{ACT} = active device yield, and Y_{PASS} = passive circuit yield. Y_{DC} measures all defects (i.e. pinholes etc.) present in the material or introduced by manufacturing. We use the Price Model for DC yield, which has been derived previously as

$$Y_{\text{DC}} = Y_0 \left(\frac{1}{1 + D_0 A} \right)^N$$

where Y_0 = possibility that a portion of the wafer is totally unacceptable no matter how small it is,

D_0 = defect/cm^2 for the process,

N = number of critical processing steps,

A = device area, for both passive and active components.

Since the active components and passive components are not related, each yield may be treated separately. The active components consist of field effect transistors and therefore the FET yield is also denoted as Y_{ACT}. The yield for passive components includes device structures such as capacitors, resistors and inductors.

For a given FET, we can also evaluate part of the manufacturing process: i.e. the yield for making a via hole as the ith step of the process:

$$Y_i = Y_{0i}\left(\frac{1}{1 + D_{0i}A_i}\right)^{N_i}$$

This may also be extended to other steps such as ohmic contacts, where Y_j = ohmic contact yield or die attach.

The RF yield is the number of chips taken to the subsequent steps of processing, after the DC tests have been completed.

$$\text{RF Yield} = \frac{\text{number of chips meeting gain, power specs}}{\text{number of DC qualified chips}}$$

In this case, we require completely processed components or die. The RF yield may be calculated at the die level or at the packaged component level.

In addition to the DC and RF yield one considers "line" yield which is given as:

$$L_n = \frac{\text{number of wafers completing process } N}{\text{number of wafers submitted to process } N}$$

At least three critical measurement points are used and are defined as line yield breakpoints:

(a) following active layer formation,
(b) post metal,
(c) final front side processing.

The breakpoints are used for information calculation only. The total line yield would be the product of the line yield of each of a total of N process steps.

A yield/cost model can now be developed which would quantitatively relate the chip cost to the baseline process variables, and to the process yield. In order to develop a cost model, a cost analysis of each of the process steps must be made. In general, the cost has three components, fabrication, assembly and testing.

$$\text{Unit Cost} = \text{Fabrication Cost} + \text{Assembly Cost} + \text{Test Cost}$$

The fabrication cost may be expressed in terms of yield:

$$C_F = \frac{[C_{\mathrm{wafer}} + F(C_N, L_N)]}{Y_{\mathrm{PR}}C_{\mathrm{MAX}}}$$

where C_F = chip fabrication cost
C_{wafer} = cost to process the wafer
C_{MAX} = total chips per wafer
Y_{PR} = process yield = $Y_{\mathrm{ACT}} \times Y_{\mathrm{PAS}} \times Y_{\mathrm{RF}}$
L_N = line yield of Nth check point

C_N = cost to process wafer to Nth check point
$F(C_N, L_N)$ = impact function of the line yield in a cumulative basis = $[(1 - L_1)C_1 + (1 - L_2)C_2 + \cdots + (1 - L_N)C_N]$
N = process or costing breakpoints

C_{MAX} accounts for edge losses for a chip of area A, and a wafer of radius R_0.

$$C_{\mathrm{MAX}} = \frac{\pi(R_0 - \sqrt{A})^2}{A}$$

The parameter C_{MAX} then takes into account wafer cracks which produce a deviation in the total chip area available.

Example A wafer processing cost including line yield = \$5000 for a 3inch wafer
10% process yield of 0.090 inch × 0.070 inch chips
C_F(chip) = \$45.00

The total MMIC cost must also include terms for assembly and test costs.

Assembly costs to package the chips depend on assembly yield Y_A, labor, and materials costs.

$$C_A = \frac{(C_{AL} + C_{AM})}{Y_A} + (1 - Y_A)C_F$$

where C_A = assembly cost
C_{AL} = labor costs
C_{AM} = material cost (package cost for MMIC)
C_F = chip fabrication cost
Y_A = assembly yield
$(1 - Y_A)C_F$ = cost of good chips which are eliminated due to the yield of the assembly operation

In this step, even though good chips have been introduced into the packaging phase, a number of them will be lost during assembly.

Example 85% assembly yield using \$10 package, \$10 labor and \$45.00 chip.
then C_A = \$34.00

After the ICs pass the packaging stage, the final yield loss occurs during testing. Losses from testing arise from effects such as electrical overstress or electrostatic discharge. Human error also plays a large role in test yield effects.

Test cost directly depends on the complexity of the unit to be tested. We must also determine the final test yield Y_T in order to calculate testing costs. The total cost in terms of total wafer yield is given as:

$$C_T = \frac{C_{TL}}{Y_T} + (1 - Y_T)(C_F + C_A)$$

C_T = test cost
C_{TL} = test labor
Y_T = final test yield

$(1 - Y_T)(C_F + C_A)$ = cost of assembled chips which do not meet test specifications

Example 90% test yield, \$10.00 test labor,
\$34.00 assembly cost, \$45.00 chip
C_T = \$19.00 can be calculated by substituting the above numbers

The total cost of the quality assured MMIC is:

$$C_{\text{Total}} = C_F + C_A + C_T = \$45.00 + \$34.00 + \$19.00 = \$98.00$$

This number is a baseline cost which does not include profit or costs from research and development effort.

3.3 RELIABILITY INDICATORS IN MICROELECTRONICS

Reliability prediction is fundamental to system design. It involves the quantitative calculation of system reliability prior to development. Reliability predictions provide baseline values for reliability growth and demonstration testing, maintainability, supportability and logistic cost studies. Up-front reliability prediction in the electronic design process can result in a reduction in the design lag time and a savings in lifecycle cost of the product [1]. It is therefore important that the predictions be precise enough so that design tradeoffs and design decisions can be reasonably made. If the prediction of the mean time to failure is too low, then overdesign can occur, resulting in higher costs. On the other hand, an optimistically high evaluation can result in disastrous consequences.

Occasionally, we need to be reminded of the real meaning and use of the reliability prediction process. Reliability predictions in general have been criticized by a few industry groups as statistical manipulations which are unnecessary, serve no useful purpose, and are not indicative of the system's actual reliability. These critics fail to realize that a reliability prediction is just a part of the overall reliability program. By itself, it represents an estimate of inherent reliability of the design based on historical failure data, and should be interpreted as such. There is no element of statistical confidence in the prediction, and the resulting "mean time between failure" should not be blindly compared to operational reliability [2]. The prediction is first and foremost a design tool which should be used to compare design options and to identify overstressed parts. It may also serve as an input to failure mode, effects and criticality analyses, effectiveness analyses, logistic support analyses and the spares provisioning process.

Reliability indicators (RIs) are being investigated by many researchers as a means of evaluating true reliability of a system or product. They have been defined as measured parameters or a weighted set of parameters that indicate the reliability of a system or product [3]. They have also been defined as methods of detecting unreliable components by measuring their informative parameters [4]. Reliability indicators are related to life testing and reliability screens, but they differ conceptually by virtue of the fact that their measurements do not entail an application of stress to the component and involve life tests only when correlation between RI measurements and failure modes are required. RIs may also be considered as alternative test strategies. They can be used to replace voltage testing and increase fault coverage during production tests, or as additional measurements to extend the coverage of functional tests to include defects that are difficult to detect using standard techniques [5].

Traditionally, reliability and testing environments have been relatively decoupled as the test engineer has been mainly concerned with the verification of the functionality of the device, rather than the quality of the process and ensuring that the product meets specification. Reliability engineers on the other hand are concerned with such things as process control, failure mechanisms, failure analysis and product integrity, and assurance rather than production test. Reliability indicators have, however, forged a link between the two disciplines for two reasons. They have the potential of detecting failure modes and faults that cannot be detected by traditional digital test methods. They are able to provide indications of the quality of the process and the detection of many degradation mechanisms which may well cause failure. Secondly they are capable of detecting infant mortalities from a batch of devices which otherwise would involve expensive and time consuming procedures. Thus we see that the development of RIs is highly desirable in the context of component reliability.

3.3.1 Emission microscopy for reliability indication

Emission microscopy has been applied to the failure analysis of AIGaAa/GaAs HEMTS. The emission of light from these devices, when biased, has been studied, and a spatial localization of the defects has been identified. In some cases the emission of light using optical fibers was studied in order to correlate the emission spots with the detected failure mechanisms [6]. These tests have been carried out in order to evaluate the reliability of HEMTs and to determine the mechanisms responsible for their degradation.

Callesio, Marchisio, and Riva [6] carried out emission microscopy tests for the analysis of HEMTs. Prior to the start of the life tests, they carried out an accurate technological analysis; then, in order to determine the optimum temperature to be used in the accelerated tests, the thermal resistance of the devices was computed. The test fixture was developed and the samples mounted on it. The tests were interrupted during regular intervals to examine the changes in electrical characteristics and the visual signs of degradation. The damaged devices were examined in order to correlate device degradation to failure modes and mechanisms.

The equipment used was an EMMI 1600 which has a detector in the form of a S25 + photocathode sensitive to wavelengths ranging from 180 nm to 1150 nm. EMMI detects the radiation emitted by the radiative recombination of electron and hole pairs in semiconductors. The light emitted by the biased devices and detected by an optical microscope is focused on to an image intensifier. The resulting video image is processed by a computer system and the final result is displayed on a monitor. In the good devices the light emission was seen to occur in the active area of the HEMTs when they are biased at high drain voltages. In some devices, nevertheless, there was a lack of uniformity of the current flow in the gate area. This can be explained as inhomogeneity of the gate metallization due to bad technological processing. So the lack of uniformity of the current flow in the active area, detectable with the emission microscope, is an important criterion to evaluate the criticality of the device from a technical point of view, which will probably cause a malfunctioning of the device itself during its operating life.

The observation of other failed devices by EMMI revealed a gate current interruption due to some intrinsic defects induced in the semiconductor by ESD phenomena [7]. Another device failed due to overheating, detected by the EMMI by thermal emission. The current density that flowed into this region was so high as to cause thermal infrared emission. To confirm this hypothesis. a spectral analysis was carried out with the aid of wavelength bandpass filters.

Some literature exists on the use of light emission spectral analysis to identify the different failure mechanisms to correlate the observed emission to the physical processes that generate it [8, 9]. It is not easy to identify the different failure mechanisms on the basis of their emission spectra, as each single mechanism does not always exhibit a typical spectrum which maintains the same characteristics every time. The spectral analysis itself is very crucial.

With the aid of a set of filters, one can obtain the intensity of the emission in arbitrary units, but this intensity must then be normalized with the filter transmissivity and the detector's wavelength sensitivity. The errors during this correction can make the interpretation of the results difficult. Currently, investigations are going on in finding other methods to correlate spectral analysis to physical mechanisms.

3.3.2 IDDQ

The detailed monitoring of supply current (IDDQ) in CMOS VLSI circuits has been performed and interpreted by several scientists with renewed focus over the past decade. It can be used to improve fault coverage to a degree which is better than that achieved by voltage testing. This method employs a technique to derive a process production lot reliability quotient, developed by statistical analysis of IDDQ measurements from prepared life test samples. This technique has been used in detection problems such as floating gates, oxide shorts, and bridging faults [5].

IDDQ measurements are being interpreted here to extract their reliability content. In general life test results are not feasible for all production lots due to time and economic constraints. Samples may be taken from production lots, early in the manufacturing process. IDDQ signatures are extracted and then life testing is performed on the samples. Current signatures can be analyzed statistically and a reliability quotient can be derived based on average change over time. A lot related reliability quotient can then be available at the point of shipping in less time than it would take to get life test results. IDD is the term given to the positive supply current and Q denotes quiescent [10].

The technique proceeds as follows. Samples are selected from each fabrication lot after probe/assembly. The samples are tested burned in and life tested dynamically for eight hours. 100% IDDQ testing is then completed at maximum supply levels and the results stored. Vital failures at any point are analyzed, but degraded parts from the sample are returned to life testing. Life testing is then continued at 150 °C for a minimum of 160 hours with read points at various intervals. At each interval, IDDQ is recorded for the maximum number of test vectors. An average delta in IDDQ is calculated relative to zero time at each break point, together with other relevant statistics. These deltas are then plotted.

Figures 3.5 and 3.6 show typical results from IDDQ testing [5]. These signatures can be compared to a reference signature taken from a good device and an accept/reject threshold can be defined based on the electrical effects of the most probable defects and the spread of normal measurements. When using this technique, it is essential to determine whether the reference signature is in fact normal. The reject threshold should be set so that normal variations in IDDQ are not interpreted as an indication of device degradation.

A technique that has been used in the testing environment is to data log several hundred parts per design. One IDDQ measurement is taken per part and this data is presented as a distribution over the batch. The threshold can then be derived. The requirements for reliability indicator measurements will vary, depending on the defects that are causing reliability problems. It may well be the case that a number of different RI measurements may be required and the accept/reject threshold may depend on the results from more than one measurement. A statistical technique called the linear discriminant method proposed by Molcoft can be used to derive a weighted combination of measurable parameters of the component that will indicate whether the component is good or bad. Let us take an example where it is necessary to use two indicators to achieve the necessary discrimination. The combined indicator can be defined as

$$Z = a_1 X_1 + a_2 X_2$$

where Z = combined indicator
X_1 and X_2 = RI measurements
a_1 and a_2 = weighting factors

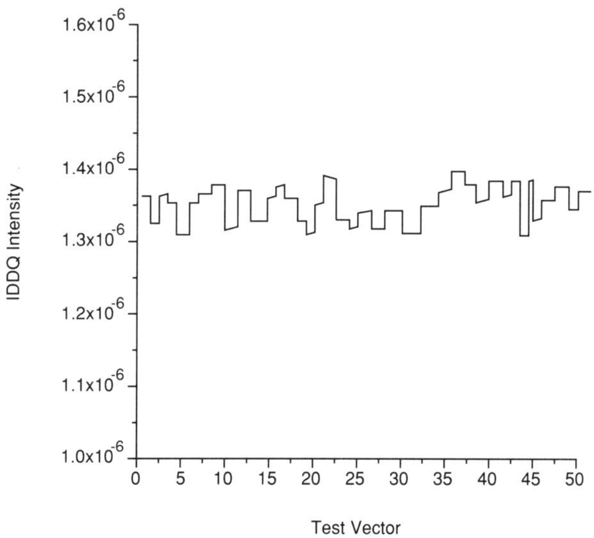

Figure 3.5 IDDQ signature of a CMOS circuit.

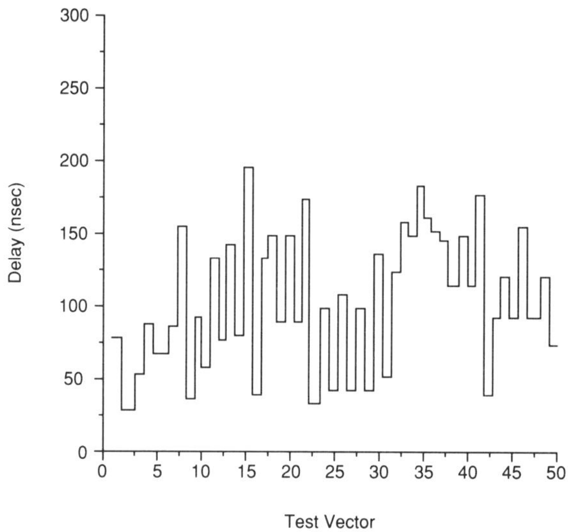

Figure 3.6 IDDQ signature of a highly degraded CMOS circuit.

To derive the weighting factors, it is necessary to perform life tests on the devices concerned to compile two distributions.

(1) Distribution of indicator measurements before stress for components passing the life test.

(2) Distribution of measurements before stress for bad components that fail the life tests.

Various mathematical techniques can then be used to derive the weighting factors.

The problems associated with testing complex VLSI devices are not simplified by assumptions that all hard faults have an equal probability of occurrence and that the derivations of the fault lists are in general layout independent. It would be prudent to concentrate future efforts in covering faults that have a realistic probability of occurrence. Inductive fault analysis techniques address this problem by examination of the device layout to generate a fault list with a ranked probability of occurrence.

3.3.3 Other reliability indicators

3.3.3.1 Marginal voltage

The minimum operating supply and signal voltage in digital circuits can indicate threshold voltage variations, identify leakage problems due to high impedance shorts and detect junction degradation in TTL circuits. A number of variations of the basic technique have been investigated. Permanent marginal voltage analysis involves applying a stream of data to the device under test (DUT) at some predetermined test vector, holding inputs steady and reducing the supply to the test value for a short period. The data is then clocked out of the device and checked to see if the drop in the supply voltage level has caused data corruption. This technique is most suited to sequential circuits. Dynamic marginal voltage analysis involves applying the full test set to the device for each supply voltage level. The supply levels are decremented until failure occurs.

3.3.3.2 Noise tests

Electrical noise (resistance fluctuations) is a very sensitive and general indicator of a lack of ideality, quality and hence reliability in electronic devices. It is many times more sensitive than simpler indicators such as resistance change. It also gives more information in its spectral form. However, it is very general, since almost any type of defect can generate the extra noise component. This noise is above any noise components which are called "fundamental," such as mobility fluctuations and quantum noise. The most detailed correlation between excess electrical noise and reliability is shown in cases of discrete bipolar transistors and in integrated circuit interconnects failing due to electromigration.

The increase in electrical noise with electrical damage is to be expected since the damaged area will be a small volume under high current density. These factors always increase noise. However, the measurements are usually carried out under conditions harsher than those of accelerated stress. It has been shown recently that the second harmonic generation in interconnects is also a sensitive indicator of damage and, hence, reliability.

3.3.3.3 Performance tests

Performance tests have been used widely by test engineers for verification of operating speed specification. Their use as RIs, however, is based on the concept that device degradation and the switching speeds of internal components are strongly related. The measurements may take the form of a maximum operating frequency measurement or delay measurements. In the case of sequential logic, measurement of the maximum operating frequency will depend upon the slowest component in the circuit. In the case of combinational logic, delay measurements are more appropriate and involve the application of test vectors that exercise the circuit in such a way as to propagate logical values from the primary inputs to a primary output via a chain of logic. The test set is designed so that each node in the device is at some time during the test sequence driven both low to high and high to low.

Defects causing changes in carrier mobility, nodal capacitance, threshold voltages and interconnect impedance will therefore affect the speed of the circuit. Approaches to delay testing are based on models of propagation delay with every gate within a specified range, and the delay between primary input and output for a sensitized path within specification, regardless of the individual gate delays. Both models have their merits. If the major concern is the maximum operating speed of a block of combinational logic, then the individual delays of internal gates are irrelevant. In this case test patterns are latched into the circuit and a second clock loads the output latch at a predetermined time after the inputs have been applied. If the logic outputs for the whole test set are true, then the device passes the test.

3.3.3.4 Transient current

When a CMOS circuit is biased, current pulses are observed on the supply rails. These arise from the charge flowing through the internal gates as they change state, and are called transient current pulses. Their height and width contain valuable information regarding the quality of the device. The pulses produced correspond to particular transistors or groups of transistors switching, and are dependent on the rise and fall times of the internal components, and the charge required by the gates to change state. Signature analysis can therefore be used to identify abnormalities in the position, height or width of the pulses. This information can then be used to identify defects, including threshold voltage shifts, oxide degradation, shorts, opens and gain factor drifts. It should be noted that this technique has the potential to be used in failure analysis for identifying the position of defects in the circuit.

3.3.3.5 Phase jitter

This technique involves the measurement of the stochastic phase variation in the output signal of a component subjected to a square wave or pulsed input signal. It has been suggested that it can be used to detect mechanisms such as corrosion, electromigration, ion mobility, whisker growth, diffusion and bad bonds, all of which are reliability hazards.

3.3.4 Algorithm to incorporate noise tests in microcircuit fabrication

ESD is regarded as the main cause of device degradation. Pre-damage of devices is caused due to low ESD and not just by the one high voltage charge which causes the final burnout. There are possibilities for device hardening by ohmic metallizations. Recent developments show that an ESD protection is possible, even in the microwave frequency, if new field emission structures are used for switching devices. However, control and monitoring of the fabrication process for useful damage detection and analysis are essential in investigating the improvements proposed by such new technologies.

Using the strong correlation of fabrication parameters, device reliability test structures are designed for the in-process monitoring of the GaAs fabrication process [11]. Basically, RF, microwave and DC measurements are used to detect ESD damage. Such a set of test structures, prepared in parallel to the microwave devices with exactly the same process parameters, additionally enables deeper insight into the reliability of the devices. Testing different fabrication levels and extracted parts of the microwave devices leads to additional information regarding device modeling.

In-process monitoring of fabrication quality based on DC, RF and RF noise measurements is an important tool to improve device technology and reliability. Based on the different structures, quality of the process step can be directly determined by electrical measurements. Every process step can be additionally investigated without correlation to the foregoing steps. This is an advantage in improving the parameters of the process step. Further, it is possible to find correlations between process step parameters and the electrical properties of the completed device. The electrical measurements during the in-process monitoring can then be augmented by further characterization of the completed devices.

The evaluation of ESD sensitivity of GaAs is important regarding devices to be used in a hard environment. DC, S-parameter and RF noise measurements are performed before and after the devices are stressed by ESD [11]. Early detection of ESD degradation is needed to find the real ESD sensitivity of the devices. Tests were carried out to evaluate the ESD sensitivity [11]. It was observed that with lower current densities the noise figure after stress remained high. With greater current densities it decreased more and more. A similar effect was observed with the GaAs microwave diode structures stressed by electrical voltage spikes (ESD). After a stress with a ESD voltage lower than 60 V, the diodes showed increased leakage current. If the diodes were pulsed a second time, even with a higher voltage, the leakage current was significantly lower. Degradation is caused mainly by a change in the ohmic contact characteristic or the semi-conductor material of the channel region.

3.3.4.1 Algorithm

The following algorithm is proposed for reliability indicators applicable to both digital and analogue circuits.

1. Develop the following test structures
   (a) Common TLM structure for specific resistance calculation.
   (b) Test pattern for investigation of lateral material movement between ohmic contacts.
   (c) Two transmission lines for S-parameter and noise measurements to detect ESD damage and to control process parameters.
   (d) Four gate-mesa crossing test patterns for investigation and analysis of the gate blow off.
   (e) Different test transistors varied in gate length and gate width.
   (f) Optimized transistor structure.
   (g) Microprobe calibration standards for the inner transistor.
2. Take different measurements of serial resistance and leakage current by using DC, Sparameter and RF noise measurements.
3. Using special transmission line test patterns, investigate the ESD sensitivity of the fate metallizations with the help of S-parameter and RF noise measurements.
4. Compare the measurements performed with the transmission line pattern to the noise measurements. This comparison gives additional knowledge as to the localization of damage caused by ESD.
5. Apply one small ESD pulse to the devices in order to achieve initiation of ESD.
6. With the help of RF noise and S-parameter measurements at the transmission lines, the best contact material and pre-handling regarding built-in ESD and electromigration hardness, can be found out.

3.3.5 Summary

The applications of reliability indicators on the production line may well depend upon factors such as equipment required, cost, speed of application and the complexities of test pattern generation. The improvements in reliability might justify the additional cost of these indicators, since they would replace the existing reliability screens and also save considerable time. Many of the indicators might find only limited use in practical applications in a commercial environment. The noise and transient voltage tests are two such examples. However, as products become more and more competitive and the demand for quality and reliability increases, reliability indicators should become an integral part of the manufacturing process.

REFERENCES

[1] Pecht, M. and Kang, W.C., "A critique of Mil-Hdbk-217E reliability prediction methods", *IEEE Transactions*, **37**, Dec. 1988, 453.

[2] Feduccia, A.J., "Reliability prediction — use it wisely", *IEEE Transactions*, **37**, Dec. 1988, 457.

[3] Jenson, F. and Moltoft, J., "Reliability indicators", *Quality and Reliability Engineering International*, **2**, 1986, 39-44.

[4] Luchino, A.I., "Rejection of unreliable items by their informative parameters", *Reliability Technology — Theory and Application*, 1986.

[5] Richardson, A.M. and Dorey, A.P., "Reliability indicators", *3rd European Symposium on Reliability of Electron Devices, Failure Physics and Analysis*, October 1992, 277-281.

[6] Gallesio, A., Marchisio, L. and Riva, D., "Emission microscopy as a tool for the reliability assessment of AIGaAs HEMTs", *3rd European Symposium on Reliability of Electron Devices, Failure Physics and Analysis*, October 1992, 161-165.

[7] Kiefer, S. and Oyler, M., "Evaluation of gate effects in GaAs MESFETs by emission microscopy", *Proc. ISTEA*, 1991, 325-334.

[8] Wallinger, T., "Characterization of device structure by spectral analysis of photoemission", *Proc. ISTFA*, 1991, 325-334.

[9] Zappe, H.P. and As, D.J., "Carrier transport in HEMTs analyzed by high field electroluminescence", *IEEE Electron Device Letters*, **12**(11), 1991, 590-592.

[10] Keane, M., "IDDQ current monitoring in CMOS circuits for testing and reliability", *3rd European Symposium on Reliability of Electron Devices, Failure Physics and Analysis*, October 1992, 285-286.

[11] Bock, K. and Hartnagel, L., "Test structures and procedures for in-process monitoring and noise measurement based detection and analysis of device damage caused by ESD", *3rd European Symposium on Reliability of Electron Devices, Failure Physics and Analysis*, October 1992, 407-410.

4

MANUFACTURING OF MICROCIRCUITS

4.1 INTRODUCTION

The manufacture of a large number of circuit elements on silicon or GaAs chips has profoundly revolutionized the information industry, the communications industry and even the entertainment industry. The industrial advances have been due to our ability to manufacture many components per circuit per chip. We have exceeded one million transistors and are approaching 100 million transistors per circuit, as shown in Figure 4.1. In 1990 we surpassed one million components per circuit, which introduced the period of ULSI (ultra large-scale integration).

Technological developments in the manufacture of integrated circuits have concentrated on increasing production yield along with the reduction of density of defects per chip. Of course, the ability to pack a large number of transistors has

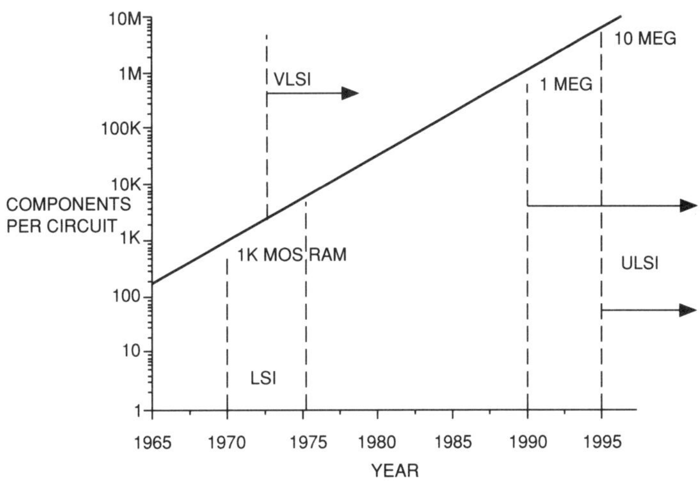

Figure 4.1 The progress of components per circuit through the years including projections up to 1995.

required chip manufacturers to reduce dimensions. This also required the reduction of defect density and the introduction of volume production along with computer integrated manufacturing with appropriate data bases. Volume production and cost track each other, and as industry's production exceeded 2000 million, the average price per unit fell below 0.50 dollars, as shown in Figure 4.2. The unit production level exceeds 20 billion circuits, and is therefore an indication of the relationship between cost and production.

Since each circuit may be reduced to a number of bits, the cost may be expressed as cost per unit shown in Figure 4.3. The cost per bit of information has declined rapidly after 1980. Therefore, the manufacture of microcircuits depends critically on the volume and density of circuit elements such as resistor, diode and transistor, and their efficiency, especially in digital applications. We will now review the basic device building blocks of an integrated circuit.

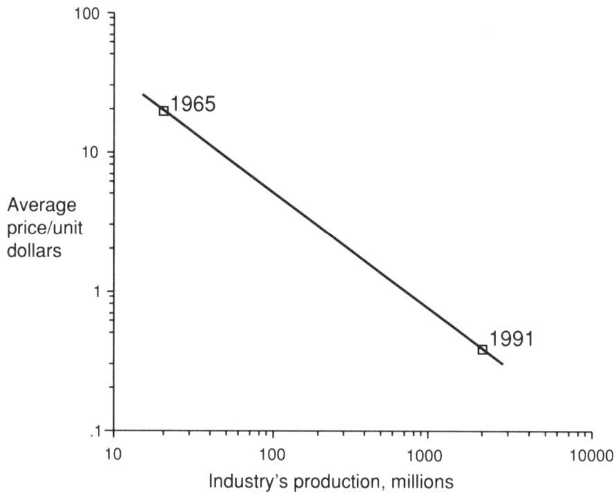

Figure 4.2 The effect of volume production on a circuit's average price per unit.

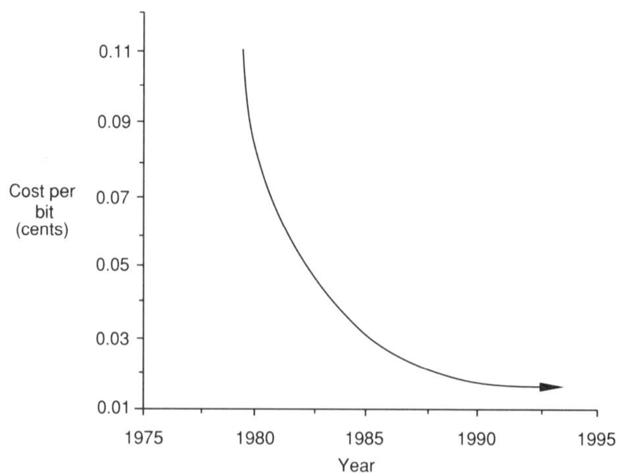

Figure 4.3 The reduction in the cost per bit.

4.2 MICROELECTRONIC CIRCUIT ELEMENTS

The microcircuit is based on three elements listed below:

(a) *Resistors* change the value of the current flow in a linear way and can shift levels of information in a linear way.

(b) *Diodes* provide asymmetry to the signal. Depending on how it is connected it can either block or transmit a signal.

(c) *Transistors* are also asymmetric but are capable of amplification.

The diode and transistor each consists of similar p-type and n-type junctions, as shown in Figure 4.4. Further clarification is shown in Figure 4.5 for both.

The semiconductor devices for microcircuits are summarized in Figure 4.5. The materials necessary for VLSI and ULSI manufacture must be discussed in terms of the energy bands. Figure 4.6 shows the basic band structure which determines the devices' electrical characteristics.

Three types of materials make up microcircuits: metals, insulators and semiconductors, as shown in Figure 4.6. The basic difference among the materials is the magnitude of the energy gap. Metallic materials form the conducting element, while the insulator is the dielectric which provides isolation between transistor elements.

The manufacture of a microcircuit is a totally integrated process, which starts with the growth of the crystal and proceeds to packaging. It is noted that yield and reliability are the driving forces for the success of any manufacturing scheme for a new technology. Yield must be maximized for each processing step while maintaining reliability in excess of 10^7 hours. The die process steps are the most complex, as shown in Figure 4.7. A typical manufacturing flow starts with the conception of a new circuit and ends with the final acceptance test. The die processing sequences then proceed from crystal growth up to the final metallization step. The oxidation process provides for isolation while lithography and plasma etching allow for the pattern to be transferred to the semiconductor. Diffusion

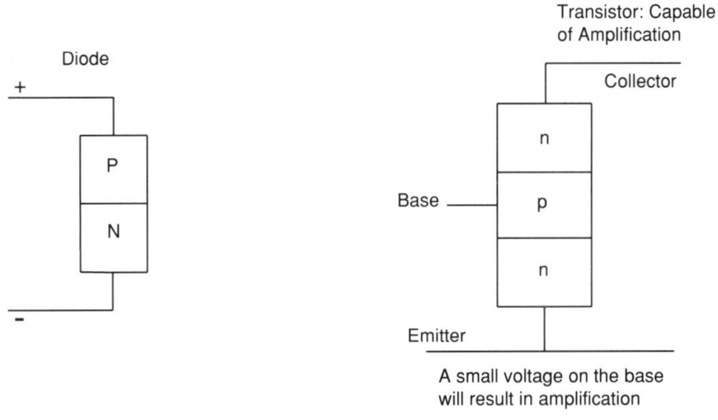

Figure 4.4 Schematic of diode and transistor elements. The transistor can either be a surface device or a bulk type depending on its fabrication.

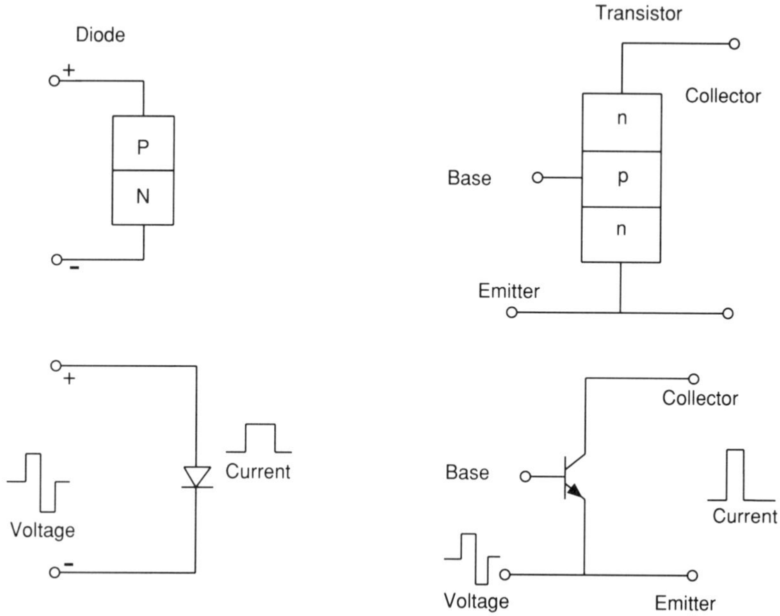

Figure 4.5 Semiconductor devices from which microelectronic circuits are constructed include the diode (left) and the transistor. Both devices are made of apposed regions of silicon doped with impurity elements; the two kinds of doped silicon are called n-type and p-type.

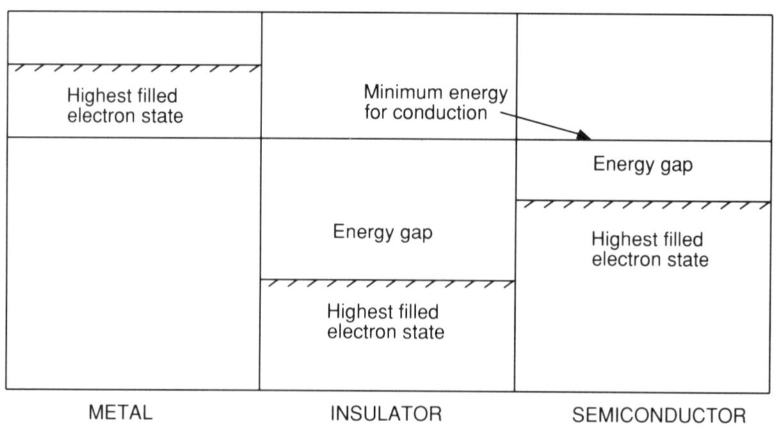

Figure 4.6 The electronic structure of metals, insulators and semiconductors determines their electrical properties. The highest possible filled electron state is shown for each type of material, along with the position of the minimum energy for conduction or current flow.

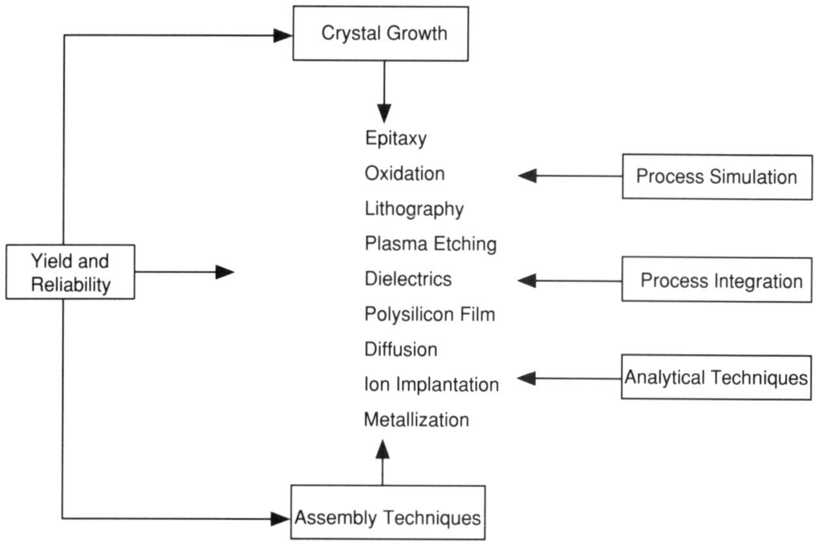

Figure 4.7 A summary of microcircuit manufacturing sequences.

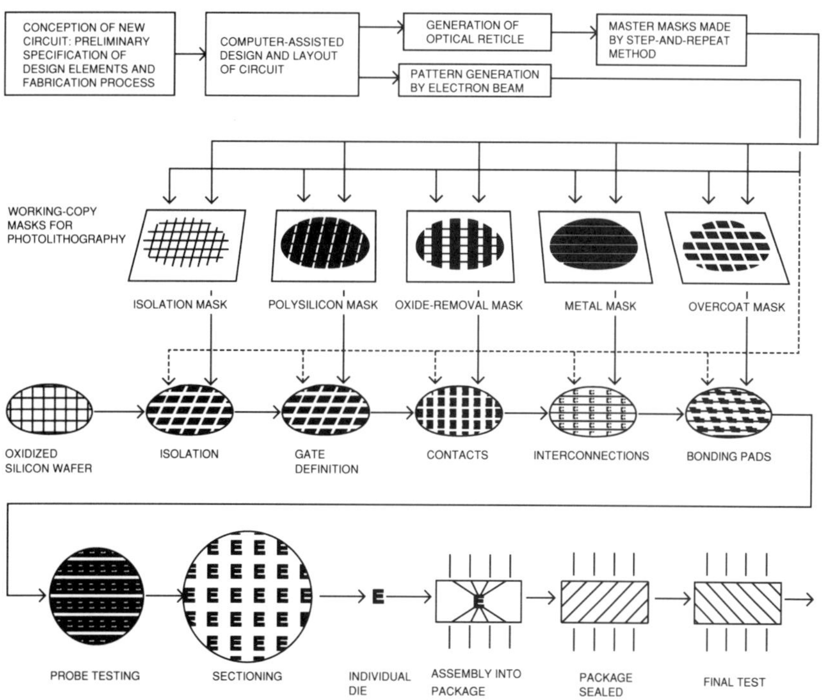

Figure 4.8 Process flow for a CMOS microcircuit (from [1]).

and ion implantation are doping techniques which allow the diode and transistor elements to become electrically active. These manufacturing processes will be discussed in the following chapters.

The overview of a manufacturing sequence indicates that a minimum of seven processing steps are required before the die is ready for final assembly and packaging. Seven process steps are required for CMOS type die, and may be increased for more complicated bipolar circuits. Figure 4.8 summarizes the manufacture process flow up to the packaging steps. Also shown in the process flow of Figure 4.8 are the required masks for photolithography. The initial process starts with the circuit and the CAD layout of the circuit.

4.3 VLSI DEVICE FUNDAMENTALS

4.3.1 Materials

Materials are classified according to their bandgap, as discussed in the previous section, and their ability to carry current as semiconductors, conductors and insulators. In semiconductors, conduction arises from: (1) thermal excitation, (2) optical excitation, (3) donor and acceptor impurities, and (4) lattice defects. The semiconductors themselves may be elemental (Si, Ge) or compound (GaAs, InP). Silicon is the dominant semiconductor presently used in industry with GaAs also being used for optical and microwave applications. These materials have the following descriptions.

Silicon The crystal structure is diamond with the distance between nearest neighbors being 2.35 Å. The silicon atom has 4 valence electrons that provide covalent bonding with the nearest-neighbor atoms. Silicon can be operated or processed up to 175 °C. In addition, Si has a thermally grown oxide which is an excellent dielectric.

Germanium The crystal structure is also diamond, however the nearest-neighbor distance is 2.45 Å. Germanium, however, does not have a stable thermal oxide for monolithic integration.

GaAs The crystal structure for GaAs is zinc blended and the energy gap is larger than for silicon, 1.43 eV to 1.12 eV.

4.3.2 Device classification

Microelectronic transistors are fabricated in a single crystal of silicon through a series of operations that require access to only one surface of the silicon chip. In the example discussed here, the entire chip is doped with a p-type impurity, then islands, of n-type silicon are formed. Smaller p-type and n-type areas are next created within these islands, in order to define the three fundamental elements of the transistor: the emitter, the base and the collector. In an n–p–n transistor, a positive voltage is applied to the base and the collector, and as a result holes flow from the base to the emitter and electrons are injected by the emitter into the base. Many of the injected electrons, however, migrate all the way through the base to reach the collector, and this emitter-to-collector current can be much larger than the emitter-to-base current. The device exhibits gain because a small signal applied to the base can control a large one at the collector. For simplicity in manufacturing, a p–n–p transistor is generally constructed according to a different plan, in which the emitter-to-collector current is lateral rather than vertical. The principles of operation are the same, except that all polarities are reversed. Connections to the circuit elements are made through aluminum conductors deposited over an insulating layer of silicon dioxide; some areas of n-type silicon are heavily doped to improve their conductivity. The large n-type islands are required to isolate the transistors. Because charge carriers of both polarities participate in the operation of these devices, they are called bipolar transistors.

Field-effect transistors, made by the metal-oxide semiconductor (MOS) technique, differ from bipolar ones in that only one kind of charge carrier is active in a single device. Those that employ electrons are called n-MOS transistors; those that employ holes are p-MOS transistors. In an n-MOS device two islands of n-type silicon, called the source and the drain, are formed in a p-type substrate. Over the channel between the source and the drain is a metal electrode, the gate, that is prevented from making contact with the semiconductor by a thin layer of silicon dioxide. In an enhancement-mode transistor, a positive potential at the drain exerts an attractive force on electrons available from the source, but the electrons cannot pass through the p-type channel, which is rich in holes. When a positive charge is applied to the gate, however, the resulting electric field attracts electrons to a thin layer at the surface of the channel and current flows from the source to the drain. In a depletion-mode n-MOS transistor there is a continuous channel of n-type silicon between the source and the drain, so that the transistor normally conducts; only when a negative voltage is applied to the gate is the current halted, since electrons are then expelled from the channel. The corresponding p-MOS devices can be constructed by forming islands of p-type material in an n-type substrate; here the charge carriers are holes. Because MOS transistors require no isolation islands, they can be packed more densely on a chip of silicon than bipolar transistors.

Complementary MOS devices include both n-MOS and p-MOS transistors on a single chip of silicon. If the circuit is fabricated in an n-type substrate, as discussed here, then a p-channel transistor can be made in the normal way. However, an n-channel device requires an island of p-type material. The need for such islands adds another step to the manufacturing process and also reduces the packing density, and thus the total number of transistors can be fitted onto a single chip. Complementary MOS devices can be arranged to achieve low power consumption. In the circuit shown here the gates of both transistors are connected to a single input; since the two transistors require signals of opposite polarity for conduction, they are never turned on at the same time and little current flows from the power supply (V+) to ground. A cross-section of a CMOS device is shown in Figure 4.9.

Electronic logic gates evaluate arithmetic and logic expressions in which binary values are represented by voltages. By convention, a binary 1 is represented by a high voltage and a 0 by a low voltage. The gates shown here are constructed from metal-oxide semiconductor transistors. The simplest is the "not" gate, or inverter. When the input to this gate is in the low state, the transistor does not conduct and only a negligible current flows from the supply voltage (V+) through the resistor and transistor to ground. As a result, there is little voltage drop across the resistor

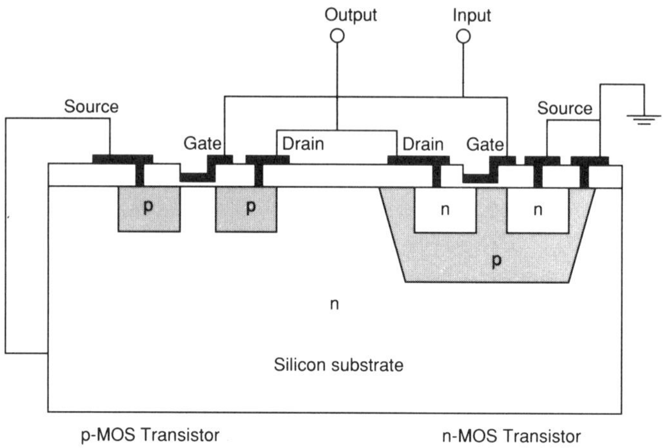

Figure 4.9 Cross section of a CMOS device where the nMOS is constructed by a p diffusion into n-type silicon (from [1]).

and the output is in effect connected to the supply voltage. When a high signal is applied to the input, the transistor conducts and the comparatively large current flowing through the circuit produces a considerable voltage drop across the resistor. The output voltage is now near ground and is therefore in the low state. The "and" gate has two input transistors connected in series; current flows through them only when both receive a high signal simultaneously. In order to restore the proper polarity of the signal, the output of the two series transistors is followed by an inverter. In an "or" gate the input transistors are arranged in parallel, so that a high signal applied to either one of them results in conduction. Again, an inverter is required to change the polarity of the output. The logic gate configuration is shown in Figure 4.10.

The equations which govern a field effect transistor describe the condition of modulating the conductivity of a semiconductor by applying an electric field normal to its surface. The control electrode is the gate, the active region is the channel,

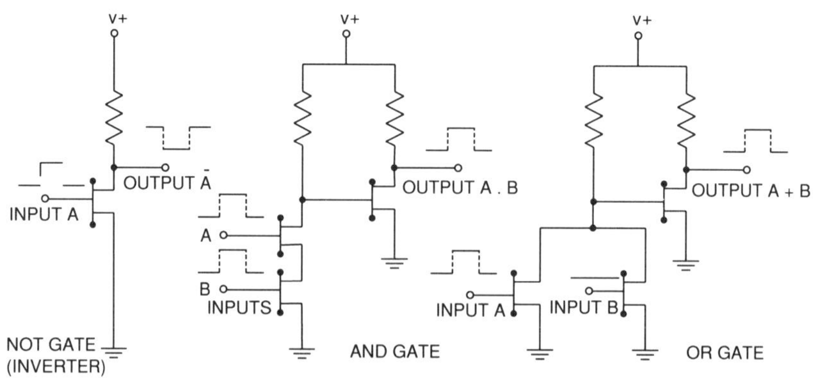

Figure 4.10 Logic gates constructed by combining transistors.

while the two terminals which connect the modulated conductor are the drain and source. The JFET was first realized in 1952; two equations describe its operation.

In the linear region (before channel pinch off),

$$I_d = K\left\{V_{ds} - \frac{2}{3}(V_{bi} - V_p)\left[\left(\frac{V_{ds} + V_{bi} - V_{gs}}{V_{bi} - V_p}\right)^{3/2} - \left(\frac{V_{bi} - V_{gs}}{V_{bi} - V_p}\right)^{3/2}\right]\right\}$$

where K = device constant,
V_{ds} = drain to source voltage,
V_{gs} = gate to source voltage,
V_{bi} = built in potential,
V_p = pinch-off voltage.

In the saturation region

$$I_{d(\mathrm{sat})} = K\left\{V_{gs} - V_p - \frac{2}{3}(V_{bi} - V_p)\left[1 - \left(\frac{V_{bi} - V_{gs}}{V_{bi} - V_p}\right)^{3/2}\right]\right\}$$

If the saturated drain current at $V_{gs} = 0$ is equal to I_{dss}, then the drain saturation current is

$$I_{d(\mathrm{sat})} = I_{dss}\left(1 - \frac{V_{gs}}{V_p}\right)^2$$

The MOS structure, on the other hand, is the basis of the MOSFET and CMOS. The key device parameters in the linear region are

$$I_d = \frac{k}{2}[2(V_{gs} - V_t)V_{ds} - V_{ds}^2]$$

In the saturated region,

$$I_d = \frac{k}{2}(V_{gs} - V_t)^2$$

where V_t is the threshold voltage.

The bipolar junction transistors are n–p–n or p–n–p and the three terminals of the transistor are called the emitter, base and collector. Unlike the FET, the BJT is a minority carrier device capable of current gain, voltage gain, power gain, and a transconductance function. The common-emitter current gain β is given by

$$\beta = \frac{I_c}{I_b}$$

The base-emitter voltage in the normal region of transistor operation is given by

$$V_{be} = V_T \ln\left(\frac{I_e}{I_{es}}\right) + I_b r_b + I_e r_e$$

The emitter current I_e is described by the Ebers equation

$$I_e = I_{es}(\mathrm{e}^{V_{be}/V_T} - 1) - \alpha_R I_{cs}(\mathrm{e}^{V_{bc}/V_T} - 1)$$

where I_{es}, I_{cs} are the emitter and collector junction saturation currents, and V_T is the thermal voltage. The key device parameters are: size, input capacitance, transit time, and threshold voltage.

4.3.2.1 Interconnect materials

The materials commonly used to provide interconnect functions are metals, refractory metals, silicides, polycides, and polysilicon.

The key parameters are linewidth, line separation, the distributed resistance and the distributed capacitance.

Polysilicon This is used as a gate material in MOS technology, because of the high sheet resistance currently available in polysilicon.

Silicides These materials are used for Schottky barriers and for interconnects.

Refractory metals These materials are used because of their ability to adhere to SiO_2. Refractory metals are also common in FET technologies.

4.3.3 VLSI device parameters

Devices, circuits, and interconnects (networks) are the basic building blocks of an IC device density, circuit integration level, circuit switching speed, IC power consumption, and IC reliability are the key figures of merit. All of these are affected by the physical dimensions.

The device dimensions may be further separated between sizes for active devices and those for interconnects. The key device dimension is the minimum feature size and the minimum device area. Minimum feature sizes are presently at the 0.5–1.0 μm level.

4.3.3.1 Interconnect dimensions

(a) *Minimum linewidth* 1–2 μm can be achieved with present technology including lithography and reactive ion etching.

(b) *Minimum separation* The primary motivation is to achieve the maximum density. The line separation is usually three to four times the linewidth.

(c) *Thickness* The thickness must be small enough so as to be able to cover steps on the wafer. Thicknesses range from 0.5 μm to 1.5 μm.

4.3.4. Device models and parameters

Circuit simulation programs are used for nonlinear DC, nonlinear transient and linear AC analysis. The common circuit simulation model for VLSI design is the SPICE program. The newest version is the SPICE 3, used by most designers. Various types of device models are utilized in SPICE.

SPICE 2 BJT model This is the basic large signal bipolar model described by Ebers-Moll. For better transient analysis, charge storage effects must be added. The BJT model based on integral charge is the Gummel–Poon model.
SPICE 2 JFET model This is the basic junction FT model which gives the transconductance. Likewise there is a basic SPICE 2 MOSFET model, which usually predicts the DC characteristics.
SPICE 3 MESFET model The MESFET model in SPICE 3 is a basic GaAs FET model which is presently available in an updated version of SPICE.

4.4 DATA-BASE MANAGEMENT SYSTEM FOR IC MANUFACTURE

The Microcircuit Data-base Management System (DBMS) used for IC manufacturing combines an existing data-base management system with a relational distributed data-base management system to promote data/information dissemination to processing personnel in the most efficient manner. This system not only improves the efficiency of data exchange among design, manufacturing and test, but also provides risk and cost assessment to increase the efficiency of the forward flow of information and enhance information feedback. An implementation plan is described which has combined existing hardware into an interactive data base among all participants of the microcircuit industry. This management system is described in order to indicate the requirements for manufacturing of microcircuits. The description is included at this point in order to understand the interrelationships between the subsequent processing steps, and the requirement to combine the processing steps in a totally integrated environment.

Efficient and cost-effective dissemination of critical data/information to microcircuit processing personnel will require a comprehensive data-base management and network system. A customized DBMS, fully implemented, is a vital factor in producing high-reliability, low-cost microcircuit chips such as VLSI, MMICs and optical interconnects.

Microcircuit manufacturing comprises various disciplines and groups interacting in such areas as system analysis, modeling, circuit design, mask layouts, materials, processing, fabrication, testing, quality control, packaging, inventory control, etc. Each group conducts a set of specific tasks; therefore, they have unique data management and data transfer needs. Many of these groups use different hardware, operating systems, and software to acquire, generate, manipulate, and manage the data. Moreover, microcircuit personnel are located at many different facilities.

A simplified model of a typical design/production cycle along with the information flow paths are shown in Figure 4.11. The cycle proceeds in a clockwise direction and encompasses various design, manufacturing, and test efforts shown at different radial distances. The solid lines are data flow paths that are vital to chip processing, while the dashed lines show paths where information/data transfer is either informal or non-existent, but is important for improving yield and performance and reducing cost. This model illustrates the inefficiency of current systems.

In contrast, the model shown in Figure 4.12 illustrates how information flow can be improved for manufacturing intensive IC programs. This production cycle involves the same steps as in Figure 4.11, and includes implementation of a logically unified (although physically distributed) data base. The manual or physical

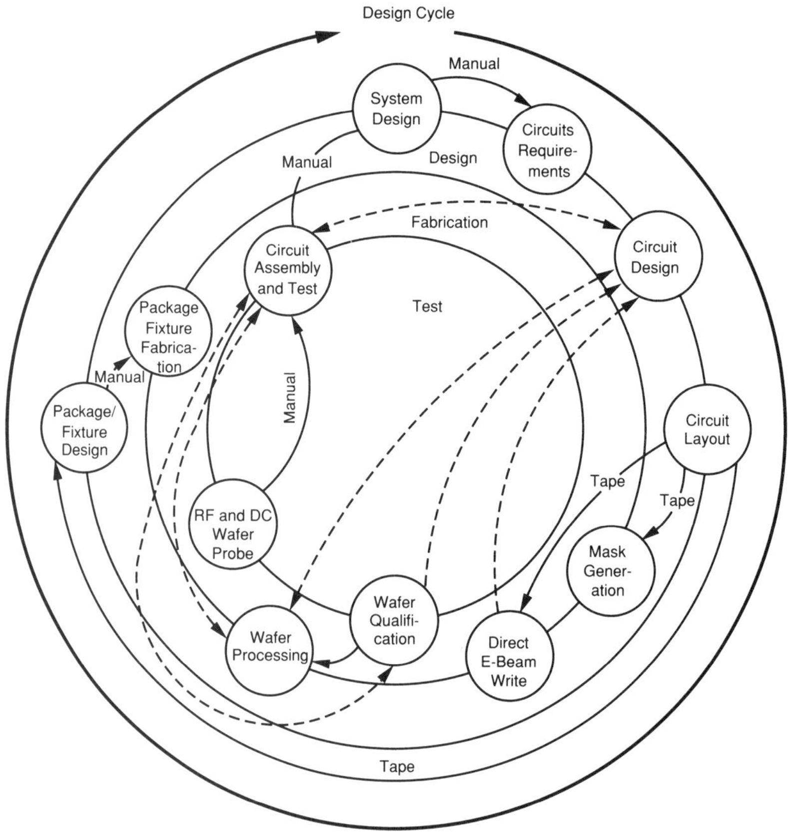

Figure 4.11 Design/production cycle and information flow path.

transfer of information along the forward flow paths has been replaced by more cost-effective electronic data transfer. This model also provides feedback paths for:

(a) Developing and verifying processing and qualification models.

(b) Isolating the cause(s) of parameter variations within and between wafers and from wafer-to-wafer and boule to boule.

(c) Indicating whether changes in circuit performance are due to processing equipment anomaly, inadequate circuit element models, variations in the active layer, changes in the material properties, etc.

The DBMS forms a part of a more comprehensive/computer integrated manufacturing (CIM) system that will be critical for providing a state-of-the-art design and manufacturing capability for microcircuits in the 1990s and beyond. A detailed CIM architecture development must be undertaken to prevent the growth of isolated islands of automation of DBMS, CAD, computer-aided manufacturing (CAM), computer-aided testing (CAT), material resource planning (MRP), etc. A detailed CIM architecture development provides coherent, rational, and timely implementation of the individual components of CIM.

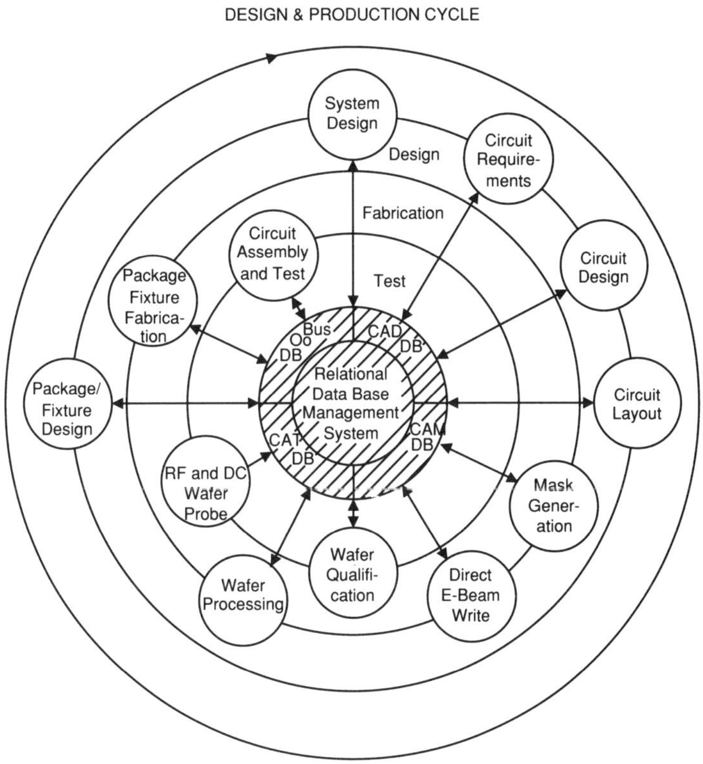

Figure 4.12 A typical improved microcircuit integrated data access system. The local DBMS (outer ring) is combined with a relational data base (inner ring).

4.4.1 Computer integrated manufacturing

The development of an overall CIM architecture provides the long-term roadmap for implementing computer-aided automation and data-base management approaches. This development integrates microcircuit design, manufacturing, packaging, testing, and business management operations. A critical part of any CIM development and implementation plan is a close examination of the procedures, information processes, and nature of the business enterprise. Benefits of CIM are many. For example, rapid access to timely, accurate data will allow process engineers to control manufacturing operations, provide designers with accurate data, and provide managers the tools they need to efficiently allocate resources, pinpoint troubled operations, and minimize manufacturing costs.

The detailed CIM architecture will provide a roadmap for rational, timely implementation of individual islands of automation, such as the DBMS, CAD, CAM, CAT, and MRP. Following the roadmap will ensure that the implemented islands of automation are fully integrable at a later date. Automation of the microcircuit production operations can be planned in phases. The areas with the most pressing need for coordination and data transfer (such as design, manufacture and test) should be integrated first (unshaded areas of Figure 4.13). IDEF functional

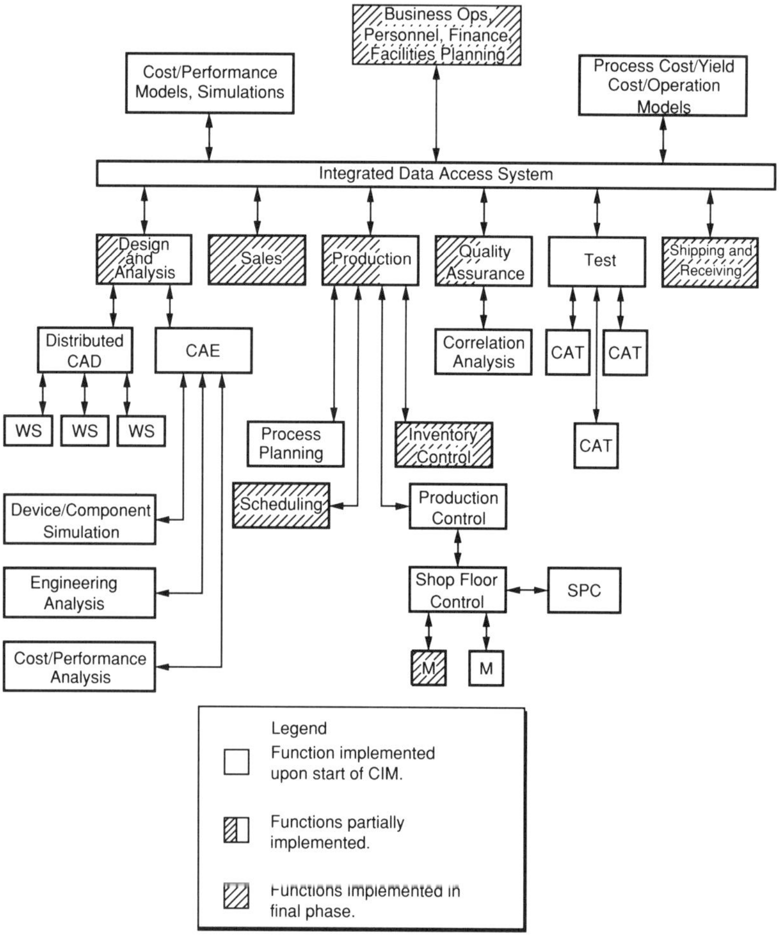

Figure 4.13 CIM architecture for microcircuit manufacture. M denotes shop floor machine and WS denotes workstation.

methodology may be used to aid in analyzing the microcircuit business and method of operations. The costs, benefits, and technical problems of fully implementing the microcircuit CIM environment may be analyzed.

Based on payoffs from full CIM implementations in other industries, we can ultimately expect 30 to 60% reduction in lead times, 40 to 60% reduction in planning activities, productivity increases of from 5 to 30%, and a 30 to 50% reduction in inventories. These CIM approaches for microelectronics are being fully implemented in the 1990s and will become the focal point for quality improvement and cost reductions.

4.4.2 Data-base management system and data correlation

The microcircuit integrated data access system, developed and demonstrated by many companies along with state-of-the-art networking, is often undertaken for

Table 4.1 Microcircuit correlation analysis (based on FETs).

Type of correlation analysis	How is the analysis used?	Data involved	Correlation output	Where will the analysis be performed?
Layout design	To determine the effect of active and passive device/circuit geometrical layout on electrical performance	Geometrical parameters from layout design	FET parasitic capacitances and high-frequency performance Characteristics of transmission lines	Mask layout design station
		Gate length of FET Pad dimensions and separations in a FET	Gain, maximum operation, and noise figure capacitances affecting maximum frequency of operation	
		Microstrip line width	Characteristics impedance of a microstrip transmission line	
Packaging	To determine effect of package design on device/circuit performance	Package and fixture design Heat sink design	Electrical performance of the packaged circuit power performance measures	Packaging
Design	To determine the effect of circuit design on electrical performance	Device characteristics and circuit elements; CAD data and procedures	Simulation results and design predictions	CAD design stations
Material properties	To characterize the influence of starting material on device/circuit performance	Epitaxial layer doping and mobility	Pinch-off voltage of FET Gain and noise figure	Material station
		Trap density	Flicker noise of FET and phase noise of an oscillator	

continued overleaf

Table 4.1 *(continued)*

Type of correlation analysis	How is the analysis used?	Data involved	Correlation output	Where will the analysis be performed?
Processing parameters	To determine the effect of various processing steps on device/circuit performance	Processing for a given circuit design	Device/circuit parameters required to obtain a given function	Processing station
		Wafer thickness	Characteristic impedance of transmission lines	
		Wet and dry etch rate and timing	Pinch-off voltage of FET	
		Ohmic metallization and annealing	Parasitic resistances of a FET	
		Ion implantation, dose energy and uniformity	DC and RF performance of FET	
		Via hole diameter	Ground inductance	

implementation during manufacturing to promote efficient data/information dissemination among microcircuit fabrication components.

DBMS ties together the individual CIM elements. In an initial implementation stage, the investigation, development and linkage of both local and global database management systems should be undertaken. This will link team members of manufacturing together in a common data network. The increased efficiency in data/information transfer between fabrication components results in rapid quality improvement.

The microcircuit integrated data access system combines local data bases and a commercially available relational distributed data-base management system (RDDBMS) to form one logical data base. Figure 4.12 depicts this approach. The data generated in various microcircuit operations, such as design, production and test will be resident on one or more local data bases. The RDDBMS will contain summarized and synthesized data from all steps and disciplines. The RDDBMS will ensure the management of data exchange among CAD, CAM, and CAT operations either at one location or over geographically distributed sites, as shown in the CIM architecture of Figure 4.13. The elements of the CIM system are grouped by an integrated data access system.

4.4.3 Correlation analysis

The CIM approach identifies correlations among the design, production and test parameters to improve the yield, reliability and performance of microcircuits. A partial list is shown in Table 4.1. The correlations are in the area of design, packaging material parameters and processing parameters. The correlations are then used at critical stages in order to optimize yield and to ensure the manufacturability of a product. In the chapters which follow, the manufacturing technology of each microcircuit element is presented up to and including packaging. A chapter on robotics for IC manufacturing reviews the application of computer aided (CAM) manufacturing techniques to assembly.

REFERENCES

[1] Meindl, J.D., *Microelectronics*, Scientific American, Freeman and Co., San Francisco, 1977 Sept., pp. 12–23.

[2] Sze, S.M., *VLSI Technology*, McGraw-Hill, New York, 1988, pp. 422–462.

5

CRYSTAL GROWTH AND SUBSTRATE MECHANISMS

5.1 SILICON STARTING MATERIAL

In this chapter, defects present in the semiconductor substrate and which affect yield, performance and reliability are evaluated. Polycrystalline, high purity silicon is used as the initial starting material in the production of single crystal silicon. The initial reaction involves the removal of carbon which is present in the starting material in the form of a carbide.

$$\mathrm{SiC} + \mathrm{SiO_2} \longrightarrow \mathrm{Si(liquid)} + \mathrm{SiO} + \mathrm{CO} \tag{5.1}$$

The melting reaction shown in equation (5.1) results in molten silicon and gases SiO and CO. The subsequent reaction is between silicon and HCl to form a silane and a carrier gas of hydrogen, as shown by equation (5.2).

$$\mathrm{Si(solid)} + 3\mathrm{HCl(gas)} - \mathrm{SiHCl_3(gas)} + \mathrm{H_2(gas)} \tag{5.2}$$

The carrier gas is H_2 and the reaction with a solid surface results in the solidification of silicon.

$$2\mathrm{SiHCl} + 2\mathrm{H_2(gas)} \longrightarrow 2\mathrm{Si(solid)} + 2\mathrm{HCl} + 2\mathrm{H_2(gas)} \tag{5.3}$$

A slim rod of silicon is used to nucleate the reaction. The major change during growth is from liquid to solid, resulting in a pure solid form of polycrystalline silicon.

During the growth of single crystal silicon, three types of defects are incorporated in the silicon material. These defects are: point defects (vacancies), dislocations, and stacking faults. The concentration of vacancies is given by

$$N_d = A\exp(-E_a/kT)$$

where $E_a = 2.6$ eV to form a vacancy in silicon. The activation energy of 2.6 eV indicates that a significant concentration of vacancies may be quenched only by rapid solidification from temperatures in excess of 1000 °C.

5.2 CRYSTAL GROWTH THEORY

The solidification of silicon from the molten state is based on heat transfer and the latent heat of fusion. The liquid–solid monocomponent growth system is described in Figure 5.1, indicating in a schematic the temperature gradients across the solid-liquid interface.

One initially considers the heat transfer across each interface in order to obtain a solidification velocity. The heat transfer across the interface is given by equation (5.4):

$$L\frac{\mathrm{d}m}{\mathrm{d}t} + k_l\frac{\mathrm{d}T}{\mathrm{d}x_1}A_1 = k_s\frac{\mathrm{d}T}{\mathrm{d}x_2}A_2 \tag{5.4}$$

where L = latent heat of fusion,
$\mathrm{d}m/\mathrm{d}t$ = mass solidification rate,
k_l = thermal conductivities of liquid,
k_s = thermal conductivities of solid.

If $\mathrm{d}T/\mathrm{d}x_1 = 0$ then $V_{\max} = k_s/L\rho_{Si}(\mathrm{d}T/\mathrm{d}x)$ represents the maximum solidification rate, or the crystal pull velocity, where ρ_{Si} = density of silicon.

Figure 5.2 shows the schematic representation of the crystal pulled at the velocity $V_{\max}$. The maximum velocity for the moving interface is also the pull rate given by:

$$V_{\mathrm{Max}} = \frac{k_s}{L\rho_{Si}}\frac{\mathrm{d}T}{\mathrm{d}x} = \text{pull rate}$$

The following critical points must be noted concerning the pull rate:

(a) V_{Max} (pull rate) influences the incorporation of impurities, especially vacancies.

(b) The growth rate influences dislocations and defect density, i.e., 2 mm/min eliminates vacancies. A fast growth rate may result in voids and inclusions.

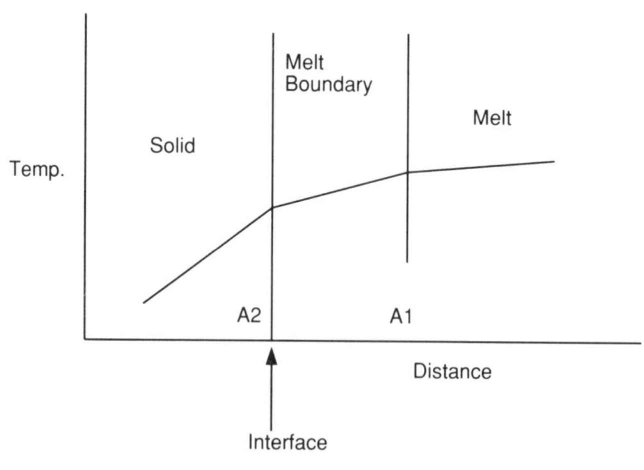

Figure 5.1 Temperature gradients across the solid–liquid interface during solidification of silicon.

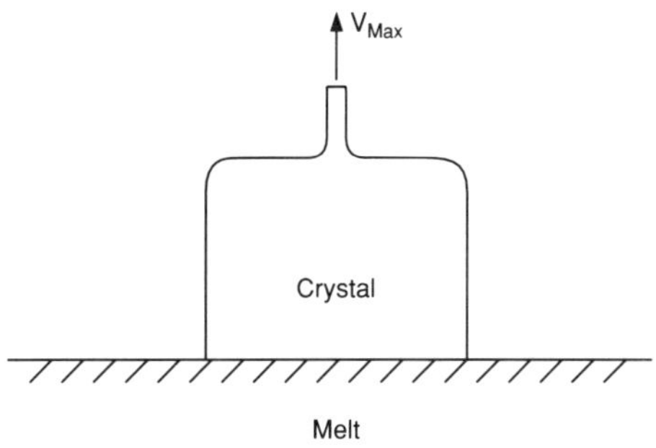

Figure 5.2 Silicon crystal pulled from the melt as a single crystal.

(c) Remelt must be suppressed since this produces defects and redistributes impurities. Remelt occurs when V_{max} is smaller than the equilibrium rate.

Distribution of impurities (dopants) is determined by the segregation coefficient k_0. This coefficient describes how defects leave a solid solution, and may be determined from the following equation:

$$c_s = k_0 c_0 (1 - x)^{k_0 - 1} \tag{5.5}$$

where x = fraction of the melt solidified,
c_0 = initial melt concentration,
c_s = solid concentration,
k_0 = segregation coefficient.

The practical aspects of crystal growth may be reduced to selecting the proper growth rate based on the material's latent heat, density and thermal conductivity. Previous observations for silicon have indicated that growth rates of 2–3 mm/min result in vacancy clusters.

The segregation or movement of solute atoms results in a decrease in c_s from the initial value of c_0. The value of k_0 is that which each particular solute atom has in solution with silicon. For, instance, oxygen and carbon tend to have a large k value and hence may easily form clusters and inclusions. Figure 5.3 shows the segregation coefficients and their effect on concentration as a function of solute

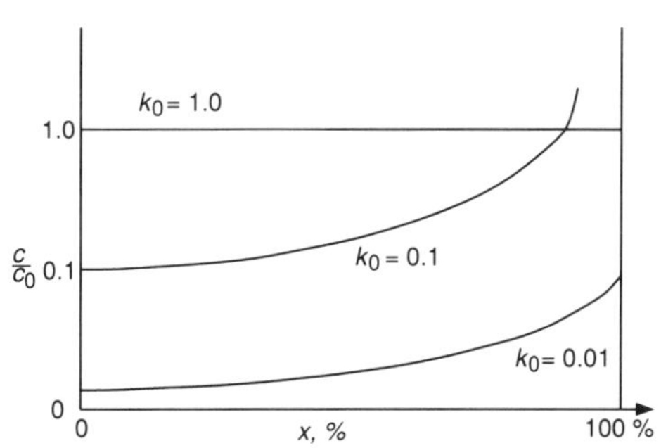

Figure 5.3 Typical segregation coefficients for solutes in silicon.

concentration. The low k value is shown to play a major role in solute atom segregation for dilute solutions.

5.3 CRYSTAL CHARACTERISTICS AND EVALUATION

After the formation of the single crystal silicon, one must quickly evaluate the crystal's physical properties. The routine evaluation carried out involves the measurement of: (1) resistivity, (2) crystal perfection and (3) mechanical properties. The resistivity is measured at both ends of the boule and is given by:

$$\rho = \frac{V}{I} 2\pi s (\Omega.\text{cm})$$

The parameter s is the distance between probes in a four-point probe measurement while V and I are the voltage and current drop across the inner probes. From measurements of resistivity, the dopant density may be determined from previous calibration.

Crystal perfection is determined by either X-ray or electron beam diffraction techniques. These techniques are quite complex and will not be discussed in the present section. After characterization of the crystals and the determination of crystal character, an identification method must be developed in order to have a reproducible method of placement of wafers in processing or fabrication machines. The exact placement of the wafer is critical with respect to its orientation since fabrication always involves a diffusion method which is highly directional.

Wafer shaping involves two operations necessary for manufacturing:

(a) Cutting with diamond in order to determine an identifying direction on the wafer.
(b) Identifying flat determination and placement on the wafer in order to prepare for manufacturing. The flats are either primary or secondary, as shown in Figure 5.4. The wafer shaping operation may result in edge cracks which would tend to decrease wafer manufacturing yield.

The placement of the primary and secondary flats quickly allows the processing engineer to identify both the orientation and doping polarity of the wafer. Table 5.1 shows typical wafer specifications which may be used for procurement. The placement of the primary flat is also used to align microcircuit elements, such as gates for MOSFETs. In various cases, the primary flat is also used to predict and identify preferential etching directions. The primary flat may also be used for

Table 5.1 Typical silicon wafer specifications.

Diameter (mm)	150±0.5
Primary flat (mm)	55.60
Secondary flat (mm)	35–40
Thickness (mm)	0.65–0.7
Bow (μm)	60
Total thickness variation (μm)	50
Surface orientation	$(100)^\circ \pm 1^\circ$

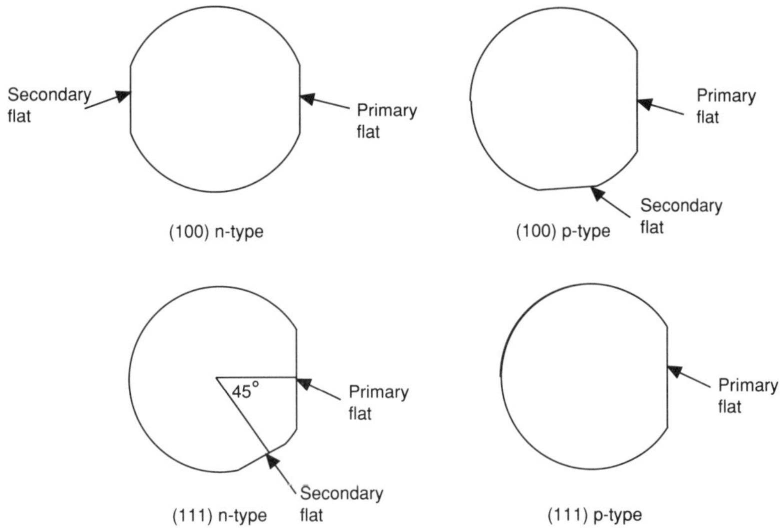

Figure 5.4 The location of primary and secondary flats in wafers.

the placement of wafers in plasma etching systems and is particularly important for wafer placement in cassette to cassette manufacturing where robotic motions can automatically identify wafer orientation and placement within a cassette. Orientation of wafers is also critical in ion implantation operations, where the effect of ion channeling is highly dependent on wafer orientation. Following the wafer flat placement, subsequent manufacturing steps include wafer etching and polishing.

Etching is the process which follows the identifying flat preparation and is intended to remove damaged and contaminated zones from the wafer. These zones may be as much as 10 μm from the surface. In addition, regions with small microcracks from the edges must be removed, since they reduce yield, especially for devices processed at the wafer periphery. Figure 5.5 shows surface and edge removal. The process necessitates the removal of damaged regions by chemical etching using mixtures of HF, nitric, and acetic acids. The semiconductor etching is an oxidation–reduction process. The acid reactions are as follows:

(1) HNO_3 (nitric) oxidizes and
(2) HF removes the oxidized products.

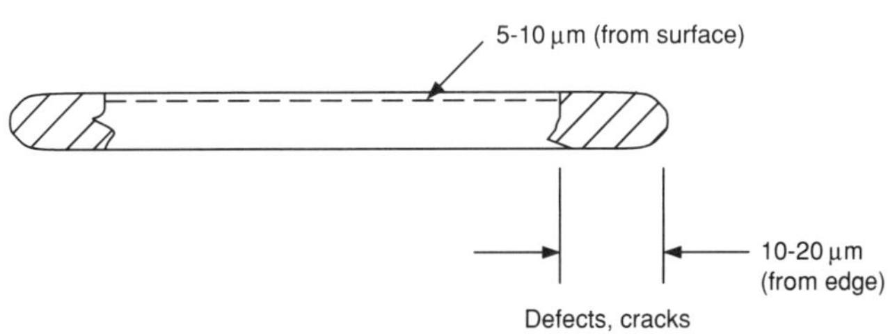

Figure 5.5 Wafer cross-section showing the effect of etching.

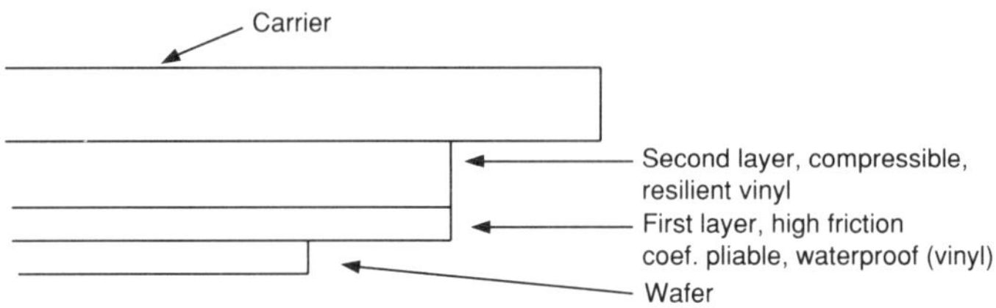

Figure 5.6 Mounting of the wafer on a carrier for polishing.

The combined action results in a newly exposed silicon surface free of microcracks and surface damage.

Polishing is the final step in producing a smooth, specular surface. The surface must be flat with a minimum of local slope. Wafers must be mounted so that no breakage occurs during polishing. A typical mounting procedure uses a two-layer vinyl system, where the first layer is of high friction material, while the second layer is resilient as shown in Figure 5.6. The effect of the mounting procedure is to reduce the stress transferred to the wafer.

5.3.1 Processing preparation

Following the wafer preparation stages, preparation for processing involves the removal of impurities using two techniques

(a) Chemical cleaning: mainly organic contaminants.
(b) Gettering treatments by wafer damage.

Both techniques are effective methods for the removal of potential failure mechanisms in bipolar transistors and MOS devices. The most common way of removing impurities is through gettering. In order to carry out a successful gettering process one must understand the location of impurities in the lattice. We note that impurities are located at interstitial or substitutional lattice sites and are recombination centers for carriers, therefore their removal is critical for good IC performance. The process of gettering consists of the following two steps:

(a) Gettering = process of removing impurities from damaged areas.
(b) Pregettering = process which provides wafer with sinks to absorb impurities.

The most common treatment creates dislocations in order to trap diffusing species at dislocation sites. Impurities are trapped in the near surface of a wafer in order to maintain the clear flat surface for processing. Two techniques are commonly used for gettering:

(1) Mechanical abrasion of wafer backside
(2) High power laser beam to create micromachined spots

The mechanical abrasion placement is precisely controlled, in order to minimize further wafer damage. In addition, excimer laser irradiation may create micromachined spots for further gettering. The processing preparation demands precautions necessary for the removal of thermal stress effects due to uneven cooling of the boule from the molten state.

During high temperature processing, the wafer is raised to temperature T. When removed from the furnace, the temperature difference results in a thermal stress gradient across the surface of the wafer given by:

$$\sigma = \alpha E \Delta T$$

α = coefficient of thermal expansion,
E = Young's modulus,
ΔT = Temperature difference between the edge of the wafer and its center.

The residual stress in the silicon wafer may be removed by annealing at 90 °C for 30 minutes.

5.3.2 Yield strength of silicon

The yield strength of silicon is affected by two factors:

(1) Precipitates in silicon, which are stress concentration areas.
(2) Temperature.

The yield strength of silicon is typically 2×10^9 dynes/cm^2 at 23 °C. The effect of temperature is related to impurity diffusion and the movement of oxygen precipitates within the silicon wafer. In the case of GaAs, mobility of carriers is also affected by changes in yield strength, but indirectly through the presence of impurities in the lattice. As shown in Figure 5.7, the critical shear strength of silicon varies with temperature and with impurity concentrations.

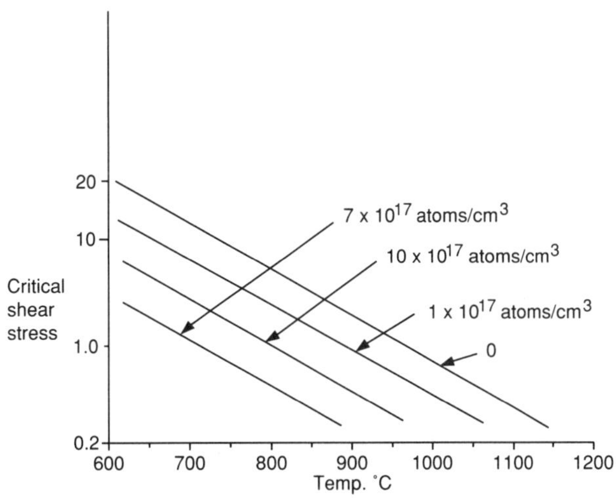

Figure 5.7 The variation of critical shear stress with temperature.

5.4 EPITAXIAL GROWTH ON SEMICONDUCTOR SUBSTRATES

Thin ordered layers on the substrate must be grown in order to achieve the low background density which is required for the manufacture of high performance VLSI circuits. The purpose of epitaxial growth is to completely remove the substrate polished surface from the active device region. Figure 5.8 shows typical dimensions of the epitaxial layer in comparison with the silicon substrate thickness.

In the case of epitaxial growth on GaAs or InP substrates, the epitaxial layer must consist of other semiconductor materials which are lattice matched to the substrate. These materials may be GaAlAs, InGaAs or InAlAs depending on the specific application for the material system. The prevalent growth techniques for the epitaxial layers consist of vapor phase epitaxy or molecular beam epitaxy. Figures 5.9 and 5.10 show plane view and cross-sections of epitaxial interfaces on compound semiconductors.

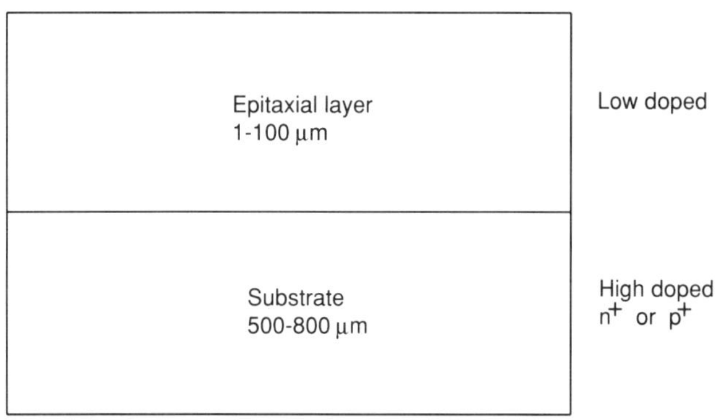

Figure 5.8 Schematic of the epitaxial layer deposited on a silicon substrate.

Figure 5.9 TEM plane view of epitaxial interfaces on compound semiconductor grown on silicon.

5.4.1 VPE of silicon

In the VPE technique silicon vapors and reaction gases flow over the silicon surface and are deposited on the substrate. By convection, the flux of atoms onto the surface is:

$$J = D\frac{\mathrm{d}n}{\mathrm{d}y} = \frac{D(n_g - n_s)}{y} = \text{molecules/A/unit}$$

where n_g = gas concentration,
n_s = surface reactant concentration.

Under steady state conditions, the flux of the reactant on the surface is equal to the chemical reaction rate given by:

$$J = k_s n_s$$

where k_s is the reaction rate constant. Equating the two fluxes allows us to solve for the surface concentration:

$$k_s n_s = \frac{D(n_g - n_s)}{y}$$

$$\text{and} \quad n_s = \frac{n_g}{1 + ky/D} = \text{Surface concentration}$$

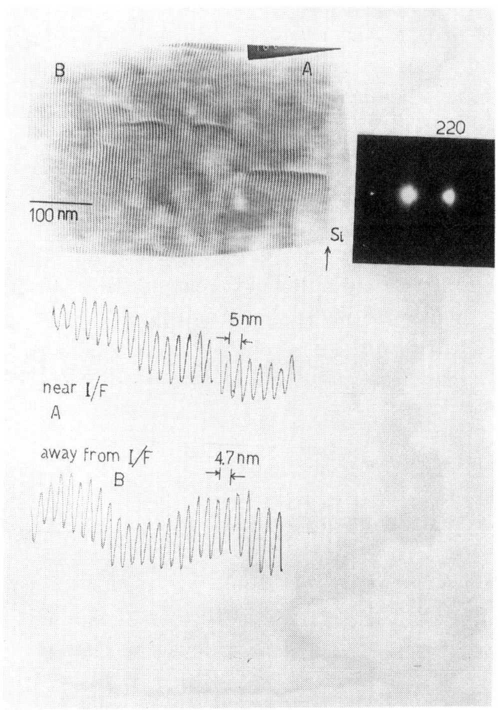

Figure 5.10 Cross-sectional TEM image of epitaxial interfaces on compound semiconductor grown on silicon.

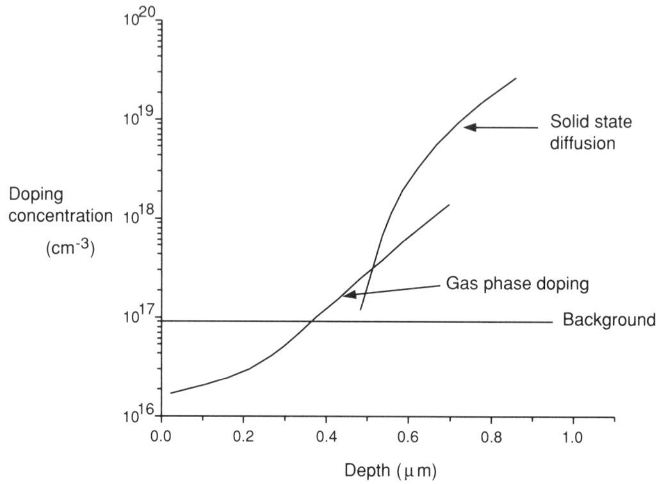

Figure 5.11 Doping concentration after dopant redistribution.

The main reaction used industrially includes silane and hydrogen:

$$\text{Main reaction: } SiCl_4(\text{gas}) + 2H_2(\text{gas}) \longrightarrow Si(\text{solid}) + 4HCl(\text{gas})$$

The above method may be extended to include the doping of the epitaxial layers. The dopant atoms are included by the same technique, first mass transport followed by the surface chemical reactions at 800–900 °C.

Usually hybrids of impurities are the source of dopant, such as arsine or phosphine:

$$2AsH_3(\text{gas}) \longrightarrow 2As(\text{solid}) + 3H_2(\text{gas})$$

In the arsine case shown above, the arsenic is incorporated in the growing epitaxial surface. Further redistribution of dopant atoms occurs by diffusion. The comparison of doping concentrations is shown in Figure 5.11.

The growth or diffusion velocity is equal to $2\sqrt{D/t}$, where t is the diffusion time in minutes, and D is the dopant diffusion coefficient.

5.4.2 Advanced CVD methods: MBE

Advanced epitaxial methods utilize Knudsen cells to produce Ga and As beams. The incident flux is As_2 in the precursor state. The GaAs surface, which is conducive for MBE growth, is the Ga stabilized surface. The incident beam is As_4 which desorbs to As_2. The sticking coefficient of As_2 is very close to one, while the sticking coefficient for As_4 is less than 0.5. The flux is given by:

$$J = \frac{1.118 \times 10^{22} PA}{l^2 (MT)^{1/2}} \text{ molecular cm}^{-2}\text{ s}^{-1}$$

where P = source pressure,

A = cell area,

l = source to substrate distance.

The flux therefore is controlled by the pressure and temperature of the Knudsen cell. A typical growth rate is one micron per hour for both doped or undoped films.

5.4.2.1 MOCVD

The technique of metallorganic chemical vapor deposition is a basic CVD technique using metallorganic gases. The reaction is as follows:

$$\mathrm{R_sGa + AsH_3 \longrightarrow GaAs + 3RH}$$

where:

$$R = \mathrm{CH_3 \text{ or } C_2H_5}$$

The substrate temperature for MOCVD growth is approximately 700 °C. The growth rate may be controlled through pressure, flow rates, and substrate temperature. In MOCVD, the kinetics must be expressed in terms of the gas-phase mass transfer coefficient. If N_0 is the concentration on the surface, then the flux toward the surface is $J_1 = h_G(N_G - N_0)$, where h_G is the gas phase mass transfer coefficient. The reaction on the surface is $J_2 = k_s N_0$. Under equilibrium conditions $J = J_1 = J_2$ or $N_0 = (h_G/h_G + k_s)N_G$

The flux during MOCVD growth is

$$J = \frac{k_s h_G}{h_G + k_s} N_G$$

and the growth rate is $\mathrm{d}y/\mathrm{d}t$ given by:

$$\frac{\mathrm{d}y}{\mathrm{d}t} = J/N = \frac{k_s h_G}{k_s + h_G}\frac{N_G}{N}$$

The surface reaction rate coefficient is $k_s(T) = B\mathrm{e}^{-(E_G/kT)}$, where E_G is the activation energy for the surface reaction. The activation energy is typically less than 1.0 eV for most of the semiconductor surfaces.

5.4.2.2 Growth mechanism for MOCVD

In order to prevent the formation of a stagnant boundary layer of gas across the surface, it is necessary to tilt the substrate with respect to the direction of gas flow, as shown in Figure 5.12. This is important, since mass transport to the growing film through this layer is the rate determining step in the growth process. For temperatures between 600 and 800 °C the growth rate is independent of temperature and is linearly dependent on the concentration of metallorganic constituent provided excess arsenic is present. Figure 5.12 shows the positioning of the substrate within the MOCVD reactor. In the following example, we examine the growth rate as a function of temperature.

Example:

Calculate the growth rate of Si from $SiCl_4$ at 1250 °C, given that:

$$h_G = 5\mathrm{cm/s}$$

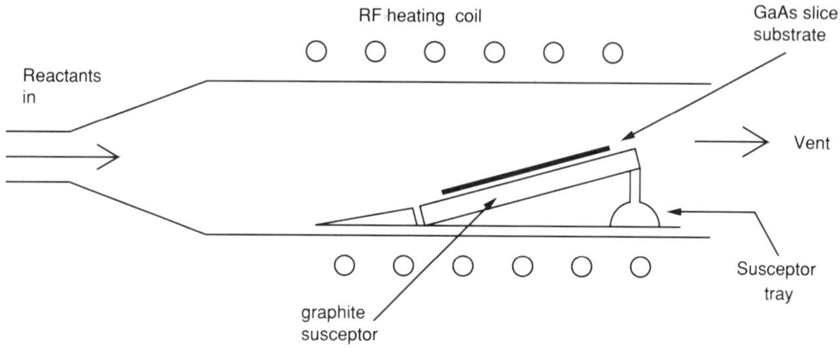

Figure 5.12 Typical MOCVD reactor.

$$k_s = 1 \times 10^7 \exp(-1.9\ \text{eV/kT})\text{cm/s}$$

$$\text{at} \quad T = 1250\,^\circ\text{C}, \quad N_G = 3 \times 10^{16}\text{atoms/cm}^3$$

(a) At $T = 1250 + 273 = 1523$ K

$$k_s = 1 \times 10^7 \exp\left[-\frac{1.9\ \text{eV}}{8.62 \times 10^{-5}(\text{eV/K}) \times 1523}\right] = 5.1836\ \text{cm/s}$$

$$N = 5.0 \times 10^{22}\ \text{atoms/cm}^3 \text{ for silicon}$$

$$V_y = \frac{\text{d}y}{\text{d}t} = \frac{5.1836 \times 5.0}{5.1836 + 5.0}\frac{3 \times 10^{16}}{5 \times 10^{22}}$$

$$= 1.527 \times 10^{-6}\ \text{cm/s}$$

$$= 0.916\ \mu\text{m/min}$$

(b) What is the change in growth rate if the reaction temperature is increased by 1%?

$$N_0 = \frac{h_G}{h_G + k_s} N_G$$

differentiate V_y with respect to k_s

$$\frac{\text{d}V_y}{V_y} = \left[\frac{h_G}{h_G + k_s}\right]\frac{\text{d}k_s}{k_s} = 0.491\frac{\text{d}k_s}{k_s}$$

Also

$$\frac{\text{d}k_s}{\text{d}T} = \frac{\text{d}}{\text{d}T}\left[B \exp\left(\frac{Ea}{KT}\right)\right]$$

$$\frac{\text{d}k_s}{k_s} = 14.47\left(\frac{\text{d}T}{T}\right)$$

Then

$$\text{d}V_y = 0.491 \times 14.47\frac{\text{d}T}{T} = 7.106\frac{\text{d}T}{T}$$

If the temperature changes by 1%, the growth rate changes by 7%.

5.5 SUBSTRATE MECHANICAL FAILURES AND RELIABILITY

The failures of silicon and GaAs substrates during manufacturing is the major reason for the reduction of yield and for the introduction of microcracks and dislocations during processing. Mechanical failures such as brittle fracture and thermal fatigue crack propagation in the integrated circuit substrate reduce the reliability levels of many electronic products. In this section we will develop the methodology for calculating the mechanical properties of GaAs substrates, noting that the discussion also applies to silicon substrates. Since the prevalent integrated circuit for the GaAs technology is the microwave integrated circuit, the present discussion pertains to monolithic microwave integrated circuits called MMICs. In this section, we will analyze the mechanical stresses of GaAs substrates due to the typical power and temperature cycles, and address the fracture failure mechanisms and models which can be used in substrate reliability assessment. The same methodology may be extended to silicon substrates.

Mechanical failure of a MMIC substrate occurs when the local mechanical stress, induced by various environment conditions, exceeds the strength of materials. The mechanical reliability problems usually start at the GaAs wafer handling, dicing, chip attach, and other manufacturing stages, resulting in various defects, voids, and microcracks. These voids or cracks due to the processing and manufacturing damage produce very high local stresses because of the stress concentration under applied thermal-mechanical loading. Fracture of a GaAs substrate during field operation is generally an extension of a pre-existing flaw subject to mechanical and thermal stress. Such flaws, often initiated during the processes of crystallizing of single crystal ingot, wafer cutting, or handling, therefore control the mechanical strength of a packaged product. Crack extension through junctions may cause a junction short or parameter change. The causes of GaAs substrate failures can be summarized into the following three categories: weak material properties, the damage induced in wafer processing and package manufacturing, operational thermal and mechanical stresses.

The material properties of the GaAs or silicon substrate, which are directly related to the mechanical reliability of a MMIC device include:

- dislocation structure,
- defects density,
- fracture toughness,
- modulus of rupture,
- modulus of elasticity,
- Poisson's ratio, and
- coefficient of thermal expansion.

The fracture toughness is the most critical macrostructural property for the mechanical failure of the GaAs substrate. The damage induced in wafer processing and package manufacturing includes:

- etched vias,
- gate recess,
- surface scratches from polishing or lapping,

- chip slicing and dicing damage, and
- chip bonding damage.

The thermal or mechanical stresses in the device operation which are responsible for mechanical failures include:

- normal microwave pulse operation,
- power cycle,
- temperature cycle,
- thermal shock stress, and
- mechanical shock and vibration.

Experience shows that instead of excessive temperature and mechanical shock, the dominant failure mechanisms of IC substrates are due to coefficient of thermal expansion mismatches under cyclic temperature changes. These stresses increase with the size of the substrate and may induce the fatigue failure. In particular, when the defects exist in the substrate, the local cyclic stresses may dramatically increase, driving the propagation of the microcracks into the active device, and inducing final failure.

The mechanical reliability of GaAs devices may increase by improving the material properties, adjusting processing and manufacturing procedures, or derating the applied stress. A correct reliability analysis and design must focus on all three aspects and their interrelationships. Material properties of the substrate play an important role in reliability improvement, since the determination of both the manufacturing procedure and the allowable applied stress should be based on these. For example, a wafer ("pick and place") handling machine may not damage a silicon wafer but may damage a GaAs wafer because of different mechanical properties. Similarly, a microcrack in a GaAs die may propagate to the active transistors and induce final failure of the device. However, an identical crack in a silicon die may not propagate under the same environmental stress conditions, because their mechanical properties are different.

On the other hand, the stress analysis and failure mechanism identification of a substrate are also very important. The major mechanical failures of device substrates in electronic packages were found to be induced by thermal stress-related mechanisms. Therefore, information on critical stresses, their distribution in the substrate, and the values of the maximum stresses, are all critical for a reliability analysis and design. Thermal stress analysis also provides a basis for further determination of the local stress state for defect and crack analysis. Because the manufacturing defects are unavoidable, therefore, the concepts of fracture mechanics must be included in reliability assessment modeling and design methodology.

5.5.1 Thermal-mechanical stresses in semiconductor substrates

Valuable insight into the thermal stress in multi-layer structures was provided by Timoshenko [3], Chong *et al.* [4], Williams [5], Hussein *et al.* [6, 7], and others on the basis of the theory of elasticity methods. The obtained solutions are, however, so complex that they cannot be directly used in substrate analysis. Recently, various simplified two-dimensional thermal stress analysis models for bonded assembly have been proposed, based on elementary methods of strength of materials and structure mechanics, with various assumptions. The accuracy of the results, therefore, largely depends on the correctness of these assumptions for a particular assembly. Chen and Nelson [8] first considered the stress distribution in bonded materials used in electronic devices. This model can help determine the magnitude and distribution of the normal stress in the GaAs substrate, and the shear and peeling stresses in the bottom surface of the substrate. However, it requires the solution of a sixth order differential equation with solutions not in explicit form. Simpler models based on shear lag theory have been used for shear stress analysis of die attach [9–12]. However, the accuracy cannot be guaranteed, and some development based on this model gives unreasonable conclusions. Suhir [13–14] proposed a practical model which enables one to determine the shear and peeling stresses in the bottom surface of the GaAs substrate, and the normal stress in the substrate. This model has been used to analyze the influence of attachment material selection on thermal stress by Suhir [15], who assumed that the temperature in the assembly changed uniformly and the die material was isotropic.

Suhir's model could be considered as the appropriate model for GaAs substrate stress analysis, since it is a trade-off between accuracy and complexity. In the following sections, we will briefly introduce Suhir's model and use it in failure analysis and design in the rest of the chapter. Suhir's thermal stress analysis model is based on the following assumptions:

- Shear stresses in the attachment vary across its thickness, but they are identical at the interfacial surfaces with the die and the substrate.
- Due to the constraint of the attachment material, the die and the substrate are subject to either tensional or compressive force and moments, which result in a linearly distributed stress throughout the thickness. The attachment is subjected to a moment, also.
- The peeling stresses do not affect the determination of other stress components.
- The effect of constraints in a second direction in the plane of the die is small.
- The contribution of the shear stress to the longitudinal deformation should be considered, since all the layers have the same radius of curvature.
- There is no slippage in the two interfacial layers.

The free body diagram shown in Figure 5.13 depicts an element of unit width and length dx with the stress components, including the normal forces P_d and P_c, in the device substrate and the case; the moments M_d and M_c in the device substrate and the case; shear stresses τ_d and τ_c in the bottom surface of the device substrate and the top surface of the case. The last assumption implies displacement compatibility, which requires that the displacements of the lower extreme fiber of the die are equal to those of the upper extreme fiber of the attachment, and the displacements of the lower extreme fiber of the attachment are equal to those of the upper extreme fiber of the substrate. In the determination of the surface fiber displacements, the contributions from the temperature change, the tensile or compressive force, the shear stress, and the bending are taken into account. The displacement compatibility at two interfacial surfaces gives two equations; moment–curvature relations in three

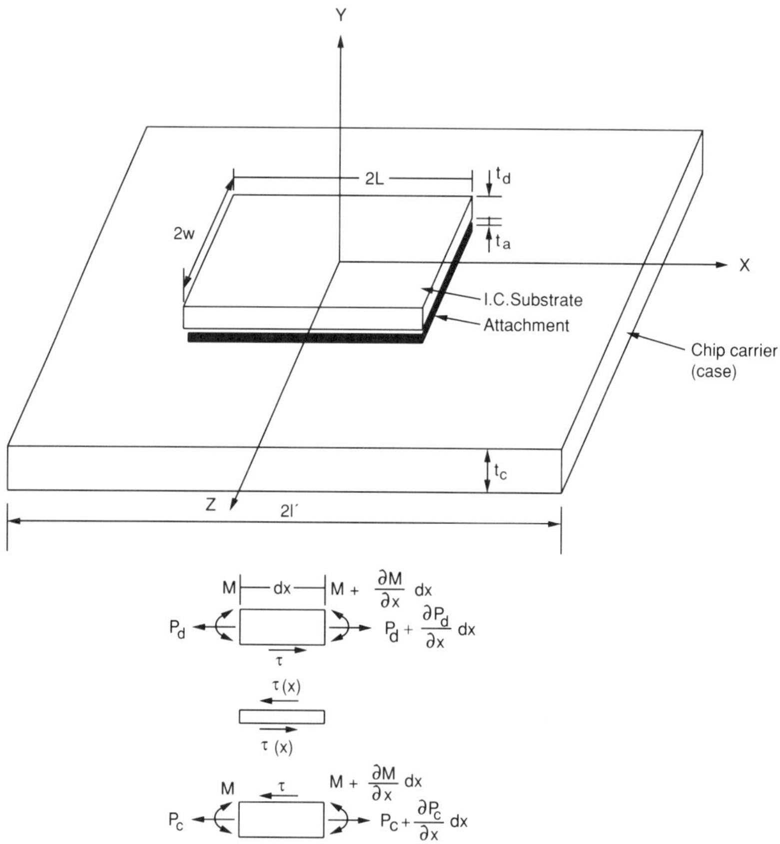

Figure 5.13 Free body diagram for stress analysis.

layers give three equations; equilibriums of the horizontal forces in three layers and equilibrium of the total moments give four equations. These nine equations could be used to solve for nine unknowns: two tensile forces, three shear stresses, three moments, and the radius of curvature. Because of the third assumption, the peeling stress is not included in these equilibrium conditions. Solving for the tensile forces in the die and the substrate yields a coupled simultaneous integral equation:

$$\left(\frac{t_d}{3G_d} + \frac{t_d}{3G_a}\right)\tau_d(x) - \left(\frac{1-\nu_d}{E_d t_d} + \frac{1-\nu_a}{E_a t_a} + \frac{(t_d+t_a)^2}{4D}\right)^x \int_o P_d(\xi)\,\mathrm{d}\xi$$

$$+ \left(\frac{1-\nu_a}{E_a t_a} - \frac{\sqrt{(t_d+t_a)^2 + (t_a+t_s)^2}}{4D}\right)^x \int_o P_c(\xi)\,\mathrm{d}\xi = (\alpha_a - \alpha_d)\Delta T \quad (5.6)$$

$$\left(\frac{t_d}{3G_a} + \frac{t_a}{3G_c}\right)\tau_c(x) - \left(\frac{1-\nu_a}{E_a t_a} + \frac{1-\nu_c}{E_c t_c} + \frac{(t_a+t_c)^2}{4D}\right)^x \int_o P_c(\xi)\,\mathrm{d}\xi$$

$$+ \left(\frac{1-\nu_a}{E_a t_a} - \frac{\sqrt{(t_d+t_a)^2 + (t_a+t_s)^2}}{4D}\right)^x \int_o P_d(\xi)\,\mathrm{d}\xi = (\alpha_c - \alpha_a)\Delta T \quad (5.7)$$

where τ_d and τ_c are shear stresses applied to the bottom surface of the device substrate and the top surface of the package case. G_d, E_d, ν_d, α_d; $G_a E_a$, ν_a, α_a; and $G_c E_c$, ν_c, α_c are shear and tensile moduli of elasticity, Poissons ratios, and coefficients of thermal expansion (CTE) for device substrate, attachment, and case materials, respectively. t_d, t_a, and t_c are the thicknesses of the device substrate, the attachment, and the case, respectively. ΔT is the temperature change. The flexural rigidity, D, is defined by

$$D = \frac{E_d t_d^3}{12(1-\nu_d^2)} + \frac{E_a t_a^3}{12(1-\nu_a^2)} + \frac{E_c t_c^3}{12(1\nu_c^2)} \tag{5.8}$$

P_d and P_c are the normal tensile or compression forces applied horizontally on the substrate and the case, which can determined by

$$P_d(x) = \int_{-L}^{x} \tau_d(\xi)\,\mathrm{d}\xi \quad P_c(x) = \int_{-L}^{x} \tau_c(\xi)\,\mathrm{d}\xi \tag{5.9}$$

Since the shearing stresses of the three-layer assembly must be anti-symmetric with respect to the origin, the solutions to equations (5.6) and (5.7) have the following form:

$$\tau_d(x) = C_1 \sinh(A_1 x) + C_2 \sinh(A_2 x) \quad \tau_c(x) = C_3 \sinh(A_1 x) + C_4 \sinh(A_2 x) \tag{5.10}$$

For die and substrate attach problems, the thickness and the elastic modulus of the attach material, t_a and E_a, are so small that these two interfacial shear stresses, τ_d and τ_c, are assumed to be identical. Therefore, Suhir simplified the solution by considering a single eigenvalue only, which gives the following solution for shear stresses by substituting equation (5.9) into equations (5.6) and (5.7):

$$\tau(x) = \tau_d(x) = \tau_c(x) = \left[\frac{1-\nu_c}{E_c t_c} + \frac{1-\nu_d}{E_d t_d} + \frac{(t_d+t_c)^2}{4D}\right]^{-1/2} \times \left[\frac{t_d}{3G_d} + \frac{2t_a}{3G_a} + \frac{t_c}{3G_c}\right]^{-1/2} \frac{\sinh(Ax)}{\cosh(AL)}(\alpha_c - \alpha_d) \tag{5.11}$$

where

$$A = \sqrt{\frac{\dfrac{1-\nu_c}{E_c t_c} + \dfrac{1-\nu_d}{E_d t_d} + \dfrac{(t_d+t_c)^2}{4d}}{\dfrac{t_d}{3G_d} + \dfrac{2t_a}{3G_a} + \dfrac{t_c}{3G_c}}} \tag{5.12}$$

When $x = L$, the shear stress, reaches its maximum value which is

$$\tau_{\max} = \left[\frac{1-\nu_c}{E_c t_c} + \frac{1-\nu_d}{E_d t_d} + \frac{(t_d+t_c)^2}{4d}\right]^{-1/2} \left[\frac{t_d}{3G_d} + \frac{2t_a}{3G_a} + \frac{t_c}{3G_c}\right]^{-1/2} \times \tanh(AL)(\alpha_c - \alpha_d)\Delta T \tag{5.13}$$

From the horizontal force equilibrium along the die length direction we have (see Figure 5.13):

$$\frac{\partial P_d(x)}{\partial x}\mathrm{d}x + \tau(x)\,\mathrm{d}x = 0 \tag{5.14}$$

The tensile force in the device substrate, $P_d(x)$, can be determined by integrating equation (5.8) with the boundary condition that $\sigma_d(x) = 0$ at $x = L$. Substituting equation (5.8) into equation (5.14) gives the normal stress in the device substrate

$$\sigma(x) = \frac{P_d(x)}{t_d} = \frac{\dfrac{1}{t_d} + 3(t_d + t_c)\dfrac{E_d t_d}{12(1-\nu_d^2)D}}{\dfrac{1-\nu_d}{E_d t_d} + \dfrac{1-\nu_c}{E_c t_c} + \dfrac{(t_d+t_c)^2}{4D}}\left[1 - \frac{\cosh(Ax)}{\cosh(AL)}\right](\alpha_c - \alpha_d)\Delta T \tag{5.15}$$

When $x = 0$, the normal stress reaches its maximum value which is

$$\sigma_{\max} = \frac{\dfrac{1}{t_d} + 3(t_d + t_c)\dfrac{E_d t_d}{12(1-\nu_d^2)D}}{\dfrac{1-\nu_d}{E_d t_d} + \dfrac{1-\nu_c}{E_c t_c} + \dfrac{(t_d+t_c)^2}{4D}}\left[1 - \frac{1}{\cosh(AL)}\right](\alpha_c - \alpha_d)\Delta T \tag{5.16}$$

In the displacement compatibility conditions, equations (5.6) and (5.7), the contribution of the peeling stress is not included. However, the peeling stress on the bottom surface of the substrate is estimated by moment equilibrium after obtaining the shear stresses:

$$p(x) = -\left[\frac{t_c t_d^3 E_d}{12(1-\nu_d^2)} - \frac{t_d t_c^3 E_c}{12(1-\nu_c^2)}\right]\left[\frac{t_d}{3G_d} + \frac{2t_a}{3G_a} + \frac{t_c}{3G_c}\right]^{-1} \times \frac{\cosh(Ax)}{2D\cosh(AL)}(\alpha_c - \alpha_d)\Delta T \tag{5.17}$$

When $x = 0$, the normal stress reaches its maximum value, which is

$$p_{\max} = -\left[\frac{t_c t_d^3 E_d}{12(1-\nu_d^2)} - \frac{t_d t_c^3 E_c}{12(1-\nu_c^2)}\right]\left[\frac{t_d}{3G_d} + \frac{2t_a}{3G_a} + \frac{t_c}{3G_c}\right]^{-1} \frac{1}{2D}(\alpha_c - \alpha_d)\Delta T \tag{5.18}$$

From above analysis the following conclusions can be drawn:

- As long as the thickness and the modulus of elasticity are small compared to those of the substrate and case material, the coefficient of thermal expansion of the attachment material does not affect the thermal stresses in the device substrate.
- The shear and peeling stresses on the bottom surface attain maximum values at the edge of the substrate ($x = L$). The normal tensile stress in the substrate attains its maximum value in the middle of the substrate ($x = 0$).
- The maximum values of critical stresses in the device substrate are linearly related to the temperature maximum and minimum from the temperature cycle and the difference of the CTEs between the device substrate and the case.
- The maximum normal stress in the substrate is independent of the size of the substrate. The dependence of the shear and the peeling stresses is weak when the size of the substrate is large. Suhir [14] has shown that the maximum shear stress is independent of the substrate size when $AL > 3$, and the maximum peeling stress is independent of the substrate size when $AL > 6$.

Figure 5.14 contains plots of the distributions of the normal stress in the device substrate, and the shear and peeling stresses on the bottom surface of the device substrate for a GaAs MMIC device (EG6023 power amplifier test vehicle) made by Texas Instrument Co. The dimensions and the material properties of the substrate, attachment and case are listed in Table 5.2, and the range of the temperature cycle is 345 °C. Calculations give 7.36 MPa maximum tensile stress in the center of the substrate, 1.8 MPa maximum shear stress and 0.22 MPa maximum peeling stress at the edge of the substrate.

In the above example, if a bonding temperature $T_b = 375\,^\circ\mathrm{C}$ is assumed, then the temperature range is the difference between the bonding and the ambient temperature which is assumed to be $T_a = 30\,^\circ\mathrm{C}$, i.e.

$$\Delta T = T_b - T_a = 375\,^\circ\mathrm{C} - 30\,^\circ\mathrm{C} = 345\,^\circ\mathrm{C} \tag{5.19}$$

The residual thermal stress may induce the substrate crack when a MMIC device operates at a low temperature environment. The residual stresses can be calculated from equation (5.18).

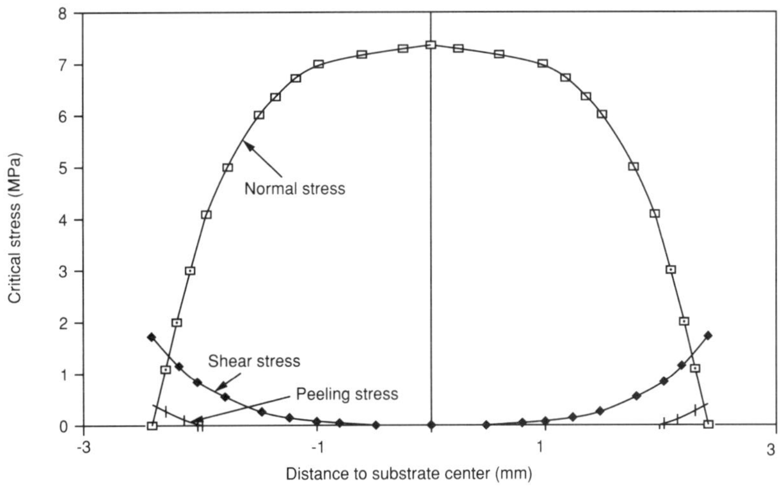

Figure 5.14 Stress distribution in GaAs substrate (EG 6023 MMIC device).

Table 5.2 Modulus of elasticity measured by vibration and static methods.

Temperature (°C)	$E_{(\mathrm{Vibration})}$ (GPa)	$E_{(\mathrm{Static})}$ (GPa)	Difference (%)
−100	126.2	122.0	3.3
21.5	123.9	119.7	3.4
100	121.8	117.1	3.9
200	119.1	113.8	4.5
300	116.3	106.2	8.7
400	113.2	(89.4)	(12.9)
500	110.5	(98.3)	(19.1)

In the reliability analysis, the temperature change due to the power cycle is needed to determine the mechanical stress. The temperature cycle range can be estimated by a first-order approximation:

$$\Delta T = T_J - T_a \tag{5.20}$$

where T_J is the estimated junction temperature, which can be determined by Fourier's law of heat transfer by conduction.

Fourier's law states that the rate of heat flow equals the product of the area normal to the heat flow path, the temperature difference along the path, and the thermal conductivity of the material. That is

$$\frac{\mathrm{d}Q}{\mathrm{d}t} = -kA\frac{\mathrm{d}T}{\mathrm{d}x} \tag{5.21}$$

where Q is the heat energy, $\mathrm{d}Q/\mathrm{d}t$ is the power expressed in Watts, k is the thermal conductivity, and T is the temperature. Denoting $\mathrm{d}Q/\mathrm{d}t$ as P, and integrating equation (5.21) we have

$$\int_0^x \frac{1}{kA}\,\mathrm{d}x = -\int_{T_1}^{T_2} \frac{1}{P}\,\mathrm{d}T + C \tag{5.22}$$

The constant of integration, C, vanishes because the temperature change is zero when the length of the heat path is zero. Assuming that the power dissipation of the device is independent of the junction temperature, and defining the thermal resistance of the system, θ, as follows:

$$\theta = \int_0^x \frac{1}{kA}\,\mathrm{d}x \tag{5.23}$$

we have

$$\theta = -\frac{1}{P}\int_{T_2}^{T_1} \mathrm{d}T = \frac{T_1 - T_2}{P} \tag{5.24}$$

In the MMIC package, the path of heat from the device junction to the case is not short compared to the heat source dimensions, and heat will also spread out laterally. This lowers the thermal resistance compared to a straight-line path. To include this thermal fan-out spreading, the surface area over which heat is conducted at different cross-sections, A, in the above equation must be considered as a function of x. The accurate determination of this function is very complicated. A reasonable simplification is to assume that the heat spreads at a 45° angle, corresponding to a truncated pyramid [16]. Therefore the thermal resistance in each layer of the die attach assembly can be determined by the following first-order approximation:

$$\theta = \frac{1}{K_i}\int_0^{h_i} \frac{\mathrm{d}x}{A_i(x)} = \frac{1}{K_i}\int_o^{h_i} \frac{\mathrm{d}x}{(L_i + 2x)(W_i + 2x)}$$
$$= \frac{1}{2K(L_i - W_i)}\left[\ln\frac{L_i(W_i + 2h_i)}{W_i(L_i + 2h_i)}\right] \tag{5.25}$$

where L_i, W_i, h_i are the lengths, widths, and thicknesses of each layer, and k_i is the thermal conductivity of each layer. The $2x$ term is based on the assumption that the heat spreads at 45 degrees. To change the angle between the vertical to the heat flow line from 45° to α, the $2x$ term in the denominator should be replaced with $2x\tan(\alpha)$.

The total thermal resistance of the MIMIC package system is the sum of the thermal resistances of each layers:

$$\theta = \sum_{i=1}^{3} \theta_i = \theta_d + \theta_a + \theta_c \tag{5.26}$$

Then according to equation (5.24) we have:

$$\sum_{i=1}^{3} \theta_i = \frac{T_J - T_A}{P} \tag{5.27}$$

where T_J and T_A are junction and maximum ambient temperatures, and P is the power dissipated on the surface of the device.

From equation (5.17), the range of temperature cycle due to power dissipation can then be estimated by:

$$\Delta T = T_J - T_a = (\Sigma\theta_I)P \tag{5.28}$$

The critical stresses in the substrate can be determined by substituting equation (5.28) into equations (5.13) (5.16), and (5.18).

The stress ranges due to the ambient temperature cycles can be obtained by substituting $\Delta T = T_{\max} - T_{\min}$, i.e. the difference between the maximum and the minimum ambient temperatures, in equations (5.13), (5.16), and (5.18). The combination of the stresses due to the power cycle and the ambient temperature change depends on the mission profile. The evaluation of it is usually difficult.

5.6 FAILURE MECHANISMS AND DAMAGE MODELS OF THE SEMICONDUCTOR SUBSTRATE

Microstructural defects and microcracks in the substrate may form during the different processing steps such as single crystal semiconductor growth and slicing, wafer scrubbing and polishing, and substrate cutting and bonding. In the actual package, the thermal-mechanical properties, the overall void concentration or defect density, and the shape and position of defects determine the failure mechanism of the substrate. The most critical failure mechanisms of the device substrate due to both thermal stress and existing defects include brittle fracture of the substrate, fatigue crack initiation, and fatigue crack propagation in the substrate.

The phenomenon of substrate fracture can be described by an appropriate fracture mechanics model, which assumes that all the materials have defects of different shapes and sizes, and that the failure is due to propagation of a critical crack. The phenomenon of fatigue crack propagation in the GaAs substrate can be described by Paris's law. The rate of fatigue crack propagation per cycle is determined by the cyclic change in the stress intensity factor, a measure of the stress intensity around the crack tip. By integration of Paris's law, the number of cycles to failure can be estimated if the material constants in the equation of crack growth rate are known, the initial crack depth is assumed, and the maximum nominal thermal stress cycles are calculated according to the thermal stress models. In this section, we introduce the phenomena and practical models for the failure mechanisms of the substrate. These models must be employed in the reliability analysis of the substrate as well as in the package design.

5.6.1 Brittle fracture

Local micro-scale cracks can cause sharp stress concentrations in brittle substrate materials which exhibit little yielding and inelasticity before fracture. Excessive stress levels in brittle materials can cause failure by sudden catastrophic propagation of the dominant crack. If a preexisting crack, after the manufacture of the device on the substrate, is equal to or greater than the critical crack size, which depends, of course, on the magnitude of thermal strength, the chip would fracture in the first stress cycle. Brittle fracture of the substrate can occur suddenly, without any plastic deformation, when a vertical surface crack at the center of the substrate or a horizontal crack at the edge of the substrate propagates during thermal cycling to the critical crack size. The microcracks nucleated at the top surface, developed from surface scratching of the wafer, may propagate to an active transistor site and induce device failure.

The basic fracture patterns on the substrate are vertical and horizontal cracking, as shown in Figure 5.15. The vertical crack is due to surface damage such as scratching and it may propagate under bending and tension stress, if the stress intensity factor reaches the fracture toughness of GaAs. The horizontal crack starts at the corner of the substrate at around 40° from the horizontal level, and then propagates along the horizontal plane under thermal or other mechanical stresses. As discussed in section 5.5, when the temperature rises during manufacture and thermal cycling, tensile stresses in the horizontal direction and peeling stresses in the vertical direction develop in the substrate. The horizontal tensile stress is responsible for vertical cracking, and the vertical peeling stress is responsible for

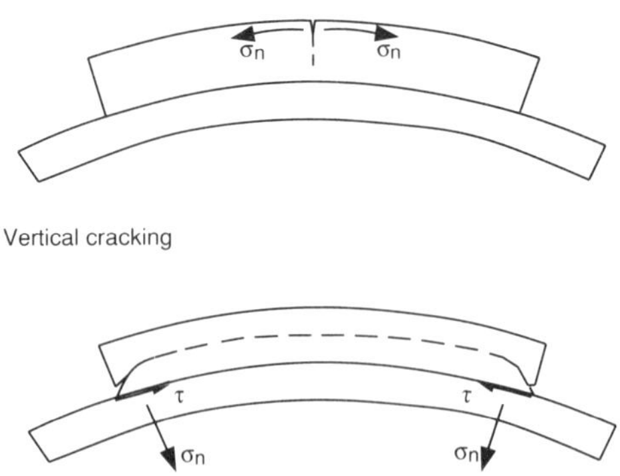

a. Vertical cracking

b. Horizontal cracking

Figure 5.15 Cracking patterns in substrate.

horizontal cracking. Beside these two types of fracture, pure horizontal cracking is also possible. Since the wire-bondings are still well connected, devices with this type of fracture could pass a full functional test without any sign of this internal failure problem. The primary cause of this type of fracture is due to "saw-and-break" chip cutting. In this cutting process, the chips are sawed to a depth of three fourths of the thickness by a diamond saw, and then the dies are separated by rolling the back of the wafer. During this process, cracks are initiated on a discontinuity plane at the bottom of the wafer, and later, the crack at the edge of the substrate propagates horizontally.

A traditional failure criterion is based on the comparison of the predicted local stress, which is the product of nominal stress and a stress concentration factor, with the modulus of rupture of the material. When the predicted local stress is greater than the modulus of rupture, i.e.

$$C\sigma_{\max} > \sigma_U \tag{5.29}$$

then failure occurs. Here C is the stress concentration factor, $\sigma_{\max}$ is the nominal maximum stress and σ_U is the modulus of rupture.

The mechanical property prediction of semiconductor substrates based on the measurement of the modulus of rupture is inadequate, since wafer failure is usually due to the presence of microcracks. Therefore, a fracture mechanics approach should be considered. From the viewpoint of fracture mechanics, fracture failure is related not only to the magnitude of the applied stress but also to the shape and the size of the crack. The failure criterion based on local stresses is not feasible, because analysis predicts infinite stresses at the tip of the flaw or crack, regardless of the level of the applied far-field average or nominal stress. The basic approach in fracture mechanics is therefore one of predicting the level of the far-field stress, at which the crack will locally propagate. By comparing the gross changes in the elastic energy of a structure, which occur as a sharp brittle crack grows to the energy required to produce new fracture surfaces, it is possible to calculate the far-field stress (if no crack was present) which will cause crack growth [17].

A parameter used to characterize the intensity of the crack-tip stress field is the stress intensity factor, K, defined as a function of crack type, initial crack size, geometry of the body, and applied stress. Fracture of the substrate would occur when the stress intensity factor is greater than the fracture toughness.

Three fracture modes illustrated in Figure 5.16 are recognized in fracture mechanics methodology [18]. Crack surface displacements in mode I are perpendicular to the crack plane. Tensile stress in the material opens the crack and the local stress concentrations at the crack tip cause crack propagation when the stress intensity factor reaches the fracture toughness of the material. Mode II and mode III cracks are caused by shearing forces in the crack plane. Mode II fractures are caused by an in-plane shearing force in which the crack surface slides perpendicular to the crack front. Mode III fractures are caused by an out-of-plane shearing force, in which the crack surface produces tearing displacements that slide parallel to the crack front. Observations of substrate cracks show that they are often initiated at corners and edges. They tend to grow inward and assume a quarter-elliptical or semi-elliptical shape. The application of fracture mechanics to

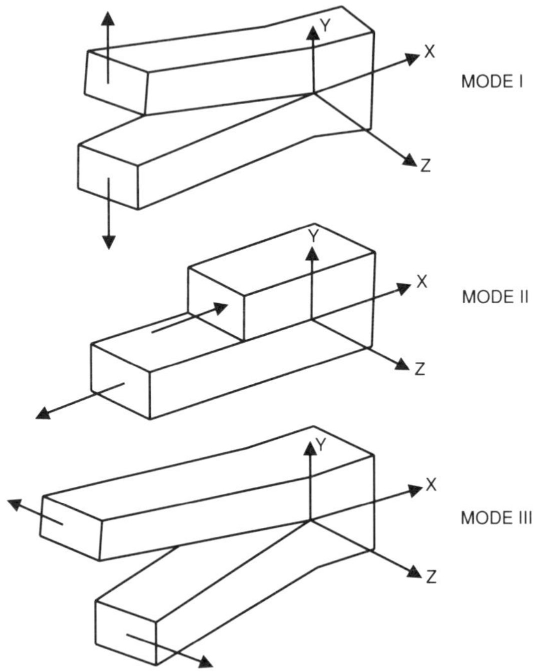

Figure 5.16 The three basic models of crack surface displacements.

these corner cracks (quarter-elliptical) and surface flaws (semi-elliptical) requires knowledge of the stress intensity factor for a crack with a curved front.

Westergaard [19] analyzed a crack in a two-dimensional infinite solid using a complex variable approach and showed that the stress intensity factor under mode I loading can be determined by

$$K_I = \sigma\sqrt{\pi a} \tag{5.30}$$

where K_I is the mode I fracture stress intensity factor, σ is the nominal far-field applied stress, and a is the half crack length. In general, the stress intensity factor for a mode I crack can be expressed by

$$K_I = \sigma\sqrt{\pi a}F \tag{5.31}$$

where F is a correction factor which is related to the geometry of the crack and the geometry of the fractured wafer.

Sneddon [20] showed that for a circular penny-shaped surface crack in an infinite three-dimensional solid, the stress intensity factor is expressed by

$$K_I = \frac{2}{\pi}\sigma\sqrt{\pi a} \tag{5.32}$$

where a is the half crack penetration depth.

Based on the stress field around an ellipsoidal cavity, Irwin [21] extended Sneddon's results to a surface flaw of length $2c$ along the surface, with a semi-elliptical shape penetrating to depth "a" below the surface, as shown in Figure 5.17, and gave the following expression for the stress intensity factor:

$$K_I = \frac{1.12}{\sqrt{Q}}\sigma\sqrt{\pi a} \tag{5.33}$$

where 1.12 is the correction for the front free surface, and

$$Q = \Phi^2 - 0.212\left(\frac{\sigma}{\sigma_Y}\right)^2 \tag{5.34}$$

where Φ is an elliptical integral of the second kind, given by

$$\Phi = \int_0^{\pi/2}\left(1 - \frac{c^2 - a^2}{c^2}\sin^2\phi\right)^{1/2} \mathrm{d}\phi$$

For a brittle solid such as GaAs, it was shown that $Q = \Phi^2$. Values for Φ can be found in mathematical tables or in graphs of elliptic functions. When $a = c$, $Q = 2.4$, and when $a = 0.4c$, $Q = 1.41$. Taking the first two or three terms of a series expression for Φ gives the following approximation [22]:

$$\Phi = \int_0^{\pi/2}\left(1 - \frac{c^2 - a^2}{c^2}\sin^2\phi\right)^{1/2} \mathrm{d}\phi \tag{5.35}$$

$$\Phi = \frac{3\pi}{8} + \frac{\pi}{8}\frac{a^2}{c^2}$$

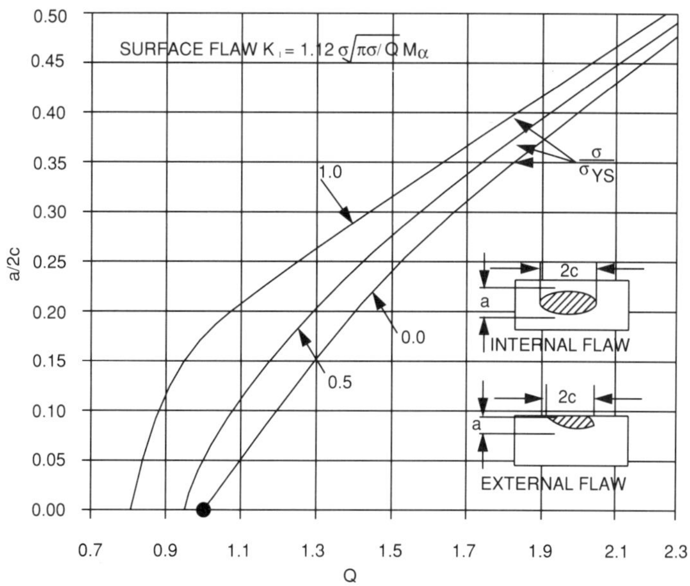

Figure 5.17 Effect of *a/2c* ratio and σ/σ_{ys} ratio on the flaw-shape parameter Q.

or

$$\Phi = \frac{\pi}{2}\left\{1 - \frac{1}{4}\frac{c^2 - a^2}{c^2} - \frac{3}{64}\left(\frac{c^2 - a^2}{c^2}\right)^2\right\} \tag{5.36}$$

To account for the effect of free surface back plane in the thickness direction, Osgood [23] introduced a factor M_K, and obtained the following expression for the stress intensity factor:

$$K_I = \frac{1.12M_K}{\sqrt{Q}}\sigma\sqrt{\pi a} \tag{5.37}$$

where

$$M_K = 0.953 - 2.369\left(\frac{a}{t}\right) + 2.74\tan\left(\frac{a}{t}\right) \tag{5.38}$$

and t is the thickness of the cracked body.

For a surface crack on the top or bottom of a GaAs substrate die, the stress intensity factor can be estimated by substituting equations (5.36) and (5.38) into equation (5.37):

$$K_I = \frac{2.85\left[0.953 - 2.369\left(\frac{a}{t_d}\right) + 2.74\tan\left(\frac{a}{t_d}\right)\right]}{3 + \frac{a^2}{c^2}}\sigma\sqrt{\pi a} \tag{5.39}$$

Equations (5.33) to (5.39) can be used to solve the fracture problem with a uniformly distributed tensile stress perpendicular to the crack propagation plane. For a MMIC packaged circuit, the nominal stresses in a GaAs substrate are not uniformly distributed. The shear and tensile stresses are applied simultaneously. However, for engineering design purposes, equation (5.38) can be used to model the problems related to cracks in the substrate, provided that the approximation gives conservative results, i.e., the upper bound of the stress intensity factor.

As an example, assuming the crack depth is only one tenth of the substrate thickness, i.e., $a = 0.1\ h_d$, and the ratio of the crack depth to the half crack length is 0.4, then equation (5.39) reduces to:

$$K_I = 0.87\sigma\sqrt{\pi a} \tag{5.40}$$

where σ is the tensile stress estimated by equation (5.16).

If the crack length is known or can be estimated, the following criterion can be used for the substrate mechanical design, in order to prevent the brittle fracture failure of the substrate in the first thermal or power cycle

$$K_I < K_c \tag{5.41}$$

where K_c is the fracture toughness of the substrate. On the other hand, the fracture criterion can be given in terms of crack size by substituting equation (5.31) into equation (5.41):

$$a < \frac{K_{IC}^2}{\pi(\sigma F)^2} \tag{5.42}$$

The right-hand side of equation (5.42) can be defined as the critical crack size, a_c, i.e.,

$$a_c < \frac{K_{IC}^2}{\pi(\sigma F)^2} \tag{5.43}$$

and the criterion of the brittle fracture failure of the substrate can be represented by:

$$a < a_c \tag{5.44}$$

If the existing crack is equal to or greater than the critical crack size determined by fracture toughness, the substrate will fail. A common design approach involves a fracture mechanics analysis based on assuming a characteristic flaw to be located at the highest stressed region in the component. According to this criterion, the allowable initial crack depth on the surface of the substrate can be determined if the fracture toughness of the substrate is known. The maximum nominal thermal stress is calculated according to the thermal stress model, and the type of crack is specified. This theoretically allowable initial crack length serves as a specification which can be used for quality control of the substrate.

5.6.2 Fatigue crack initiation and propagation

When repetitive cyclic stresses are applied to a material, fatigue failure can occur at stresses below the ultimate strength of the material due to the accumulation of damage. Fatigue failure is a wear-out type of failure mechanism which begins with the initiation of a crack. The crack develops typically at a point of discontinuity or defect in the material, which can cause local stress and plastic strain concentration.

The fatigue properties of a brittle material can be characterized by the stress–life curve. This curve represents the equation which empirically relates the stress amplitude $\Delta\sigma$ to the cycles-to-failure N_f

$$N_f = C(\Delta\sigma)^m \tag{5.45}$$

where C and m are the material constants.

Once a crack has initiated, even though it may not be of critical size at the applied service stress, it will grow to critical size gradually by stable fatigue propagation during power and temperature cycling until it becomes unstable under the applied stress cycles.

In 1961, Paris *et al.* [24] proposed a power law equation to predict fatigue crack propagation based on the stress intensity factor at or near the crack tip in the plane of propagation. As the stress varies during the thermal-mechanical cycling, K will proportionally vary as follows:

$$\Delta K = K_{\max} - K_{\min} = (\sigma_{\max} - \sigma_{\min})\sqrt{\pi a}F \tag{5.46}$$

Then the rate of fatigue crack propagation, $\mathrm{d}a/\mathrm{d}N$, will be given by

$$\frac{\mathrm{d}a}{\mathrm{d}N} = A(\Delta K)^n \tag{5.47}$$

where a is the crack size, N is the number of cycles, and A and n are the material properties.

Assuming that equations (5.14), (5.16), and (5.18) express the magnitude of the stress range for completely reversed loading, and equation (5.47) describes the factor K, then equation (5.46) becomes:

$$\frac{da}{dN} = A(\Delta\sigma\sqrt{\pi a}F)^n \tag{5.48}$$

Integration of equation (5.48) gives the number of cycles to failure:

$$N_f = \int_0^{N_f} dN = \int_{a_i}^{a_f} \frac{da}{(A\Delta\sigma\sqrt{\pi a})^n} = \frac{1}{\left(1 - \frac{n}{2}\right) A^n \sigma^n \pi^{n/2}} \left[a_f^{1-(n/2)} - a_i^{1-(n/2)}\right] \tag{5.49}$$

where a_i is the initial crack size and a_f is the final crack size. The final crack size is identical to the critical crack size a_c defined by equation (5.43) for the given stress range, or the allowable length of the crack which penetrates into the active area on the substrate surface.

The crack propagation rate, da/dN, becomes non-zero beyond a threshold value of the cyclic range of the stress intensity factor, ΔK. Crack propagation is initially characterized by a propagation rate whose slope decreases with increasing ΔK, followed by a steady-state secondary phase where the propagation rate is linearly proportional to ΔK on a log-log scale. During the tertiary and final phase, the propagation rate increases dramatically until catastrophic failure occurs. For design purposes, the steady state or secondary region is usually most important.

Equation (5.47) holds for completely reversed cyclic loading. If the loading is not completely reversed, Foreman's correction factor is used, and $\Delta\sigma$ is replaced by $(\sigma_f - \sigma_{\text{mean}})$, where σ_{mean} is the mean of the applied cyclic stress, then the rate of crack propagation becomes:

$$\frac{\partial a}{\partial N} = A\Delta K^P/[(1 - R)(K_{IC}) - (K_{\max})] \tag{5.50}$$

where A and p are material constants, and R is given by

$$R = \sigma_{\min}/\sigma_{\max} \tag{5.51}$$

This fatigue law can also be used for variable history loading, as long as the load history is split into blocks, such that the mean stress and the stress amplitude are constant within each block. Fatigue damage due to each stress block can be evaluated separately and the cumulative damage can be estimated by using damage superposition models, such as Miner's law of simulative fatigue damage [25].

5.6.3 Mechanical properties of GaAs wafers

Compared with elemental semiconductor materials such as silicon and germanium, little is known about the mechanical properties of GaAs. Swaminathan and Copley [26] investigated the stress–strain behavior of Si-doped GaAs single crystals, deformed in compression parallel to the < 100 > and < 110 > directions, at temperatures ranging from 250 to 550 °C. In both orientations, slopes occurred only on the {111} < 110 > close-packed slip system and the specimens typically broke into several columnar fragments with their long dimension parallel to the stress axis. Laister and Jenkins [27] performed bending and tensile tests at temperatures ranging from 25 °C to 1000 °C. Surface defects and impurity segregation were shown to be dislocation sources, with strains of 30% at high temperatures. Yonenaga *et al.* [28] investigated compressive mechanical properties of GaAs crystals, where the deformation characteristics were considered as functions of the deformation temperature, strain rate, and initial state of the crystal. The effective stress was determined by strain-rate cycling tests. Chen [29] analyzed cracking in silicon and GaAs wafers utilizing fracture mechanics approaches to determine the minimum wafer thickness for the opening mode crack propagation subject to bending. He predicted the critical crack size for both opening and tearing fracture. To evaluate the strength of GaAs solar cells, Chen [30] measured the modulus of elasticity, Poisson's ratio, and the fracture toughness of single crystal GaAs bulk material in different crystal orientations at room temperature. However, for the GaAs wafer material commonly employed in integrated circuits applications, information about the temperature dependence of fracture toughness and other mechanical properties is not available.

The mechanical properties of the substrate are usually orientation dependent because semiconductors are anisotropical materials. The critical stresses in the substrate and the attachment, estimated from simplified stress analysis models, however, are related only to the material properties in a direction parallel or perpendicular to the substrate cutting direction. The material properties involved in the determination of the stresses include the modulus of elasticity, Poisson's ratio, and the coefficient of thermal expansion. Semiconductors are very brittle materials with low mechanical strength. For simplified analysis, the strength of the substrate is represented by the modulus of rupture. However, the modulus of rupture of semiconductor material may be controlled by a more intrinsic material property — fracture toughness. The fracture toughness of the substrate characterizes the material resistance to fracture, and is, therefore, critical for reliable IC design.

To analyze the failure mechanisms and predict the reliability of microelectronic packages, the immediate parameters in need of investigation for GaAs wafer material at different temperatures include the modulus of elasticity, E, over the temperature range of −125 °C to 500 °C [31]. The tests for determining the modulus of rupture, σ_U, the critical value of the stress intensity factor, K_c, and the coefficient of thermal expansion, α, were conducted over the temperature range of −100 °C to 200 °C. The test specimens were cut along the substrate cutting mark in the (100) direction. In the following sections, the testing conditions and the test results are presented.

5.6.4 Modulus of elasticity

The modulus of elasticity, or the Young's modulus, is the proportionality constant between axial stress and strain for low strains, which can be determined from either a static stress–strain test or from a dynamic mechanical vibration or sonic velocity test [32, 33]. The results from vibration techniques are about 10% greater than those from static testing [34]. Employing the relationship between modulus of elasticity and sonic velocity, Chen [30] conducted testing at room temperature for a bulk GaAs single crystal with an ultrasonic pulse-echo technique. The results calculated from the measured sonic velocity indicate that the modulus of elasticity of GaAs is dependent on the crystalline orientation. The average values are $E_{<100>} = 89.3$ GPa, $E_{<111>} = 144$ GPa.

Both the mechanical vibration method and static stress–strain slope method were applied to three-point bending specimens over the temperature range from $-125\,^{\circ}\mathrm{C}$ to $500\,^{\circ}\mathrm{C}$. Tests were conducted on a Rheometrics solid analyzer (RSA-2). The test specimens were gripped by a fixture between a linear servo motor and a machine transducer. The motor applies a strain to the specimen and the transducer measures force. The force from the servo-motor is transmitted through the specimen to the transducer. The inner volume of an oven chamber and the fixtures in contact with the specimen were cooled down to $-150\,^{\circ}\mathrm{C}$ by the boiling of liquid nitrogen and heated to $600\,^{\circ}\mathrm{C}$ by a controlled AC voltage applied to the convection gun heater. The testing temperature was controlled within $5\,^{\circ}\mathrm{C}$. The process of temperature and loading increase was monitored by a microcomputer.

Test specimens were cut from 50.8 mm diameter and 0.45 mm thick GaAs wafers using a diamond saw. The wafers are chemically polished on one side, with an edge cutting mark which indicates the cutting direction of dies. Parallel to the substrate cutting direction, the specimens were cut in the middle of the wafers with the nominal dimensions of 6×50 mm, as shown in Figure 5.18. The span length of the testing fixture was 48 mm for the bending tests.

The modulus of elasticity from the vibration testing was calculated by:

$$E = \frac{\Delta P}{4W\Delta\delta}\left(\frac{L}{B}\right)^3 \tag{5.52}$$

where ΔP is the amplitude of applied loading, $\Delta\delta$ is the amplitude of deflection, L is the span length of three-point bending. W is the width and B is the thickness of the specimen.

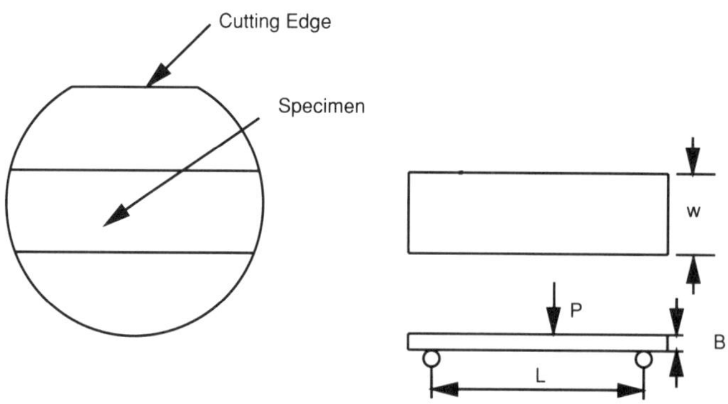

Figure 5.18 Three-point bending specimen cut from a GaAs wafer.

The measured modulus of elasticity from the strain controlled vibration testing for a temperature ramp mode is plotted in Figure 5.19 as a solid line. The vibration frequency was 1 Hz; the strain amplitude at the bottom fiber of the specimen was 0.01%; and the static precompression load was 0.3 N. During the tests, the temperature was increased continuously from −125 °C to 500 °C at a rate of 10 °C per minute. It is shown that the modulus of elasticity decreases linearly with an increase of temperature from 126.0 GPa at 125 °C to 110.6 GPa at 500 °C. The square marks on Figure 5.19 are the modulus of elasticity measurements of the slopes of the static stress versus strain curves at different temperatures. The slopes were determined using a least squares regression method. The modulus of elasticity measured from static tests is within 9% to that measured from vibration tests over the temperature range from −125 °C to 300 °C. However, when the temperature is above 300 °C, the values of modulus of elasticity measured from static tests are lower than those from vibration tests. Table 5.2 lists the measured values at various temperature levels and their differences.

The differences between the modulus of elasticity measured by the vibration method and by the static method at temperatures above 300 °C are due to the nonlinear behavior of the GaAs material. When the temperature is above 300 °C in static bending, the specimens can sustain large plastic deformations. In fact, a 2 mm irreversible deflection at the middle of the bending specimen was observed. From the load versus deflection behavior of the specimen shown in Figure 5.20, it is clear that the GaAs wafer is no longer purely elastic at temperatures above 300 °C, and the measured data from modulus of elasticity from the static method are not valid. However, over a service temperature range from −55 °C to 125 °C, GaAs wafers are elastic. Since the change in the modulus of elasticity over this temperature range is only 3.2%, the modulus of elasticity of the GaAs wafer in the substrate cutting direction is considered to be temperature independent for design purposes.

The data shown in Figure 5.19 and Table 5.2 were recorded from specimens of width 6.12 mm. Tests conducted on other specimens had widths of 6.06 mm and

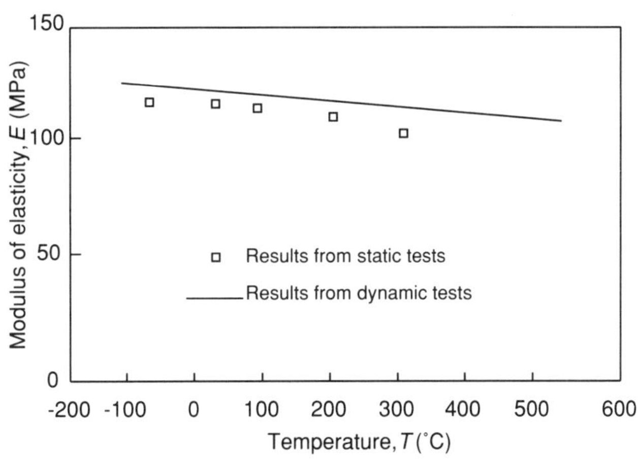

Figure 5.19 Measured modulus of elasticity.

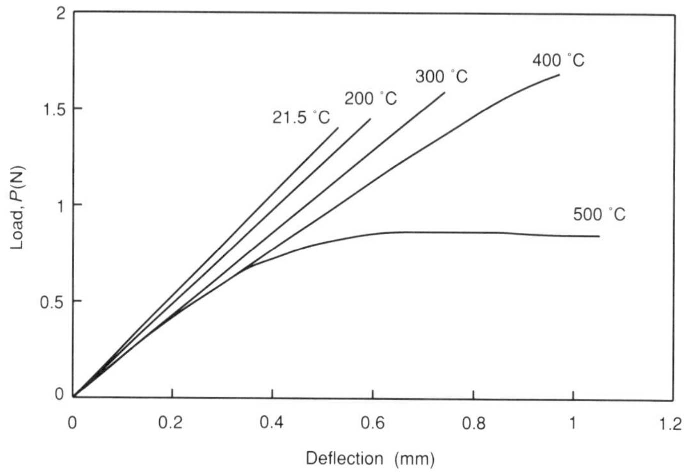

Figure 5.20 Load versus deflection behavior at different temperatures.

5.88 mm and gave the same results. The average value of modulus of elasticity obtain from the three specimen is 123.8 GPa at 23 °C.

5.6.5 Modulus of rupture

If has been demonstrated that GaAs material is linearly elastic over the temperature range of −150 °C to 300 °C. Fracture normally occurs at the surface under a tensile stress caused by flexure. The ultimate stress corresponding to failure, termed the modulus of rupture, affords a convenient parameter for determining the handling, manufacturing, and operating requirements of wafer materials. It also indicates the quality of wafers and the consistency of this quality.

The modulus of rupture is influenced by variables associated with the procedure used for its measurement, including the rate of application of stress, the test environment, the area of the specimen subject to stress, and the surface condition of specimens. ASTM C158, F417, and F394 specify the standard test methods for the modulus of rupture for glass, electric-grade ceramics, and ceramic substrates [35–37]. However, no standard test method exists for semiconductor materials. In this study, the modulus of rupture of GaAs was measured by threepoint bending tests and the effects of temperature and specimen dimensions on test results were investigated.

To determine the effect of the thickness of specimens, two different thicknesses of 0.45 m and 0.635 mm were tested. The specimens were nominally 5 × 50 mm, 7 × 50 mm, and 10 × 50 mm. The precise dimensions of each specimen were measured using a micrometer prior to testing (Tables 5.3 and 5.4). The specimens with a thickness of 0.45 mm were cut from a GaAs wafer with a diameter of 50.8 mm and the specimens with a thickness of 0.635 mm were cut from GaAs wafers of diameter 76.2 mm. Similar to the specimens for the modulus of elasticity tests, the specimens were all cut along the substrate cutting direction. To investigate

Table 5.3 Modulus of rupture for different specimen widths at 23.5 and 105 °C.

	T = 23.5 °C	
B (mm)	W (mm)	σ_U (MPa)
0.45	1.513	125.71
0.45	4.979	74.75
0.45	7.120	73.22
0.635	5.114	75.67
	T = 105 °C	
B (mm)	W (mm)	σ_U(MPa)
0.45	1.752	105.49
0.45	5.143	78.12
0.45	7.321	77.63

Table 5.4 Temperature effect on modulus of rupture (B = 0.45).

Temperature	σ_U (MPa)	W (mm)
−100 °C	70.87	5.124
23.5 °C	74.45	4.979
100 °C	78.12	5.143
200 °C	82.43	5.111

the effect of surface conditions on the modulus of rupture, three-point bending specimens with one side polished were tested, with the polished surface once in tension (polished face downwards) and once in compression (polished face upwards). All tests were conducted on a Rheometrics solid analyzer (RSA-2).

The modulus of rupture in bending is calculated by

$$\sigma_U = \frac{3P_{\max}L}{2WB^2} \tag{5.53}$$

where $P_{\max}$ is the maximum load applied to a specimen before breaking, L is the span length of three-point bending, W is the width, and B is the thickness of the specimen.

Table 5.3 lists the measured modulus of rupture at room temperature (23.5 °C) and 105 °C for the different dimensions of the specimens. Significant effects of the dimensions on the test results were observed. The narrower the width of the specimen, the higher the measured modulus of rupture. The dependence of the modulus of rupture on specimen size is expected, because the modulus of rupture is controlled by a more intrinsic material property — fracture toughness and the microcracks of the specimen. For thick and wide specimens, the possibility of a critical section containing more voids and microcracks is larger, and the modulus of rupture as a global strength property is smaller.

Table 5.4 lists the modulus of rupture for different temperatures obtained from specimens with similar widths. The values of measured modulus of rupture increase

Table 5.5 Polish effect on modulus of rupture ($T = 23.5\,^\circ$C).

Polished surface	W (mm)	σ_U (MPa)
Tension	4.979	74.45
Tension	7.120	73.22
Compression	5.121	75.55
Compression	6.878	73.31

with temperature. The modulus of rupture above 200 °C was not tested because specimen deflections were out of the range of the test machine.

The data listed in Tables 5.3 and 5.4 were obtained with the chemical-polished surface on the tension side. Table 5.5 compares the data for the polished surface in tension with that in compression. No significant differences were observed.

5.6.5.1 Fracture toughness

For brittle materials, the stress intensity factor, K, characterizes the magnitude of the stress near a crack tip as a function of the type and magnitude of the applied loading, and the type and size of a crack for a given body geometry [38]. Elastic fracture occurs when K reaches a critical value, K_c, for the maximum constraint (plane–strain) constant K_{IC}. The plane–strain fracture toughness, K_{IC}, is a material property which characterizes the crack resistance. Experimental determination of K_{IC} is possible only when specimens exceed a required thickness which depends on the crack tip plastic zone size. However, K_c under plane–stress or a transition condition between plane–strain and plane–stress can also be a characterization parameter and is used to predict failure for designs with the same material thickness and constraints [39].

The critical value of stress intensity factor determines the onset of crack extension and directly relates to fracture strength, thermal shock resistance, and fatigue behavior of chips. ASTM has developed a standard test method [40] for plane–strain fracture toughness, K_{IC}, of metallic materials. However, no standard method exists to cover non-metallic fracture toughness. Chen [30] measured plane–strain fracture toughness of single-crystal GaAs bulk material at room temperature by the Vickers diamond indentation method. The test results of different crystalline orientations are $K_{IC\{100\}} = 0.43$ MN/m$^{3/2}$, $K_{IC\{110\}} = 0.31$ MN/m$^{3/2}$, $K_{IC\{111\}} = 0.45$ MN/m$^{3/2}$.

The fracture toughness of a GaAs wafer for a mode I fracture was tested at different temperature levels [31]. The disc-shaped compact tension [DC(T)] specimen recommended by ASTM E399 [40] was adopted at room temperature, and single-edge-notched tension [SE(T)] specimens were used over the range of temperatures. The dimensions of the specimens are shown in Figure 5.21(a). The DC(T) specimen width was $W = 37.79$ mm, and the thickness was $B = 0.45$ mm. The dimensions of the SE(T) specimen were $W = 6$ mm and $L = 30$ mm (Figure 5.21(b)).

Accordingly, the conducted fracture tests were under fully plane–stress conditions. A fine crack cutting technique on the GaAs wafer was performed to generate a sharp crack tip. To investigate the initial crack length effect, specimens with different initial crack sizes were prepared.

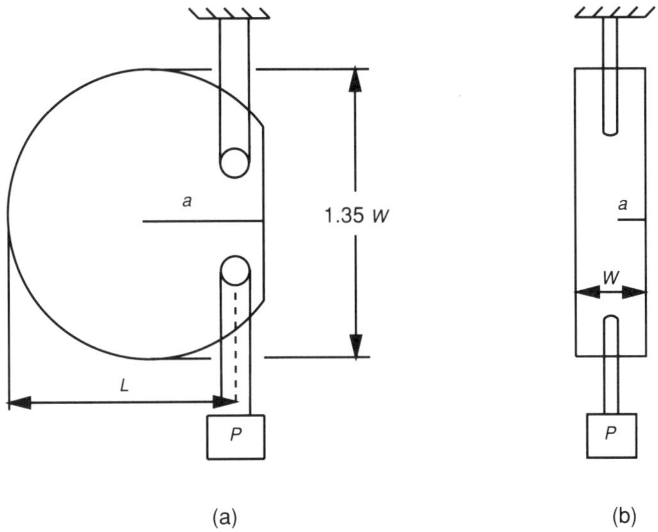

Figure 5.21 Specimen geometry of fracture testing.

For the DC(T) specimen, the critical stress intensity factor, K_c, is calculated from the following expression in units of $N/m^{3/2}$ (or ksi $in^{1/2}$)

$$K_c = \frac{P_{\max}}{B\sqrt{W}} F_1\left(\frac{a}{W}\right) \tag{5.54}$$

Here $P_{\max}$ is the maximum load applied to the specimen, B is the specimen thickness, W is the specimen width, a is the crack length, and

$$f_1\left(\frac{a}{W}\right) = \frac{2 + a/W}{(1 - a/W)^{2/3}}\left[0.76 + 4.8\frac{a}{W} - 11.58\left(\frac{a}{W}\right)^2 + 11.43\left(\frac{a}{W}\right)^3 - 4.08\left(\frac{a}{W}\right)^4\right] \tag{5.55}$$

For the SE(T) specimen, the critical stress intensity factor is calculated from the following expression in units of $MN/m^{3/2}$ (or ksi $in^{1/2}$)

$$K_c = \frac{P_{\text{Max}}\sqrt{a}}{BW} F_2\left(\frac{a}{W}\right) \tag{5.56}$$

where

$$f_1\left(\frac{a}{W}\right) = 01.99 - 0.41\frac{a}{W} + 18.7\left(\frac{a}{W}\right)^2 - 38.48\left(\frac{a}{W}\right)^3 + 53.85\left(\frac{a}{W}\right)^4 \tag{5.57}$$

In order to investigate the specimen geometry effect, the fracture toughness for a mode I fracture was tested at room temperature, using DC(T) and SE(T) specimens. Figure 5.21 shows that the DC(T) and SE(T) specimens in mode I fracture tests were hung by a ribbon structure with the load applied vertically at room temperature. The loading was increased by continually adding weights. The fracture testing at higher or lower temperatures was conducted on a Rheometrics solid analyzer with a thermal chamber. The process of temperature and loading increase was monitored by a microcomputer.

Table 5.6 lists the measured fracture toughness from a 0.45 mm thick specimen at room temperature (22 °C) with the different specimen dimensions and types. Compared with the fracture test data of silicon reported in [41], the fracture toughness values are only about one third that of silicon. Table 5.6 also compares the fracture toughness for different specimen thicknesses. The average value of K_C, $K_{C,\mathrm{Av}}$, is the sum of all values divided by the number of specimens. The overall average value of the fracture toughness at room temperature is 0.385 $\mathrm{MN/m^{3/2}}$.

Table 5.6 Plane-stress fracture toughness at room temperature (23 °C) (B = 0.45 and B = 0.635 mm).

Specimen type	B (mm)	W (mm)	a_0/W	K_c ($\mathrm{MN/m^{3/2}}$)	$K_{C,\mathrm{Av}}$ ($\mathrm{MN/m^{3/2}}$)
SE(T)	0.45	5.34	0.31	0.413	
SE(T)	0.45	6.21	0.33	0.301	
SE(T)	0.45	7.14	0.43	0.344	
SE(T)	0.45	7.45	0.51	0.515	0.393
DC(T)	0.45	37.8	0.51	0.475	
DC(T)	0.45	37.8	0.52	0.303	
DC(T)	0.45	37.8	0.71	0.344	0.374
SE(T)	0.635	11.22	0.27	0.542	
SE(T)	0.635	11.22	0.30	0.438	
SE(T)	0.635	12.61	0.49	0.538	
SE(T)	0.635	12.61	0.54	0.438	0.495

Table 5.7 Temperature effect on plane-stress fracture toughness (B = 0.45 mm).

Temperature (°C)	W (mm)	a_0 (mm)	a_0/W	K_C ($\mathrm{MN/m^{3/2}}$)	$K_{C,\mathrm{Av}}$ ($\mathrm{MN/m^{3/2}}$)
−100	6.56	2.121	0.32	0.264	
−100	5.15	2.232	0.43	0.343	
−100	7.11	3.139	0.44	0.332	0.313
23	5.34	1.655	0.31	0.413	
23	6.21	2.409	0.33	0.301	
23	7.14	3.070	0.43	0.344	
23	7.45	3.799	0.51	0.515	0.393
100	7.12	2.124	0.30	0.567	
100	5.15	1.761	0.34	0.551	
100	9.89	5.132	0.52	0.434	0.551
200	5.38	1.197	0.22	0.743	
200	5.23	2.133	0.40	0.652	
200	7.18	3.091	0.43	0.814	0.736

The temperature effect on fracture toughness was investigated by testing the SE(T) specimens at −100 °C, 0 °C, 23 °C, 100 °C, and 200 °C. Table 5.7 lists the fracture toughness of the 0.45 mm thick SE(T) specimens at different temperatures. The average value of the measured fracture toughness increased with temperature.

5.6.6 Coefficient of thermal expansion and summary

The coefficient of thermal expansion, i.e. change in length per unit length per thermal degree, of a GaAs wafer was investigated in orientations parallel and perpendicular to the cutting edge over the temperature range of −75 °C to 325 °C. Square-shaped specimens with nominal dimensions of 7 mm × 7 mm × 0.45 mm were used. Tests were performed on a Perkin-Elmer mechanical analyzer (DMA-7) using quartz fixtures. The rate of temperature increase was 5 °C per minute. A typical curve of dimensional change as a function of temperature for the GaAs wafer is shown in Figure 5.22. The least squares regression of the data in the temperature range of −50 °C to 275 °C gives the coefficient of thermal expansion shown in Table 5.8 for different specimens. The coefficient of thermal expansion measured the same in both the directions. The average of the measured coefficient of thermal expansion is $5.58 \times 10^{-6}/\,^{\circ}C$.

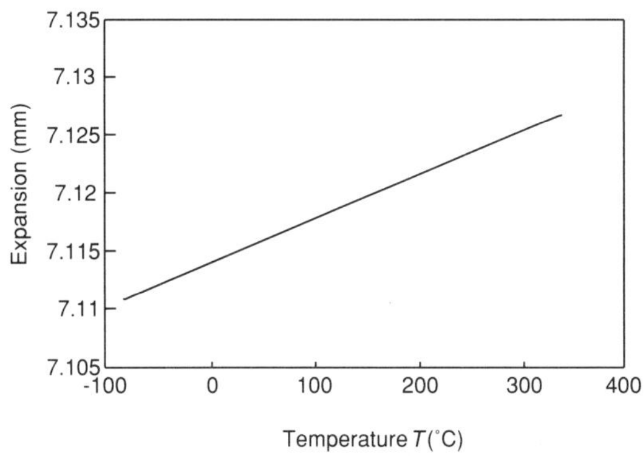

Figure 5.22 Thermal expansion of GaAs wafer material.

Table 5.8 Coefficient of thermal expansion in directions parallel to and perpendicular to die cutting edge.

$\alpha(\times 10^{-6}/^{\circ}C)$	
Parallel	Perpendicular
5.68	5.69
5.36	5.41
5.67	5.64
Average: 5.57	5.58

The above results provide very important information for GaAs devices, package reliability analysis and reliability design. The following conclusions have been drawn:

- A GaAs wafer is an elastic material over the temperature range of −125 °C to 300 °C. However, elastic–plastic behavior was observed at temperatures above 300 °C.
- The modulus of elasticity of the GaAs wafer in the substrate cutting direction is a negligible linear function of test temperature. Over the range of service temperatures of the majority of semiconductor devices in an electronic package, the modulus of elasticity changes from 126 GPa at 55°C to 121 GPa at 125 °C. Since this change is only 3.2% within the range of service temperature, it could be considered as a temperature independent constant for design purposes.
- The modulus of rupture of the GaAs wafer increases with increasing test temperature and decreasing specimen width. The effect of chemical polishing of the wafer on the modulus of elasticity was not significant. If the modulus of rupture is to be a strength parameter, the specimen geometry requirement should be specified.
- The average fracture toughness of a GaAs wafer material measured from 0.45 mm thick specimens is 0.385 $MN/m^{3/2}$ in the substrate cutting direction, which is only one third of the fracture toughness value of silicon. Temperature has a major effect on fracture toughness.
- The coefficient of thermal expansion of the GaAs wafer measured in directions parallel and perpendicular to the cutting edge gave consistent values of $5.58 \times 10^{-6}/{}^{\circ}C$.

5.7 APPROACH FOR MECHANICAL RELIABILITY DESIGN

Reliability is defined as the probability that a product will perform its intended function without failure under stated conditions for a given period of time. The concept of reliability as a probability implies that any attempt to quantify it must involve the application of statistical methods, typically in assessing the mean time to failure and the standard deviation of the lifetime. In fact, estimates of the failure rates of electronic components based on empirical models fit to field data are available in military handbooks [42]. The failure models used empirically often incorporate physical and operating parameters, but the equations are typically not derived from or based on any physics or mechanics of failure. As such, they do not give the designer any insight into or control over the actual failure mechanisms. Empirical field failure based methodologies generally assume and compute a constant hazard rate for the components. This is conceptually problematic, since the hazard rates of most electronic related failure mechanisms are known to be modeled only by a wear-out distribution (i.e. Weibull, log normal). Thus, when designing a new product or a product with new technologies where little or no field-history exists, extrapolation from field failure data of old designs may be inappropriate and misleading.

For the design of the device substrate and the substrate attachment assembly, current specifications and standard reliability test methods fail to evaluate fracture properties due to lack of knowledge of the physics of failure mechanisms. For example, MIL-STD-883C [43] specifies the acceptable initial crack length based on visual tests, without considering the fracture toughness of the semiconductor material. However, from the viewpoint of fracture mechanics, the allowable crack length must be calculated by:

$$a_c = \frac{K_c^2}{\pi(\sigma F)^2} \tag{5.58}$$

where F is a shape factor depending on the geometry of the device. Assuming the applied stresses imposed on a GaAs substrate and a silicon substrate are the same with identical geometries of the devices, the allowable crack length in GaAs is:

$$a_{c,\mathrm{GaAs}} = \frac{K_{C,\mathrm{GaAs}}^2}{K_{C},\mathrm{Si}^2} a_{c,\mathrm{Si}} = \frac{1}{9} a_{c,\mathrm{Si}} \tag{5.59}$$

where, $K_{c,\mathrm{GaAs}}$ and $K_{C,\mathrm{Si}}$ are the fracture toughness of GaAs and silicon materials, respectively. From MIL-STD-883 notice 4, the crack length in silicon substrate must be less than 0.0762 mm (3 mil). Since the fracture toughness for GaAs is about one third of that for silicon die, then, in order to prevent the GaAs substrate fracture, the allowable crack in the GaAs substrate should be less than 0.006 86 mm (0.27 mil), which is much smaller in size and much more difficult to inspect than in silicon.

Realistic reliability estimation modeling techniques are required to provide guidelines for reliability design. The goal is to provide the designer with a method to proactively incorporate reliability in the design process. A useful approach for this purpose is mechanostochastic modeling proposed by Pecht *et al.* [44, 45], which because it is based on actual physics of failure mechanisms, has the advantage of being useful for the design of new structures, using new materials and technologies. The mechano-stochastic process steps include:

- Identification of the appropriate failure mechanisms for a design assembly,
- Identification of the input parameters in the failure models, including those associated with material characteristics, damage properties, relevant geometry at failure sites, manufacturing flaws and defects, and environmental and operating loads,
- Identification of the variabilities and statistical distribution functions in each of the parameters,
- Computation of the effective reliability function.

The design is considered acceptable only if the estimated time dependent reliability function, $R(t)$, meets or exceeds a threshold value, $R^*(t)$ over the required time periods.

By understanding the failure modes and mechanisms, engineers can select materials, geometries, and manufacturing technologies to minimize the probability of failure. Of course, designers often have to make trade-offs based on functionality, availability, manufacturability, testability, yield, cost, as well as reliability. The design constraints and requirements differ with the application: in a locomotive, weight and volume generally would not be terribly important but in a spacecraft they would. Dollar cost in a military fighter airplane could be less important than high reliability, but in an automobile the opposite may be true, and so forth. Fortunately, engineers are often furnished with information such as sensitivity factors to aid in their decision process.

5.7.1 Design variables and constraints

The objective of the reliability design of the MMIC substrate or any type of substrate is to prevent the brittle fracture of the substrate and the fatigue crack propagation in the substrate during the mission life of the IC devices. The objectives of the design are to determine the geometry of the substrate and package, and to select the materials of attachment and package case so that the above mentioned failure mechanisms are prevented. The following are the design variables:

- device substrate thickness (t_d),
- device substrate dimensions (L_d, and W_d),
- attachment thickness (t_a),
- case thickness (t_c),
- attachment material properties (E_a, ν_a, α_a, k_a,), and
- package case material properties (E_c, ν_c, α_c, and k_c,).

The following variables are pre-specified requirements prior to the design. As the design constraints, they will not be changed during the design loop:

- device substrate material properties (E_d, ν_d, α_d, $K_{C,d}$, A_d, n_d, and k_d),
- device power dissipation (P),
- maximum value of operational temperature range (ΔT),
- maximum value of operational temperature ($T_{\max}$),
- minimum value of operational temperature ($T_{\min}$),
- required number of operational temperature cycles (N_f),
- life cycle mission profile (combinations of temperature cycles),
- manufacturing capabilities, and
- costs of materials and manufacturing processes.

The designed substrate and package must pass the specified functioning requirements and tests. Sufficient mechanical strength of the substrate is essential not only for the life of the device operation but also during the manufacturing. The substrate must survive the accelerated mechanical, thermal and chemical stresses that may be imposed on the circuit during processing, screening, and testing. The following variables should be included in the substrate design:

- maximum bonding temperatures of various bonding techniques ($T_{\max}$),
- maximum bonding force of various bonding techniques ($F_{\max}$),
- maximum testing and screening temperature range ($\Delta T_{\max}$), and
- the subjected temperature cycles during circuit testing and product screening (N_u).

REFERENCES

[1] Christou, A., "Reliability problems in state-of-the-art GaAs devices and circuits", *Quality and Reliability Engineering International*, **5**, 1989, 37–46.

[2] Christou, A., Hu, J.M., and Anderson, W.T., "Reliability of InGaAs HEMTs on GaAs substrate", *29th Proceedings of Reliability Physics*, IEEE Reliability Society, 1991.

[3] Timoshenko, S., "Analysis of bi-metal thermostats", *Journal of Optical Society of America*, **11**, 1925, 233–255.

[4] Chong, K.P., Engenm K., and Hartsock, J.A., "Thermal stress and deflection of sandwich panels", *Journal of the Structure Division*, ASCE, **103**, 1977, 35–49.

[5] Williams, H.E., "Asymptotic analysis of the thermal stresses in a two-layer composite with an adhesive layer", *Journal of Thermal Stress*, **8**, 1985, 183–203.

[6] Hussein, R., Ha, K., and Fazio, P., "Thermal stresses in sandwich plates", *Journal of Thermal Stress*, **12**, 1989, 333–349.

[7] Hussein, R., Ha, K., and Fazio, P., "Thermal stresses in sandwich panels with inter-layer slips", *Journal of Thermal Stress*, **12**, 1989, 191–207.

[8] Chen, W.T. and Nelson, C.W., "Thermal stresses in bonded joints", *IBM Journal of Research and Development*, **23**, 2, 1979, 179–188.

[9] Seraphim, D.P., Ladky, R.C., and Li, Che-Yu (eds.), *Principles of Electronic Packaging*, McGraw-Hill Book Company, New York, pp. 168–171, 1989.

[10] Westinghouse Electric Corporation, "Reliability analysis/assessment of advanced technologies" *Technical Report*, No. N-7-5503, Rome Air Development Center, New York, December 1989.

[11] Bolger, J.C., "Polyimide adhesive to reduce thermal stresses in LSI ceramic packages", *14th National SAMPE Technical Conference*, pp. 12–14, October 1982.

[12] Bolger, J.C. and Mooney, C.T., "Die attach in Hi-rel P-dips: polyimides or low chloride epoxies?", *IEEE Transaction on Components, and Manufacturing Technology*, **CHMT-7** (4), December 1984.

[13] Suhir, E., "Stress in adhesively bonded bi-material assemblies used in electronic packaging", *Proceedings of Material Research Society Symposium*, **72**, 1986, 133–138.

[14] Suhir, E., "Calculated thermally induced stress in adhesively bonded and soldered assemblies", *Proceedings of International Symposium on Microelectronics*, 1986.

[15] Suhir, E., "Die attachment design and its influence on thermal stresses in the die and the attachment", *Proceedings of 37th Electronic Components Conference*, 1987.

[16] Licari, James J. and Enlow, Leonard R., *Hybrid Microcircuit Technology Handbook*, Noyes Publications, New Jersey, 1988.

[17] Shigley, Joseph Edward and Mischke, Charles R., *Mechanical Engineering Design*, McGraw-Hill Book Company, New York, 1986.

[18] Paris, P.C. and Shih G.C., "Stress analysis of cracks", *Fracture Toughness Testing*, ASTM STP 381, 1965, pp. 30–81.

[19] Westergaard, H.M., "Bearing pressure and cracks", *Journal of Applied Mechanics*, **61**, 1939, A49–53.

[20] Sneddon, I.N., "The distribution of stress in the neighborhood of a crack in an elastic solid", *P. Roy Soc.*, **A187**, London, pp. 229–260, 1946.

[21] Irwin, G.R., "The crack extension force for a part-through crack in a plate", *Journal of Applied Mechanics*, 651–654, 1962.

[22] Broek, D., *Elementary Engineering Fracture Mechanics*, Martinus Nijhoff, Boston, 1986.

[23] Osgood, C.C., *Fatigue Design*, Pergamon Press, New York, 1982.

[24] Paris, P.C., Gomez, M.P., and Anderson, W.E., "A rational analytical theory of fatigue", *The Trend in Engineering*, **13**, 1961, 9–14.

[25] Miner, M.A., "Cumulative damage in fatigue", *Journal of Applied Mechanics*, September 1954.

[26] Swaminathan, V. and Copley, S.M., "Temperature and orientation dependence of plastic deformation in GaAs single crystals doped with Si, Cr, or Zn", *Journal of American Ceramic Society*, **58**(11), 1975, 482–485.

[27] Laister, D. and Jenkins, G.M., "Deformation of single crystals of Gallium Arsenide", *Journal of Material Science*, **8**(9), 1973, 1218–1232.

[28] Yonenaga, I., Onose, U., and Sumino, K., "Mechanical properties of GaAs crystals", *J. Mater. Res.*, **2**(2), 1987, 252–261.

[29] Chen, C.P., "Prediction of critical crack sizes in solar cells", *NASA Tech. Brief*, **13**(8), Item #60, from *JPL Invention Report* NPO-17637/7135, 1989.

[30] Chen, C.P. "Fracture mechanics evaluation of GaAs: final report", *JPL Publication* 84-81, Jet Propulsion Laboratory, Pasadena, California, 1984.

[31] Hu, J.M., Pecht, P., and Barker, D., "Experimental evaluation of mechanical behavior of GaAs wafer material", *Proceedings of Spring Conference of the Material Research Society*, 1991.

[32] ASTM C623-71, "Standard test methods for Young's modulus, shear modulus and Poisson's ratio for glass and glass-ceramics by resonance", *Annual Book of ASTM Standards*, **15**(2), American Society for Testing and Materials, Philadelphia, pp. 179–184, 1990.

[33] Hearmon, R.F.S., "The elastic constants of anisotropic materials", *Reviews of Modern Physics*, **18**(3), 409–440, 1946.

[34] Smith, C.O., *The Science of Engineering Materials*, 3rd ed. Prentice-Hall, Englewood Cliffs, New Jersey, pp. 168–227, 1986.

[35] ASTM C158-84, "Standard test methods for flexural testing of glass (determination of modulus of rupture)", *Annual Book of ASTM Standards*, **15**(2), American Society for Testing and Materials, Philadelphia, pp. 19–26, 1990.

[36] ASTM F394-78, "Standard test methods for flexural strength (modulus of rupture) of ceramic substrates", *Annual Book of ASTM Standards*, **10**(4), American Society for Testing and Materials, *ASME*, **24**, 1988, 327–329.

[37] ASTM F394-78, "Standard test methods for biaxial flexural strength (modulus of rupture) of ceramic substrates", *Annual Book of ASTM Standards*, **10**(4), American Society for Testing and Materials, ASME, **24**, 1988, 322–326.

[38] Irwin, G.R., "Analysis of stress and strain near the end of a crack transversing a plate", *Journal of Applied Mechanics*, ASME, **24**, 1957, 361–364.

[39] Ewalds, H.L. and Wanhill, R.J., *Fracture Mechanics*, Edward Arnold Publishers, New York, 1984.

[40] ASTM E-399, "Standard test methods for plane-strain fracture toughness of metallic materials", *Annual Book of ASTM Standards*, **3**(1), ASTM, Philadelphia, 1990.

[41] Chen, C.P. and Leipold, M.H., "Fracture toughness of silicon", *American Ceramic Society Bulletin*, **59**(4), 1980, 469–472.

[42] "Reliability prediction of electronic equipment", *MIL-STD-217E*, Department of Defense, 1986.

[43] "Military standard test method and procedures for microelectronics", *MIL-STD-883C*, Secs. 2001.2–3015.3, Department of Defense, 1983.

[44] Pecht, M., Dasgupta, A., and Barker, D., "The reliability physics approach to failure prediction modelling", *Quality and Reliability Engineering International*, **6**, 1990, 267–273.

[45] Leonard, C.T. and Pecht, M., "How failure prediction methodology affects electronic equipment design", *Quality and Reliability Engineering International*, **6**, 1990, 243–249.

6

OXIDATION AND DIELECTRICS IN MICROCIRCUIT PROCESSING

6.1 INTRODUCTION TO OXIDATION

Oxidation of silicon and the deposition of dielectrics on GaAs is a critical part of both pattern transfer and surface isolation for silicon and GaAs technologies [1, 2]. The silicon semiconductor has a native oxide which is approximately 20–100 Å thick. This, along with the existence of an integral silicon–silicon dioxide, has resulted in the development of silicon as a viable technology. Thermal oxidation is the normal technique for the growth of an oxide. In this case gases of oxygen and/or steam are passed over the silicon substrate at a temperature of 900 to 1200 °C. The amount of silicon consumed is equal to the thickness of the SiO_2 formed. Oxide properties are summarized in Table 6.1 for steam oxidation at a pressure of one atmosphere. The other two oxidation types shown in Table 6.1 are oxidation with dry oxygen and with wet oxygen. Dry oxygen typically results in a slightly higher film density. In designing an oxidation process, one must be able to calculate the key oxidation parameters.

The principal reaction for the oxidation of silicon is the combination of silicon and oxygen at elevated temperatures, resulting in silicon dioxide.

$$Si + O_2 \longrightarrow SiO_2 \tag{6.1}$$

Table 6.1 Oxide properties.

Type	Density (g/cm^2)	Resistivity (Ω-cm)	Dielectric strength (10^6 V/cm)
Dry O_2	2.24–2.27	3×10^{15}	2
Wet O_2	2.18–2.20	$3–5\times10^{15}$	3
Steam (1 atm)	2.0 –2.2	$10^{15}–10^{17}$	6.8–9

When water vapor is included for the case of oxidation with steam, the reaction is:

$$Si + 2H_2O \longrightarrow SiO_2 + 2H_2 \tag{6.2}$$

The techniques then include dry oxygen, wet oxygen, steam, and high pressure steam exposure of a silicon substrate.

The reaction kinetics must include mass transfer from the gas phase to the solid phase, with the diffusion mechanism being the concentration difference between the gaseous and solid phases. The key parameters are defined below:

J_1 = flux incident on the wafer,
h_G = gas phase mass transfer coefficient,
N_G = concentration in the gas phase,
N_0 = oxide concentration on the surface,
$J = h_G(N_G - N_0)$.

From Henry's law, the relationship between the oxide concentration and the oxide partial pressure P_0, is shown to be:

$$N_{ox} = HP_0 \tag{6.3}$$

The oxide concentration is a linear function of the oxide partial pressure.

The oxide concentration in equilibrium with the gas is defined as N^* and, in terms of Henry's constant H, is given by,

$$N^* = Hp_G \tag{6.4}$$

where

$$p_0 = N_0kT \quad \text{and} \quad p_G = N_GkT$$

Combining equations (6.3) and (6.4), a relationship is obtained between the oxide concentration and the surface concentration:

$$N_{ox} = HkTN_0 \tag{6.5}$$

Substituting for p_G from equation (6.4) results in a new relationship for the equilibrium concentration as a function of the thermal energy kT.

$$N^* = HkTN_G$$

The incident flux is

$$J_1 = \frac{h_G}{HkT}(N^* - N_{ox}) \tag{6.6}$$

The flux in the oxide is J_2 and by definition is given by:

$$J_2 = D\frac{(N_{ox} - N_i)}{a(t)} \tag{6.7}$$

where $a(t)$ = thickness of oxide layer,
N_i = concentration at Si–SiO_2 interface.

The thickness is of course a function of growth time. At the interface of Si–SiO_2, the oxidant reacts with the silicon, and is given by the third flux, J_3

$$J_3 = k_s N_i \tag{6.8}$$

At steady state $J_1 = J_2 = J_3 = J$, and the ratio of N_i to N^* is shown to be

$$\frac{N_i}{N^*} = \frac{1}{1 + (k_s/h) + (k_s/D)a(t)}$$

The total flux is thus:

$$J = J_3 = \frac{k_s N^*}{1 + (k_s/h) + (k_s/D)a(t)} \tag{6.9}$$

The growth rate is just the equilibrium flux multiplied by the number of particles diffusing per unit volume:

$$J = \text{Growth rate}/N_v$$

The growth rate is also the time derivative of the thickness $a(t)$.

$$\frac{\mathrm{d}a(t)}{\mathrm{d}t} = \frac{k_s N^*/N_v}{1 + \left(\frac{k_s}{h}\right) + \left(\frac{k_s}{D}\right) a(t)} \tag{6.10}$$

where equation (6.9) was used.

The solution for $a(t)$ is obtained by separating variables and integrating:

$$a^2(t) + Aa(t) = Bt'$$

The parameter $t' = t + \tau$ is the shifted time which takes into account the initial oxide layer of thickness a_0 at $t = 0$. The constant A is given as:

$$A = 2D\left[\frac{1}{k_s} + \frac{1}{h}\right]$$

Two limiting cases are of interest:

$$t \gg \tau \qquad \text{long oxidation times, and}$$

$$t \gg \frac{A^2}{4AB}$$

In the case of long oxidation times,

$$a^2(t) \cong Bt \qquad \text{which is parabolic growth.}$$

For short oxidation times:

$$t \ll \frac{A^2}{4B}, \quad \text{or}$$

$$a(t) \cong \frac{B}{A}t' = \frac{k_s h}{k_s + h}\frac{N^*}{N_v} \tag{6.11}$$

which is linear growth with time.

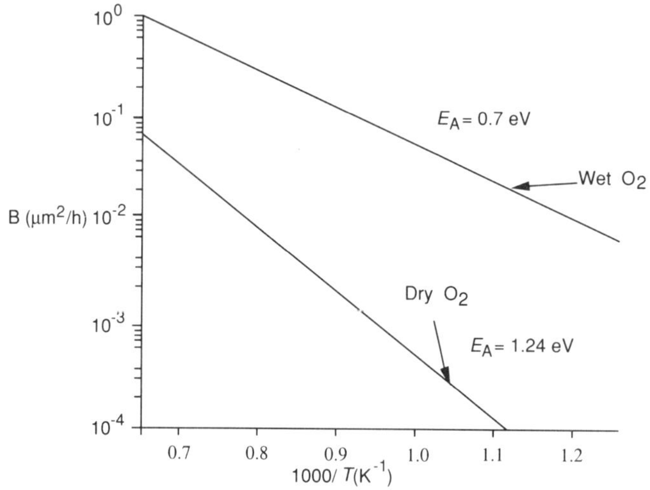

Figure 6.1 Oxidation activation energies for wet and dry oxidation.

Table 6.2 Interface concentration.

Diffusing species	N^* (cm^{-3})
O_2	5.2×10^{16}
H_2O	3.0×10^{19}

The rate constants all have a temperature dependence given in terms of an activation energy which also depends on the oxidation technique. We have:

$$k_s = k_{os}\,\mathrm{e}^{-E_A/kT}$$

$$B = B_0\,\mathrm{e}^{-E_A/kT}$$

$$A = A_0\,\mathrm{e}^{-E_A/kT}$$

The values of the activation energy for wet and dry oxidation are shown in Figure 6.1, which indicates a higher B-value for wet oxidation in comparison with dry oxidation. Typical values for interface concentration have been determined by Deal and Grove [3] and are shown in Table 6.2. These are the interface equilibrium concentrations at the Si–SiO_2 interface.

6.2 REDISTRIBUTION OF IMPURITIES DURING OXIDATION

At elevated temperatures, redistribution of impurities takes place by diffusion and will be derived in this section. In the oxide, the distribution of impurities is given by:

$$D_1\frac{\partial^2 N_1}{\partial X^2} = \frac{\partial N_1}{\partial t} \tag{6.12}$$

and in the silicon:

$$D_2 \frac{\partial^2 N_2}{\partial y^2} + V(t) \frac{\partial N_2}{\partial y} = \frac{\partial N_2}{\partial t} \tag{6.13}$$

where

$$V(t) = \frac{\sigma^2}{2ra(t)}$$

is the growth rate or velocity of the Si–SiO_2 interface and N_1 is the impurity concentration in the oxide.

For the oxide, the solution for the impurity concentration is given as:

$$N_1(x,t) = N_0 \left\{ i - \frac{mDF \operatorname{erf}\left(\frac{x}{2\sqrt{D_1 t}}\right)}{\left[mDF + \operatorname{erfc}\left(\frac{\sigma}{2r\sqrt{D_2}}\right)\right]\left[\operatorname{erf}\frac{a(t)}{2\sqrt{D_1 t}}\right]} \right\} \tag{6.14}$$

The concentration of impurities in the silicon can be shown to be:

$$N_2(y,t) = N_0 \frac{m \operatorname{erfc}\left[\frac{\sigma[(1/r) + (y/a)]}{2\sqrt{D_2}}\right]}{mDF + \operatorname{erfc}\left(\frac{\sigma}{2r\sqrt{D_2}}\right)} \tag{6.15}$$

Therefore, N_1 and N_2 are discontinuous and result in a significant redistribution of impurities directly at the interface. The four obvious cases are for buildup of impurities at the oxide side or at the silicon side. These cases are summarized in Figure 6.2.

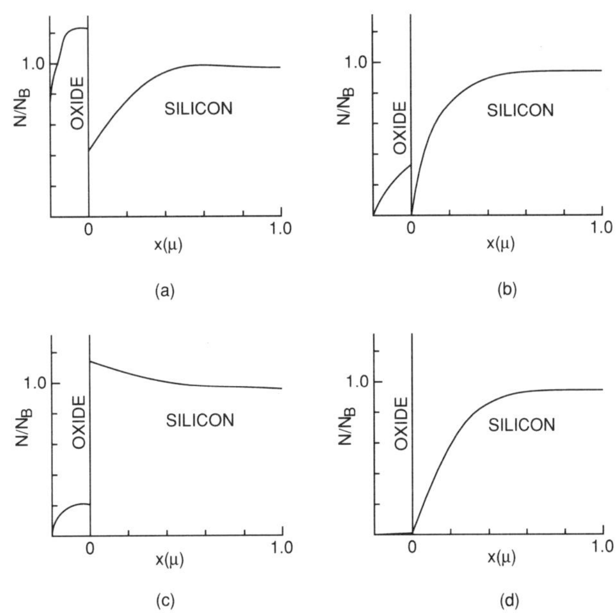

Figure 6.2 Redistribution of impurities from oxidation.

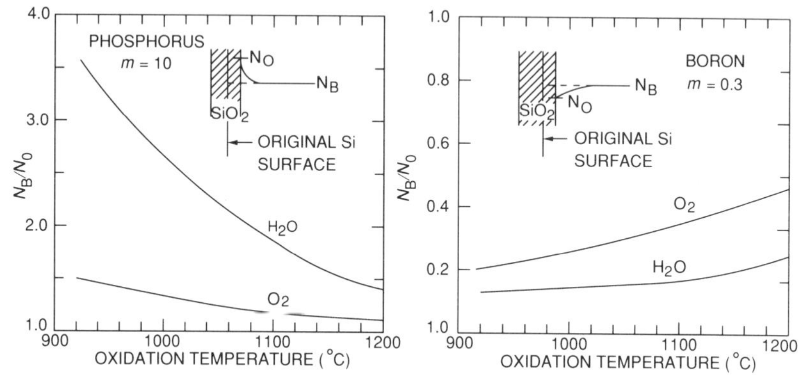

Figure 6.3 Impurity redistribution for phosphorus and boron in silicon.

In case (a), the oxide growth results in a depletion of impurities from silicon. Such a situation is expected for silicon where the m-value is less than 1. The impurity concentration may result from diffusion through the oxide as shown in case (b), where again there is a concentration depletion at the interface. When the m-factor is much larger than 1, such as the case for phosphorus, the concentration profile increases at the interface as shown in (c) and (d).

The specific cases of redistribution of phosphorus and boron are shown in Figure 6.3. Since growth kinetics of oxidation with H_2O is more rapid, the H_2O formed oxide peaks at a higher concentration for phosphorus, for which the value of "m" is taken to be ten. Likewise, when redistribution kinetics for boron result in depletion of impurities, a lower concentration is obtained for the oxides formed from wet steam at elevated temperature. It is also important to note that very little differences in oxide growth rate exist for wet and dry oxygen at temperatures in excess of 1000 °C. However, oxidation in dry oxygen at temperatures of 900 °C is much more rapid than in wet oxygen.

The only parameter which can be altered for manufacturing is thickness. The thickness is controlled by the growth temperature and the deposition technique.

6.3 THE THERMAL OXIDATION PROCESS

The oxidation rates in oxygen and water vapor have been measured by a number of exact characterization techniques. Up to 1000 Å, the oxidation process follows the following parabolic laws:

$$O_2 : X^2 = 8.4 \times 10^{10} P^{4/5} t e^{(-1.7\ q/kT)}$$

$$H_2O : X^2 = 2.54 \times 10^{13} P^{3/5} t e^{(-1.7\ q/kT)}$$

where X is the film thickness, P is the pressure, t is the time, and q is the electronic charge.

The grown films have no indicated crystalline structure. The film is essentially continuous and amorphous. If the surface impurities are not carefully removed prior to oxidation, oxide crystallites will form on the surface and will lead to non-uniform growth.

The effect of thermal oxidation is to eliminate the usual slow states observed on unoxidized silicon surfaces. The superposition of a DC bias on the AC signal does not shift the signal, indicating the absence of slow trap states. The general oxidation kinetics, as initially determined by Deal and Grove [3], show that the film thickness X can be determined by

$$X = \left[-\frac{1}{2}k_1 + \frac{1}{2}(k_1^2 + 4k_2 t)^{1/2}\right]$$

where k_1 is a surface-sensitivity parameter with units of mols/cm^3, whose value depends on temperature, crystal orientation and other surface factors. Fowkes and Hess [4] have made a detailed study of the oxidation kinetics of silicon in O_2^{2+}/Cl mixtures showing that

$$k_1 = A \times 10^4$$

$$k_2 = B \times 1.667 \times 10^6$$

Since oxygen ambients are often contaminated with trace amounts of H_2O, Irene and Van der Meulen [5] examined the effect of various amounts of H_2O contamination on the parabolic and linear rate parameters. With the introduction of small quantities of H_2O, they found an abrupt increase in the parabolic term.

In order to limit junction movement, either high pressure steam or high pressure dry oxygen oxidation is typically used. The effect of oxidant diffusion at high pressures is limited. The oxidation rate is always linear. Pressures in excess of 5 atmospheres result in highly stable shallow junctions, which are necessary for VLSI circuits.

6.4 CHEMICALLY VAPOR DEPOSITED (CVD) FILMS

Attempts to develop SiO_2 deposition techniques as alternatives to thermally grown SiO_2 were mainly driven by requirements to lower the temperature of oxidation, or to develop dielectric protective films in order to passivate the entire circuit after processing. However, with silicon device structures, thermal oxide in the active region of the device gives the best electrical performance. Thus deposited SiO_2 films are mainly used for the second layer dielectric or for protective coatings.

The most widely used method in manufacturing for depositing low temperature CVD silicon dioxide is based on the oxidation of silane (SiH_4), where the reaction is given as:

$$SiH_4(g) + 2O_2(g) \longrightarrow SiO_2(s) + 2H_2O(g)$$

The optimum temperature is in the 400–450 °C range, with the oxygen to silane ratio maintained to greater than 10. Other CVD techniques include plasma CVD, in which the reactants at relatively low pressure are excited by means of a glow discharge, so that the films can be formed at low temperatures. The CVD technique can be used to obtain other insulating films such as alumina and borosilicate glasses.

6.4.1 Sputtered dielectric films

Sputtered insulating films can be formed by either reactive sputtering from a metal or semiconductor target, or by RF sputtering from an insulating target. In the case of SiO_2, high purity silicon is used as the target material, with a sputtering ambient containing some oxygen.

When sputtered under proper conditions, SiO_2 films resemble thermally grown oxides. As expected, faster deposition rates result in lower density films; however, the effect of deposition rate is relatively small for the higher rates which are normally used for routine deposition. There are two problems associated with RF sputtered films. These are film contouring and radiation damage. The problem of contouring can be severe at metallization or semiconductor steps.

6.4.2 Silicon nitride

Silicon nitride offers an excellent barrier to the migration of ions and water and can also be used as an oxidation mask. Si_3N_4 is formed by CVD deposition in the range of 600–800 °C by the reaction of excess NH_3 with SiH_4, $SiCl_4$, or $SiBr_4$.

Silicon nitride may also be deposited at low temperatures (200–300 °C) in the presence of an RF plasma. These films, however, tend to have a lower density, but may be used as the final protective coating on some devices, especially in GaAs technology.

6.5 CHARGE STABILITY AND RELIABILITY IMPLICATIONS

Thermal SiO_2 has a resistivity greater than 10^{16} Ω-cm, a dielectric strength of 10^7 V/cm, as well as a low surface state density of below 10^{10} states (cm^2 eV). Charges occurring in the oxidized silicon structure must be minimized and controlled for a successful device product [6]. The charges present are fixed oxide charges (N_{ox}), interface state charges (N_{fs}), mobile ionic charges (N_{ion}), and radiation induced trapped charges (N_{rad}), as shown in Figure 6.4.

The fixed oxide charge has been reported to be positive in polarity, immobile and stable. However N_{ox} is found to be process dependent, and is a function of the final high-temperature treatment. The oxidation process itself introduces the fixed charge, which can be reduced by annealing in an inert ambient. Although N_{ox} is intrinsic to the oxidation process, it does not depend on the oxidation rate in dry O_2.

In summary, N_{ox} does not exchange charge when the silicon N_{ox} is independent of silicon impurity type and concentration, and it is stable. The charge is located very close to the silicon–silicon dioxide interface and is strongly dependent on temperature [7].

Other states of interest are located at the Si–SiO_2 interface and have energies within the silicon bandgap [8]. These states are in equilibrium and can exchange charge quickly with silicon when probed at very high frequencies. They are called "fast surface states" (N_{fs}) and the total charge associated with them depends on whether they are acceptor or donor states. Likewise, they are functions of the

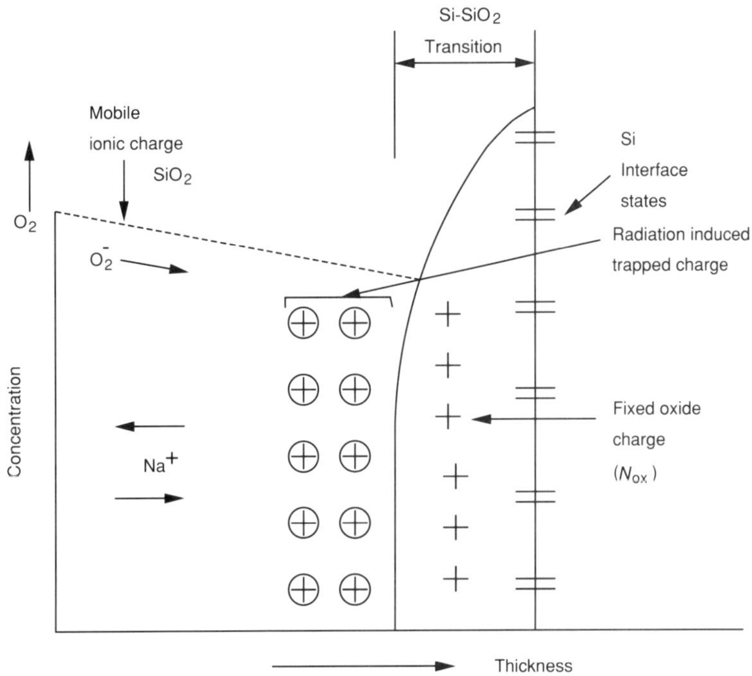

Figure 6.4 Location of charges which affect device performance and reliability.

final oxidation temperature, with the lowest N_{fs} being achieved at the highest temperature. Figure 6.5 shows the location of the donor and acceptor states within the bandgap.

The temperature dependence of the fast states affects the stability of oxidized manufactured devices. It is agreed in the literature that N_{fs} decreases upon further high temperature processing, as long as impurities are not introduced in the oxide films during annealing. The most distinguished characteristic of fast surface states is their ability to anneal out at low temperatures (300–500 °C) in the presence of oxygen. The surface state removal is due to hydrogen reacting with unsaturated silicon atoms in the interface, forming Si–H bonds [9].

A serious problem encountered in both silicon and GaAs technologies is due to instabilities caused by ionic drift in SiO_2. MOSFETs, when they contain mobile charges, have unstable C-V characteristics which shift with the application of bias. Sodium contamination is the problem, the cause of mobile charge resulting in a large positive bias instability. Therefore, the main emphasis must be in removing all possible contamination sources from the oxidation process. The utilization of water in processing can mobilize or activate normally immobile sodium by the reaction:

$$\mathrm{Si-ONa} = \mathrm{H_2O} \longrightarrow \mathrm{Si-OH} + \mathrm{Na^+} = \mathrm{OH^-}$$

The presence of mobile ionic charge is disastrous in CMOS, since charge motion within the gate dielectric will cause a substantial change in device characteristics as a function of time under operating conditions. Threshold voltage drifts and

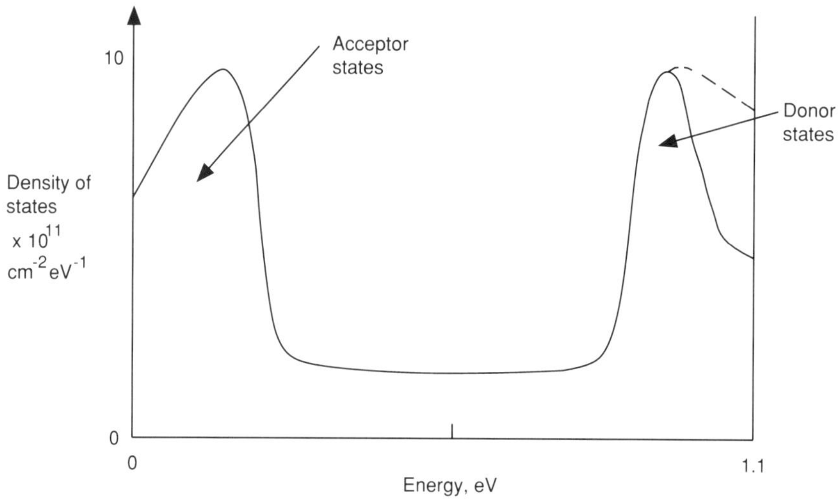

Figure 6.5 Donor and acceptor states characteristic of fast states within the bandgap of silicon.

therefore channel conductance changes and, in the case of n-channel devices, the normally off enhancement mode transistors, can be changed to normally on devices. It has also been shown by Di Stefano [17] that Na^+ ions at the $Si-SiO_2$ interface can drastically lower the barrier height for electron injection, causing increased gate leakage and eventual dielectric breakdown. The increased concentration of Na^+ at small localized sites is typically many orders of magnitude greater than the average ion sodium concentration and can result in localized dielectric breakdown.

The N_{ion} concentration is determined by the I–V characteristic technique [10]. With metal evaporation from clean sources, a mobile ion concentration of less than 10^9 mobile charges/cm^2 can result.

6.5.1 Radiation induced trapped charge

The results of irradiating MOSFETS and CMOS ICs are the following:

(1) A buildup of fixed positive charge in the proximity of the $Si-SiO_2$ interface occurs. The charge buildup increases with radiation dose and eventually saturates. The magnitude of the effective charge at saturation is dependent on the bias during irradiation. There is some evidence that positive charge may also build up near the metal–SiO_2 interface.

(2) The creation of fast surface states at the $Si-SiO_2$ interface is observed. These effects however may be eliminated by thermal annealing at temperatures greater than 150 °C or by UV exposure. What occurs, physically, is that the energy transfer at the Si–O network liberates electrons and holes, resulting in rapid electron diffusion and hole trapping or recombination. Therefore, the

concentration of trapped holes is greatest near the interfaces, but may be redistributed due to fields in the oxide. These traps at the interface do not affect CV behavior of CMOS circuits, but can be filled rapidly when hot electrons are provided, resulting in flat bond voltage shifts (V_{FB}) of several volts. The traps are usually not annealed out as a result of post-metallization anneals. In summary, all defects created during irradiation are potential electron traps and low temperature anneals do not remove any of them. The anneal treatments result in compensation.

6.5.2 Negative bias instability

CMOS integrated circuits also suffer from a dielectric induced instability at negative bias. The shift is always in the direction which indicates that positive charge is present in the SiO_2. This instability is due to "show trapping" and has an activation energy of 1.05 eV. Typically, hole injection from the silicon valence band into donor states of SiO_2 occurs. In the flat band condition, the ground states of the trap centers lie below the top of the Si valence band. Under the influence of negative bias, some of the ground states lie above the Fermi level. These active centers become positively charged by the thermal emission of electrons to the excited states, from which they can tunnel to the silicon. Figure 6.6 shows the energy diagram of the oxide centers.

There is no clear explanation of the causes or the defects responsible for the instabilities. There is a dependence of ΔV on the processing variables, strongly suggesting that the negatively biased activated centers are related to the other oxide centers introduced during manufacturing.

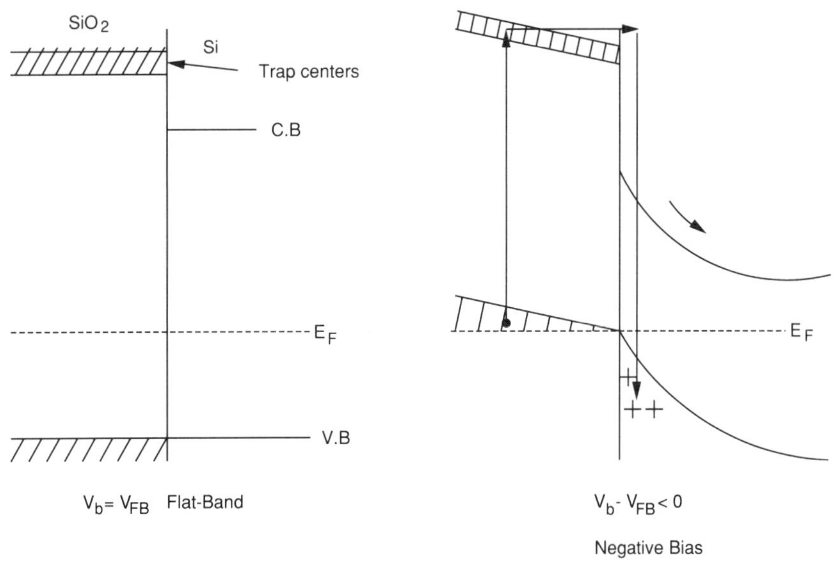

Figure 6.6 Energy band diagram showing the effect of negative bias.

6.5.3 Hot carrier trapping in SiO_2

Carriers can be injected into SiO_2 dielectrics of CMOS devices by tunneling, ionizing radiation or avalanche multiplication. These injected carriers become trapped and cause problems in integrated circuits. Positive charge can also accumulate in SiO_2 due to injection and trapping of hot holes. Threshold shifts corresponding to about 10^{12} trapped charges/cm^2 are usually achieved. Holes may be injected into SiO_2 in a variety of ways. The injection barrier is only about 0.4 eV. Once in the SiO_2, they become immediately trapped due to the large trapping cross-section.

Electron trapping after injection is particularly important since electron injection occurs easily. This is of prime concern in VLSI circuits, where hot electron effects can impose a limit on performance and transistor feature size. Electron trapping behavior is more complicated than hole trapping. With narrow channel MOSFETs, electrons can easily surmount the Si–SiO_2 barrier (3.25 eV) and enter the oxide. If there is a source of electrons in the substrate, hot electron injection may easily occur and trapping in the gate becomes important. Thermal annealing can minimize the density of some of the traps, but it is important to minimize all impurities and damage during manufacturing [11].

6.6 SUMMARY

The dielectric surface provides the first level of protection of a manufactured microcircuit and therefore may mask a series of defects prior to utilization. The dielectric is shown to serve two roles: that of surface passivation or as a key field controlling layer in a CMOS circuit. The key manufacturing variables are vacuum or system contamination level and surface properties. In general Si_3N_4 offers an excellent barrier to the migration of ions and water and may also be used in both silicon and GaAs technologies. Hot carrier trapping in SiO_2 continues to affect the reliability, performance and yield of CMOS circuits. In VLSI circuits, a majority of the yield reducing problems are related to electron injection. Thermal annealing after manufacture may minimize many of the defects.

REFERENCES

[1] Amick, J.A., Schuable, G.L., and Vossen, J.L., "Deposition technique for dielectric films on semiconductor devices", *J. Vac. Sci. Technol.*, 1977, 1053–1063.

[2] Balk, P., "Surface properties of oxidized germanium-doped silicon", *J. Electrochem. Soc.*, **118**, 1971, 494.

[3] Deal, B.E. and Grove, A. So., "General relationship for the thermal oxidation of silicon", *J. Appl. Phys.*, **36**, 1965, 3770–3778.

[4] Fowkes, F.M. and Hess, D.W., "Control of fixed charge at Si-SiO_2 interface by oxidation-reduction treatment", *Appl. Phys. Lett.*, **22**, 1973, 377–379.

[5] Irene, E.A. and Van der Meulen, Y.J., "Silicon oxidation studies: analysis of SiO_2 film growth data", *J. Electrochem. Soc.*, **123**, 1976, 1380–1384.

[6] Goronkin, H., "Origin of the fixed charge in thermally oxidized silicon", *J. Electrochem Soc.*, **124**, 1977, 314–317.

[7] Di Stefano, T.H., "Field dependent internal photoemission probe of the electronic structure of the Si–SiO_2 Interface", *J. Vac. Sci. Technol.*, **13**, 1976, 856–859.

[8] Fanet, J.M. and Poirier, R., "Charge storage in SiO_2 under low-energy electron bombardment", *Appl. Phys. Lett.*, **25**, 1974, 183–185.

[9] Kern, W., "Densification of vapor-deposited phosphosilicate glass films", *RCA Rev.*, **37**, 1976, 55–77.

[10] Ning, T.H., "Capture cross section and trap concentration of hole in silicon dioxide", *J. Appl. Phys.*, **47**, 1976, 1079–1081.

[11] Vossen, J.L., "Growth, synergism, and advances — The American Vacuum Society in 1980", *J. Vac-Sci. Technol.*, **8**, 1981, 135–142.

7

RELIABLE ACTIVE LAYER DOPING BY DIFFUSION AND ION IMPLANTATION

7.1 INTRODUCTION TO ACTIVE LAYER FORMATION

In addition to photolithography which will be presented in Chapter 8, a critical manufacturing step is the doping of the active layers in order to attain the designed impurity concentration in the transistor structure. In this chapter, we will concentrate on the two available processing techniques necessary to achieve these design objectives. The physics of diffusion and ion implantation will be reviewed as it pertains to the silicon and GaAs technologies. The methodology of treating grain boundary diffusion is emphasized in the present chapter, since enhanced diffusion occurs as a result of grain boundary mechanisms. Likewise the defects introduced by ion implantation are also presented.

7.1.1 Diffusion

Electrical activity in semiconductors is induced by thermal diffusion or by ion implantation. Diffusion involves the movement of atoms at elevated temperatures. The three types of diffusion which affect semiconductor performance are: bulk, grain boundary and surface diffusion. For doping by diffusion, the key equations are given in terms of concentration $c(x, t)$ and diffusion coefficient D. The concentration may be found through solution of the well known partial differential equation which relates doping or diffusing concentration to position and time.

$$\frac{\partial c(x,t)}{\partial t} = D\frac{\partial^2 c(x,t)}{\partial x^2} \tag{7.1}$$

For a constant source, c_s, at the surface, the solution of equation (7.1) is a complementary error function shown below:

$$c(x,t) = c_s \text{erfc}\left(\frac{x}{2\sqrt{DT}}\right) \tag{7.2}$$

For a fixed source on the surface given by c_s, the impurity distribution after an implantation and subsequent anneal is:

$$c(x,t) = c_s \exp\left(\frac{-x^2}{4DT}\right) \tag{7.3}$$

$$\text{where} \quad c_s = \frac{Q_0}{\sqrt{\pi Dt}} \quad \text{and} \quad D(T) = D_0 \exp\left(\frac{-E_a}{KT}\right) \tag{7.4}$$

The constant Q_0 is the total number of ions made available for the subsequent diffusion drive, while $D(t)$ is the temperature dependent diffusion coefficient. Grain boundary diffusion can be examined by the model shown in Figure 7.1, where the grain boundary separates two crystalline regions present on each side. The grain boundary width is taken as δ, with concentration c_b and a lattice concentration of c_l. Surface diffusion will also occur at the $Y = 0$ plane.

The concentration in the crystalline region of the specimen can be found by solving the previous differential equation for c_l:

$$D_1 \nabla^2 c_1 = \frac{\partial c_1}{\partial t} \tag{7.5}$$

Likewise, the grain boundary concentration may be found from the following equation, in terms of the grain boundary diffusion D_b:

$$D_b \nabla^2 c_b = \frac{\partial c_b}{\partial t} \tag{7.6}$$

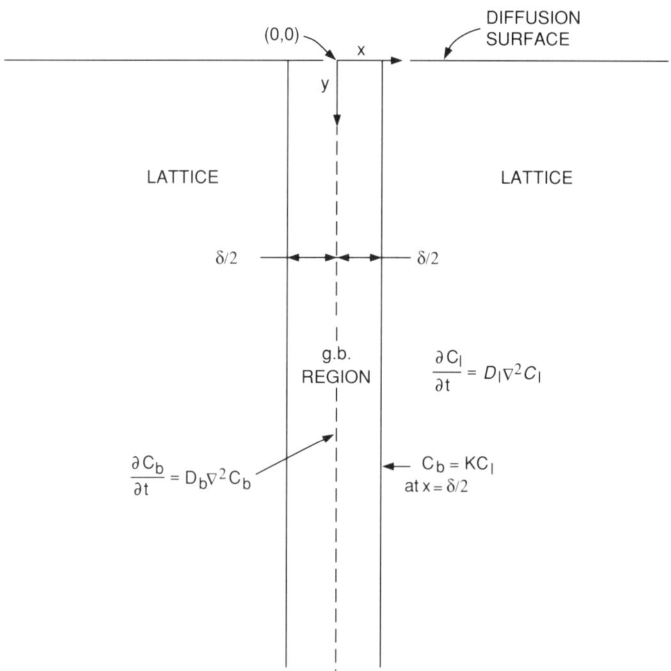

Figure 7.1 Geometric model for grain boundary diffusion in the B-kinetics regime.

The interface is characterized by continuity conditions for the concentrations and concentration gradients,

$$c_b = Kc_1$$
$$D_b\frac{\partial c_b}{\partial x} = D_1\frac{\partial c_1}{\partial x} \tag{7.7}$$

Whipple approximated the concentration within the boundary by the even function

$$c_b(x, y, t,) = c_b^1(y, t) + \frac{x^2}{2}c_b^{11}(y, t) \tag{7.8}$$

which is centered at $X = 0$ with boundary width $\delta/2$ on each of its sides. Combining equations (7.5)–(7.8), we obtain a partial differential equation for the grain boundary region.

$$D_b\frac{\partial^2 c_1}{\partial y_2} - \frac{2D_1}{K\delta}\frac{\partial c_1}{\partial x} = \left(\frac{D_b}{D_1} - 1\right)\frac{\partial c_1}{\partial t} \tag{7.9}$$

For a polycrystalline sample of average grain size d, the lattice concentration c_l is assumed to take the form:

$$c_l = c_1 + \frac{1}{d}c_2 \tag{7.10}$$

Then the boundary diffusivity is obtained by substituting equation (7.10) into (7.9).

$$KD_b\delta = \left(\frac{\partial \ln c_2}{\partial y^{6/5}}\right)\left(\frac{4D_1}{t}\right)^{1/2}\left(\frac{\partial \ln c_2}{\partial (y)^{6/5}}\right)^{5/6} \tag{7.11}$$

Equation (7.11) is the grain boundary diffusivity equation for a polycrystalline sample, in terms of D_b and D_l, as a function of distance into the boundary from the free surface.

For the case of an instantaneous source and a reflecting boundary, where the gradient of concentration at $X = 0$ is also zero, equation (7.11) may be solved by a separation of variables as shown below:

$$c(y, t) = M\delta(y)\delta(t); \qquad \frac{\partial c(o, t)}{\partial y} = \frac{\partial c(l, t)}{\partial y} = 0 \tag{7.12}$$

The intermediate concentrations c_1 and c_2 are exponentially decaying oscillating functions given by,

$$c_1 = \frac{2M}{n_0\sqrt{D_1 t}}\sum_{m=1}^{\infty}\cos(\mu_{2m}n)\exp(-\mu_{2m}^2) \tag{7.13}$$

$$c_2 = \frac{-2M}{n_0\sqrt{D_i t}}\int_1^{\Delta} \mathrm{d}\sigma \sum_{m=1}^{\infty}\cos(\mu_{2m}n)\exp(-\mu_{2m}^2\sigma)\mu_{2m}^2 \times \operatorname{erfc}\left[\frac{\dfrac{\varepsilon + (\sigma - 1)}{K\beta}}{2\left(\dfrac{\Delta - \sigma}{\delta - 1}\right)^{1/2}}\right] \tag{7.14}$$

The concentration c_2 is merely a section integral, including all section widths Δ defining a cross-section σ. Therefore c_2 may be written in terms of an integral over the entire crosssection.

$$c_2 = \frac{8M}{n_0}\int_1^{\Delta} \mathrm{d}\sigma \sum_{m=1}^{\infty} \cos(\mu_{2m}n)\exp(-\mu_{2m}^2\sigma)\mu_{2m}^2 \times [\exp(-X^2)/\sqrt{\pi} - X \operatorname{erfc} X]\sqrt{\frac{\delta-\sigma}{\delta-1}} \tag{7.15}$$

The experimentally measured profile will correspond to a sum of c_1 and c_2 as given by the previous equation, $c_l = c_1 + (1/d)c_2$, except that now c_1 refers to the solution of equations (7.13) and (7.14), and c_2 is given by equation (7.15).

We now solve equation (7.11) for a constant source with the following boundary conditions:

$$c(H,o,t) = c_0; \qquad \frac{\partial c(H,l,t)}{\partial y} = 0$$

The concentrations c_1 and c_2 in terms of the constant source may be expressed as:

$$c_1 = 1 - \frac{4}{\pi}\sum_{m=1}^{\infty}\frac{1}{(2m-1)}\sin(\mu_{2m-1}n)\exp(-\mu_{2m-1}^2) \tag{7.16}$$

$$c_2 = \frac{2c_0}{n_0}\int_1^{\Delta} \mathrm{d}\sigma \sum_{m=1}^{\infty}\sin(\mu_{2m-1}n)\exp(-\mu_{2m-1}^2\sigma)\mu_{2m-1} \times \operatorname{erfc}\left[\frac{\dfrac{\varepsilon+(\sigma-1)}{K\beta}}{2\left(\dfrac{\delta-\sigma}{\delta-1}\right)^{1/2}}\right] \tag{7.17}$$

The quantity n_0 is given by

$$n_0 = \frac{1}{\sqrt{D_1 t}}$$

and μ_{2m-1} by

$$\mu_{2m-1} = \frac{(2m-1)\pi}{2n_0}$$

The section integral in terms of n_0 takes the form:

$$c_2 = \frac{8\sqrt{D_1 t}c_0}{\mathrm{d}n_0}\int_1^{\Delta} \mathrm{d}s \sum_{m=1}^{\infty}\sin(\mu_{2m-1}n)\exp(-\mu_{2m-1}^2\sigma)\mu_{2m-1} \tag{7.18}$$

$$c_2 = \frac{8\sqrt{D_1 t}c_0}{\mathrm{d}n_0}\int_1^{\Delta} \mathrm{d}\sigma \sum_{m=1}^{\infty}\sin(\mu_{2m-1}n)\exp(-\mu_{2m-1}^2\sigma)\mu_{2m-1} \times [\exp(-x^2)/\sqrt{\pi} - X \operatorname{erfc} X]\sqrt{\frac{\delta-\sigma}{\delta-1}} \tag{7.19}$$

Table 7.1 Results of thin film diffusion measured by surface techniques and lateral spreading methods.

Diffusant†/matrix	Method of measurement	Temperature (°C)	Diffusivity (cm^2/sec)	Activation energy (eV)
Ag/Au	AES surface accumulation	30–260	$10^{-5}(D^0b)$	0.63
Ag/Au	AES sputter profiling	30	$\sim 2 \times 10^{-18}$	
Ag/Au	He ion scattering	30	$\sim 10^{-14}$	
Ag/Au	AES surface accumulation	150–260	$10^{-6}(D^0b)$	1.1
Cr/Au	AES surface accumulation	210–293	$10^{-3}(D^0b)$	1.09
Cu/Au-0.2 Co	AES surface accumulation	100 125 150	6.4×10^{-15} 3.6×10^{-14} 1.3×10^{-13}	
Cu/Au	He ion backscattering	25	$\sim 10^{-17}$	
Co/Au-0.86 Co	AES sputter profiling	100 150	3.3×10^{-18} 2.4×10^{-16}	
Cu/Al	AES sputter profiling	175	8.6×10^{-13}	
Cu/Al	Lateral spreading	150–300	$10(D^0b)$	0.81
Au/Ag	ESCA surface accumulation	141	4.3×10^{-17}	
Pt/Cr	AES surface accumulation	400–580	$10^{-2}(D^0b)$	1.69
Ag/Cu	AES surface accumulation	400	$\sim 10^{-14}$	
Pb/Cu	AES surface accumulation	125–230		0.52

Therefore, the solution of both equations results in a direct determination of c_2. Knowing c_2, the grain boundary diffusivity is determined using:

$$D_b = \frac{1}{K\delta}\left(\frac{\partial \ln c_2}{\partial y^{6/5}}\right)\left(\frac{4D_1}{t}\right)^{1/2}\left(\frac{\partial c_2}{\partial y^{6/5}}\right)^{5/6} \tag{7.20}$$

The solutions of equation (7.20) generally agree with experimental results for thin films on semiconductor substrates. Typically, from the concentration versus depth data, one can easily separate the regions of grain boundary diffusion, lattice diffusion, and dislocation enhanced diffusion.

Figure 7.2 shows the effect of enhanced diffusion through thin film structures. The change in diffusion mechanisms as a function of temperature is shown. The determination of D_b using the analytical techniques described, is shown in Table 7.1. The first 0.1 μm of the diffusion profile of Figure 7.2 shows that grain boundary diffusion of Au into MgO (polycrystalline) dominates over all of the

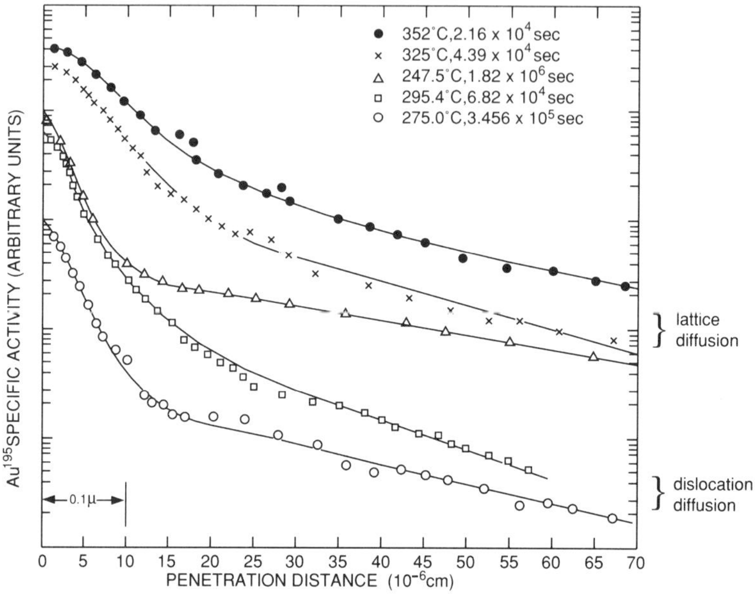

Figure 7.2 Au^{195} tracer-penetration profiles in (001) epitaxial Au films on MgO substrates. Note the sharp transition from lattice to dislocation diffusion in the lowest temperature profile.

temperature regions. Lattice diffusion clearly dominates at temperatures above 325 °C. A difference in concentration slope between lattice diffusion and dislocation diffusion is shown in Figure 7.2.

7.1.2 Diffusion of Zn $\longrightarrow$ GaAs from SiO_2

In the solution of the diffusion equations we have assumed that D is constant. We now consider the case where the diffusion coefficient is not constant. The typical diffusion temperatures are 800–900 °C. Zn diffusion produces p layers in GaAs and is used for optoelectronic components. The Zn diffusion coefficient can be determined by the Boltzmann–Matano method, where D is a function of concentration for a given temperature. The diffusion equation may be written as:

$$\frac{\partial c}{\partial t} = \frac{\partial}{\partial x}\left[D(c)\frac{\partial c}{\partial x}\right]$$

with the following boundary conditions.

$$c = c_{\text{surf}} \qquad x < 0, t = 0$$

$$c = 0 \qquad x > 0, t = 0$$

The above equation can be solved for the diffusion coefficient at concentration c_1:

$$D(c_1) = \frac{1}{2t}\frac{\int_0^{c_1} x\mathrm{d}c}{(\mathrm{d}c/\mathrm{d}x)_{c=c_1}} \tag{7.21}$$

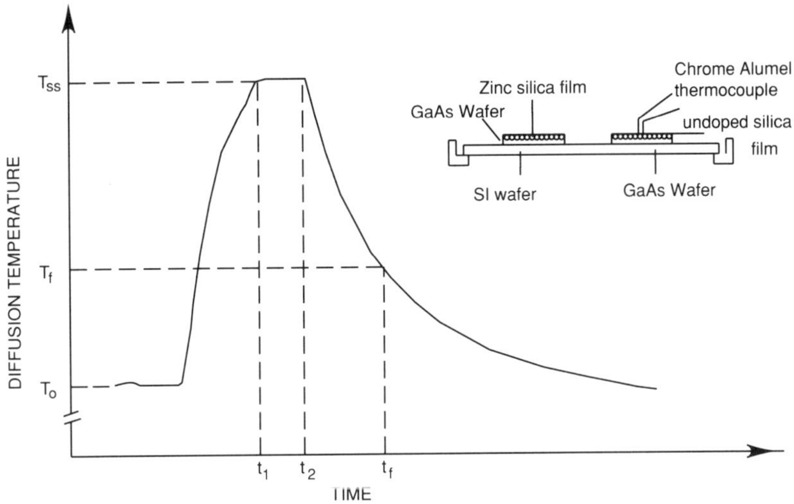

Figure 7.3 Diffusion temperature versus time for RTA drive of Zn into GaAs.

with boundary conditions now given as the integral over the surface concentration:

$$\int_0^{c_{\text{surf}}} x \, \mathrm{d}c = 0 \tag{7.22}$$

In experimentally solving equations (7.21) and (7.22), the heating cycle must be optimized in terms of duration Δt and temperature. Rapid thermal annealing (RTA) now produces a similar effect as long-term anneal, by achieving a quick increase in temperature. In such a configuration, the key times are t_1 and t_f. Figure 7.3 shows the heating cycle which must be employed in order to achieve an effective Zn doping of GaAs. The results of the Zn diffusion experiments are shown in Figure 7.4. The diffusion temperature is therefore used to control both electrical properties and sheet resistance. From Figure 7.3 the effective duration at temperature T_{ss} is $t_2 - t_1$, while a mid-temperature T_f may be also used as the approximate temperature where diffusion would no longer be significant. The diffusion temperature must also be optimized in terms of its electrical properties, such as sheet resistance and mobility, as shown in Figure 7.4. The optimum diffusion temperature in order to attain the highest p-mobility possible is shown to be 800 °C. Figure 7.5 illustrates the diffusion profile, indicating a c^2 dependence.

7.2 LAYOUT DESIGN GUIDELINES

The diffusion equation allows us to set the critical lateral and vertical dimensions. Having completed the explanation of diffusion, we may proceed to list the design guidelines for the geometric layout of integrated circuits. The layout may logically be carried out after the design of the diffusion and impurity profiles, since the decision point for proceeding to manufacturing has been met and passed. The design guidelines are listed as follows:

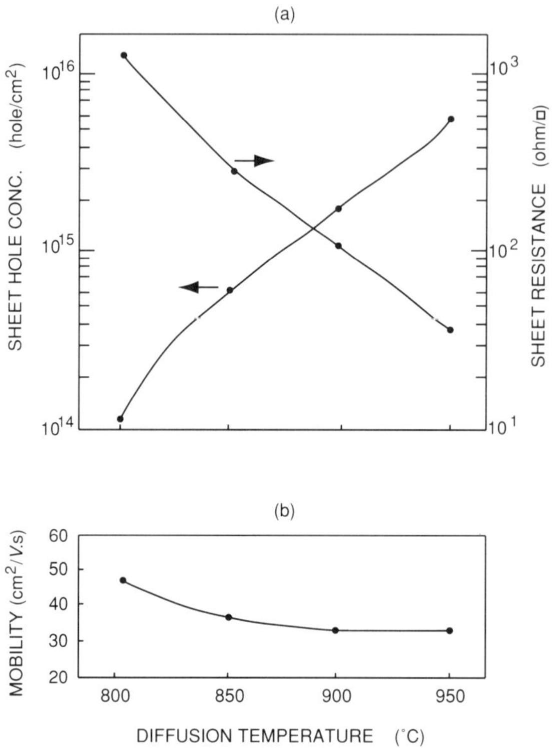

Figure 7.4 Diffusion of Zn into GaAs.

(1) Minimize the area so as to obtain a low capacitance and high yield. We must recall that yield varies inversely with chip area, and therefore both objectives may be met by designing the smallest chip possible. The low capacitance is required in order to minimize the requirements which will be placed on the packaging effort.

(2) Group components together to minimize the number of isolated regions required. The grouping optimization will further lower chip capacitance and eliminate leaky dielectric regions on the chip. Long interconnections over the dielectric region will also be avoided.

(3) Avoid large area isolation junctions in order to avert large leakage currents. Reverse polarity junctions may be used in the case of highly doped p^+ or n^+ substrates. Large area junctions will result in larger leakage currents.

(4) Minimize the number of interconnect cross-overs, since each cross-over has parasitic resistance and capacitance. In addition, reliability related problems of interconnect cross-overs such as pin-holes and first and second layer metallization interdiffusion effects will be minimized.

(5) The layout should be arranged in order to give a uniform power dissipation, and therefore eliminate chip hot-spots. Elimination of hot-spots also eliminates diffusion effects, concentration gradients, and electromigration.

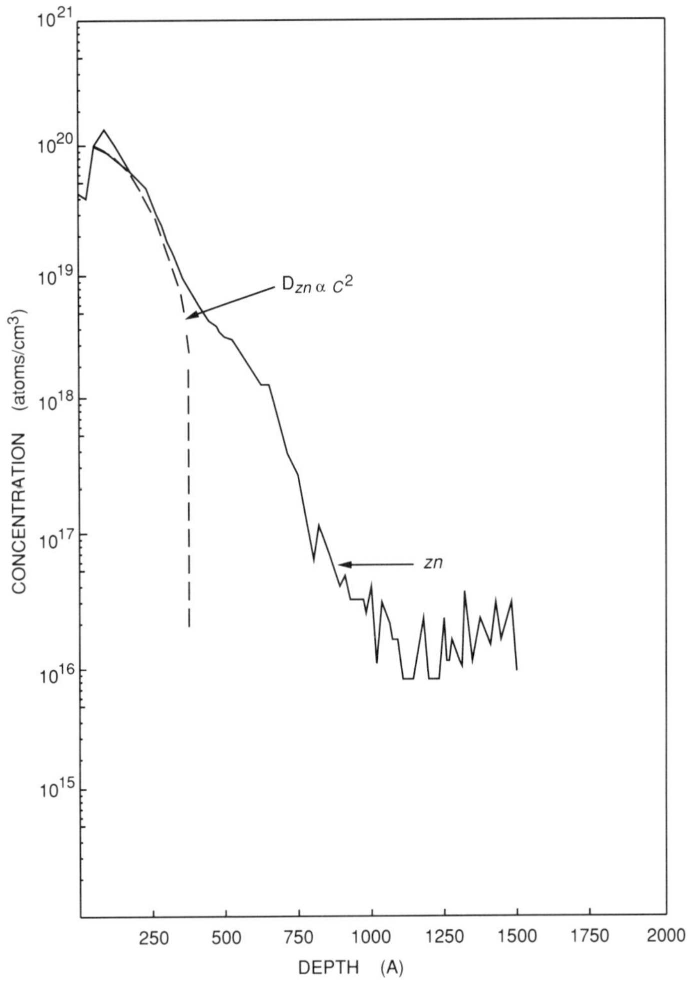

Figure 7.5 The D_{zn} fits a c^2 curve for the initial stages of diffusion.

(6) The layout is arranged so that bond pads are on the periphery. Bond pads should not be placed over buried components. In cases where the bond pads are not at the chip periphery, flip chip technologies must be used which also necessitate specialized packaging requirements.

The processing of bipolar ICs may be used to illustrate the process design guidelines. It is essentially a seven mask level process summarized in Figure 7.6. The bipolar process must include a buried layer diffusion, an isolation diffusion, a base and emitter diffusion. The seven mask processes have been designed to form the isolation regions prior to the formation of the active base and emitters.

The diffusion selection must take into account the background doping concentration. The drive-in then is calculated keeping the total impurity concentration constant, as shown in Figure 7.7. The process sequence, which requires a diffusion

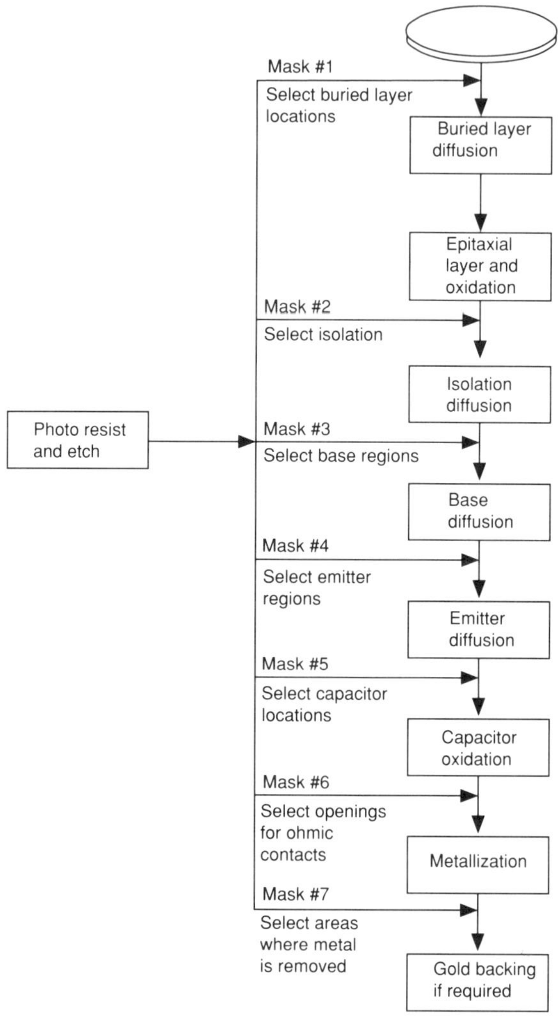

Figure 7.6 Bipolar process showing a seven-mask process.

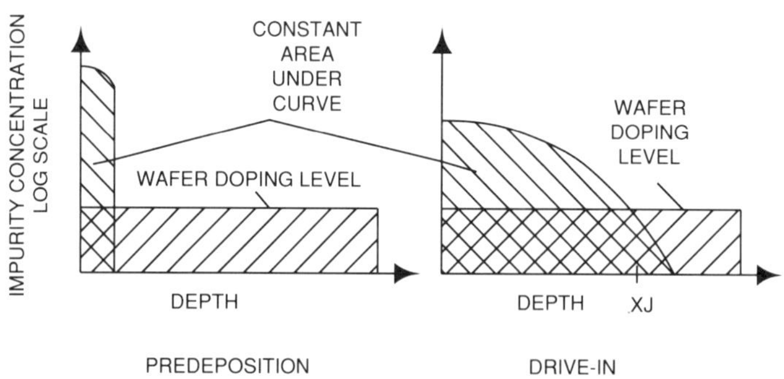

Figure 7.7 Diffusion profile and drive-in profile.

drive-in must be carried out at temperatures as high as 900 °C, so that it will not result in a low melting temperature metallization. Therefore, from Figure 7.6 the metallization steps are shown to be the final two processing steps.

7.3 ION IMPLANTATION FOR ACTIVE LAYER DOPING

Ion implantation provides exact control over impurity profiles in semiconductors, since the average depth depends on energy and the number of ions implanted depends on the ion beam current. In understanding ion implantation. We must be able to apply range distribution theory and also be able to calculate impurity distributions based on target cross-sections and stopping powers. Many of these calculations have been carried out and the data is available in tables of stepping powers.

Range distribution theory will be used to determine impurity profiles for the ion species of interest which has been implanted into a semiconductor of density ρ. Range distribution functions may be calculated by Boltzmann Transport or Monte Carlo techniques, which are based either on projectile distribution functions or on multiple scattering theory. The distribution of ions within the semiconductor may be expressed in terms of a probability for a forward scattering and and the energy transferred to the semiconductor. The basic scattering equation is known as the LSS equation and is given as:

$$\cos\theta \frac{\partial p(r, E, \theta)}{\partial r} + \frac{(1 - \cos^2\theta)}{r} \frac{\partial p(r, E, \theta)}{\partial \cos\theta}$$
$$= N \int_0^{T_m} [p(r, E - T, \cos\theta) - p(r, E, \cos\theta)]\, d\theta(T) \qquad (7.23)$$

where p = probability that an implanted ion with energy E will come to rest at elemental areas $r\,dr\,d\theta$,

T_m = maximum energy transfer

$d\sigma$ = total cross-section for energy transfer between T, $T + dT$.

7.3.1 Cross-sections and stopping powers

The cross-section for scattering may be modeled by taking into account both the type and magnitude of the individual interactions. The interactions resulting in a continuous energy loss by the incident beam of ions may be divided into scattering of electrons and scattering of the nuclei. The total cross-section is:

$$d\sigma_p(E, T) = d\sigma_n + d\sigma_e$$

The nuclei scattering cross-section is denoted as $d\sigma_n$, while the electron cross-section is given as $d\sigma_e$. The cross-section $d\sigma_n$ is modeled by assuming two-body collisions while $d\sigma_e$ is modeled also assuming two-body interactions, where the atoms exchange their loosely bonded outer electrons. Equation (7.23), also known as the LSS equation, may be solved in order to obtain the ion projected range,

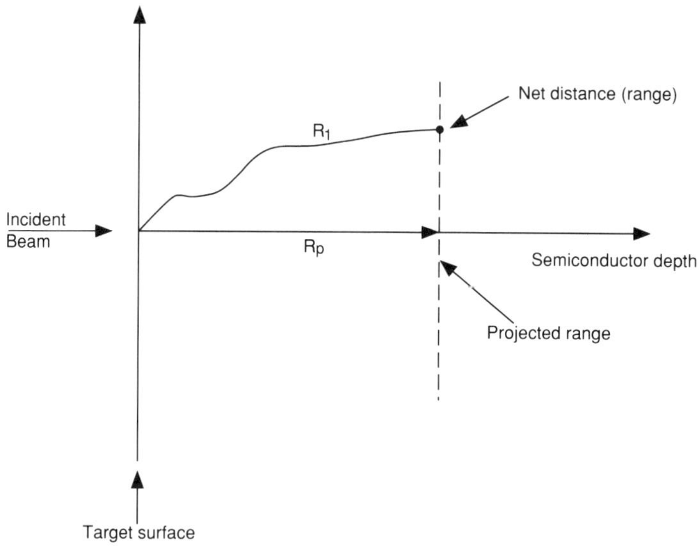

Figure 7.8 Schematic of implantation range.

assuming that the impurity distribution is a Gaussian given by:

$$n(x_p) = \frac{N_0}{\sqrt{2\pi}\sigma_p} \exp[(x_p - R_p)^2/2\sigma_p^2] \tag{7.24}$$

The number of ions as a function of projected depth, x_p is given in terms of the surface concentration N_0, the cross-section σ_p and total projected range, R_p (see

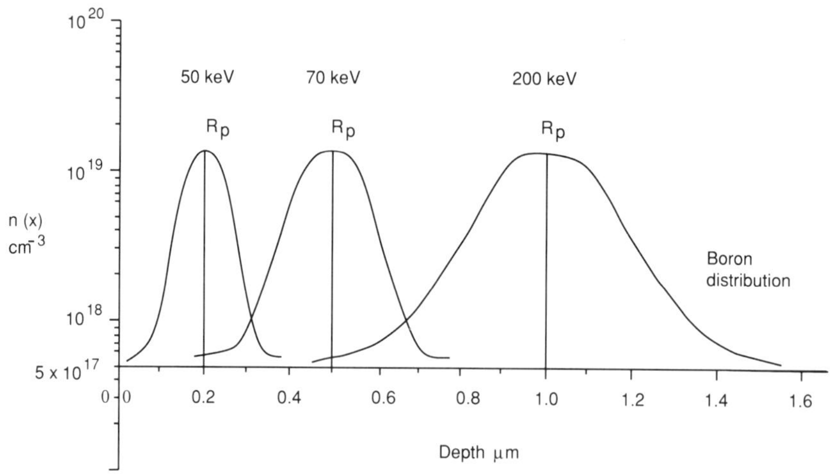

Figure 7.9 Implant distributions for three different ion stages.

Figure 7.8). The form of equation (7.24) is a Gaussian function centered around the range R_p, which spreads in width as the function of the implant depth increases. This is shown in Figure 7.9.

7.3.2 Ion channeling

The implantation of ions into simple crystal material is a highly anisotropic process. For instance, when ions are implanted into silicon and GaAs, the injected ions can channel along directions which are aligned with the beam. The result is a deeper profile, as shown in Figure 7.10, for implantation of boron along the $< 100 >$ axis of silicon.

The ion implantation profile can also be calculated for a given implant energy and dose by solving the Boltzmann transport equations. The ion distribution function as a function of depth may be approximated by the Boltzmann distribution function. The transport equation is then given by:

$$f(E, x) = \int \mathrm{d}t|V|F(E, x, t) = \text{ number of particles}$$

with the energy E, and

$$\frac{\partial f(E, x)}{\partial x} = N_s \int [\mathrm{d}\sigma(E')f(E', x) - \mathrm{d}\sigma(E)f(E)] \tag{7.25}$$

For instance, if we are implanting arsenic into silicon through a SiO_2 mask, the range distributions of As in SiO_2 or As in Si can be calculated from measurement of the total number of particles implanted. Once the profile is determined, the implant damage must be removed.

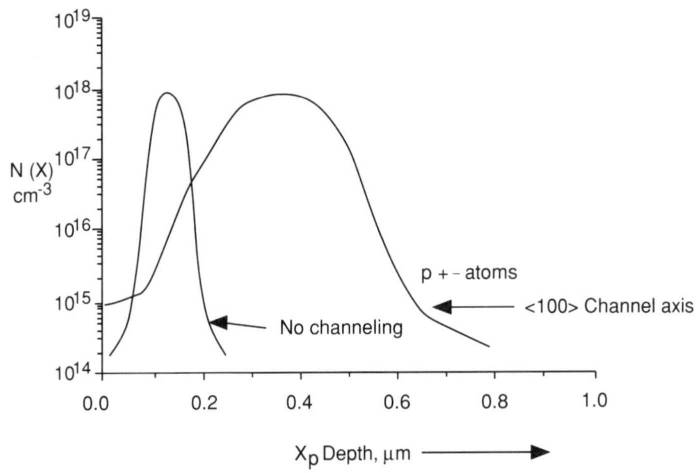

Figure 7.10 Implant distribution profile showing the effect of ion channeling.

7.3.3 Material damage by ion implantation

An ion implanted into a target with energy E_0 will lose its energy in a series of nuclear, E_n, and electronic collisions, E_e, which produce vacancies and interstitial empty sites in the host material lattice. The defects may exist as single defects, clusters, or pairs. The most common defects consisting of defect pairs are listed below.

(a) *Vacancy–interstitial pair* (Frenkel defect), is illustrated in Figure 7.11. The damage removes an atom from its site and moves it to an interstitial site. Therefore the energy loss process produces a concentration of both vacancies and interstitials, resulting in a high degree of material disorder which must be removed by annealing.

(b) *Vacancy–impurity pairs*. The incident collision creates a vacancy, and an impurity atom goes to an interstitial site, creating a bonded pair of impurity atoms attracted to a vacant site. Common vacancy–impurity pairs are: V+P (vacancy + phosphorus), V+O (vacancy + oxygen) and V+As (vacancy + arsenic).

In addition to these defects, voids and dislocations may be created by ion implantation.

A great majority of defects created can be removed by annealing at temperatures of up to 900 °C. The anneal process also recrystallizes the sample, thereby removing the damage created in the surface amorphous layer. The valency clusters may be removed by annealing at 600 °C, while the vacancy–impurity pairs may be removed by annealing between 400 °C and 600 °C.

Most of the annealing must be carried out in excess of 600 °C. A temperature of 800–850 °C is in many cases ideal. Figure 7.12 shows the typical defects introduced by ion implantation. It also indicates that some finite density of defects remains even after the elevated temperature anneal. It is noted that in the case of GaAs, the surface must be encapsulated with a dielectric such as Si_3N_4, in order to prevent surface desorption from occurring at temperatures above 400 °C. The surface damage problems may be minimized by utilizing rapid thermal annealing (RTA) at 800–900 °C for a duration of one to two minutes.

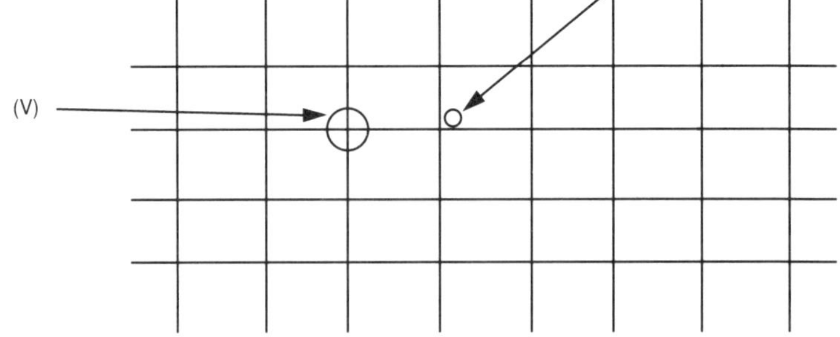

Figure 7.11 Vacancy–Interstitial pair produced by ion implantation.

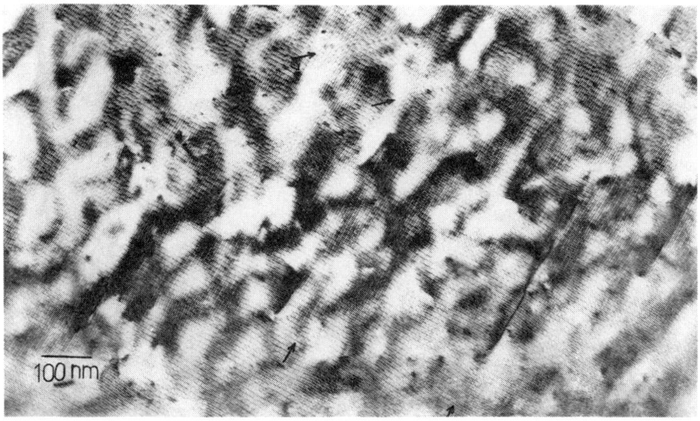

Figure 7.12 Transmission electron micrographs of 70 keV boron implanted into 100 silicon and subsequently annealed for 30 minutes at 900 °C.

7.4 SUMMARY

Active layer doping during the manufacture of microcircuits occurs by diffusion of dopant atoms and by ion implantation. The diffusion of dopants is accelerated by grain boundary effects and therefore the mathematical techniques for calculating diffusion profiles are presented. The material damage induced by ion implantation have also been discussed. These defects may be removed by annealing during the post-processing treatment. The diffusion profiles, the extent of lateral diffusion, affect the design rules and performance of the integrated circuit. Damage and residual defects affect manufacturing yield. For most instances, the implantation defects contribute to the Poisson distribution of defects from which processing yield may be determined.

8

PATTERN TRANSFER IN MICROCIRCUIT MANUFACTURING

8.1 VLSI LITHOGRAPHY

In this chapter we discuss the basics of forming a pattern on the semiconductor. The lithography operation must attain a certain resolution and the correct placement (alignment) of multiple patterns. This operation is referred to as a multistep pattern transfer process. First, the pattern is transferred to the photoresist which is a light-sensitive material; next, transfer takes place from the photoresist into the wafer surface layer. The manufacturing method is a three-step process which starts with mask alignment, proceeds to photoresist exposure and then to resist development, as shown below:

Process step	*Purpose*
Alignment and exposure	Precise alignment of mask to wafer and exposure to UV light
Development	Removal of unpolymerized resist

The optical alignment processing step relates only to optical mask aligners. Transferring the image on the mask onto the wafer surface requires ten basic steps, which may occur with many variations. Therefore, the basic manufacturing process is a ten-step process which starts with the semiconductor wafer surface preparation.

Ten-step photolithography process

(1) *Surface preparation*: etching and cleaning of surface, accomplished with wafers in cassettes.

(2) *Application of photoresist*: thin layer is applied by spinning the photoresist.

(3) *Soft bake*: carried out to promote adhesion of the photoresist to the wafer.

(4) *Alignment and exposure*: precise placement of mask to wafer and exposure to UV light.

(5) *Development*: removal of unpolymerized resist.
(6) *Hard bake*: final evaporation of solvents in order to stabilize the resist.
(7) *Inspection*: analysis of developed region.
(8) *Etch*: etching of the pattern through the exposed regions.
(9) *Photoresist removal.*
(10) *Final inspection of lithographic pattern.*

This ten step process may be automated and the wafers may be moved from process step to process step while in a protective cassette.

8.1.1 Photoresist composition

The resists contain four basic constituents which allow them to exist either in the negative or positive form. The resist components shown below vary from the main constituent, which is a polymer with various additives intended to tailor the final properties.

Component	*Function*
Polymer	• Changes structure when exposed to radiation.
Solvent	• Allows spin application, and formation of thin layers on the wafer surface.
Sensitizers	• Controls chemical reactions in the polymeric phase.
Additives	• Final properties attained in terms of viscosity.
Solid components	• Determine final resist exposure characteristics.

The exposure characteristics of the resists are discussed in the following section. The exposure differences between the resists of different chemistries result in negative and positive types. Negative resists are based on the polyisoprene polymer. Before exposure, they exist in an unpolymerized condition. After exposure, the polymers become cross-linked, and hence stable. Therefore, negative resists are not removable in the exposed regions. Positive photoresists are based on the phenol-formaldehyde polymer. After exposure, this polymer becomes more soluble, and may be removed through an acetone wash. Hence, the utilization of both positive and negative resists results in defining regions of material buildup or removal.

8.1.2 Manufacturing technology

In this section, photolithography, the highest cost process, will be included in the consideration of the present status of manufacturing. The overall measure is still the cost per functioning die shipped out of the fabrication facility. It is therefore critical that the costs pertaining to fabrication by photolithography be assessed. The total wafer costs include six factors:

- Overhead — administration and facilities.
- Materials — direct and indirect.

- Equipment for each processing step.
- Labor for each processing step.
- Yield, wafer success.
- Wafer fabrication costs.

The cost factors are shown in schematic form in Figure 8.1. Wafer fabrication costs are driven by the individual costs for the design, masks, and materials.

Typical contributions of the factors to the overall die cost are shown. The percentages will vary with the wafer, feature size, degree of automation, and number of process steps. Table 8.1 shows how each factor contributes to the total cost.

The yield is a figure of merit which depends on the product maturity. Yield is lowest at the product start date and increases to its highest value at the product maturity date, mainly due to the elimination of fabrication problems which occur during production. Yield improvement may be significantly affected by a successful photolithographic process. Figure 8.2 shows the variation of wafer yield with product maturity.

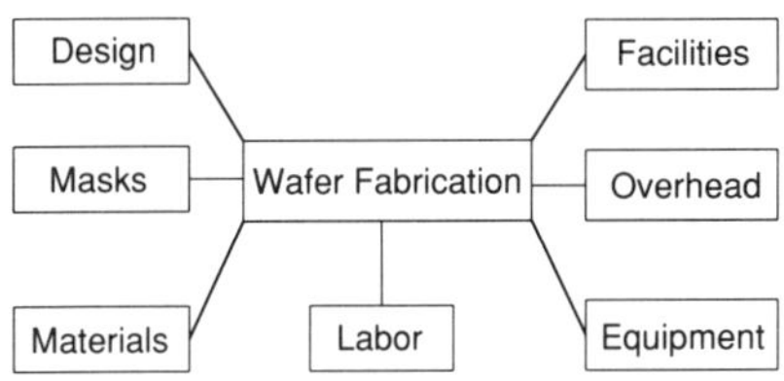

Figure 8.1 Cost factors included in the fabrication of a completed wafer.

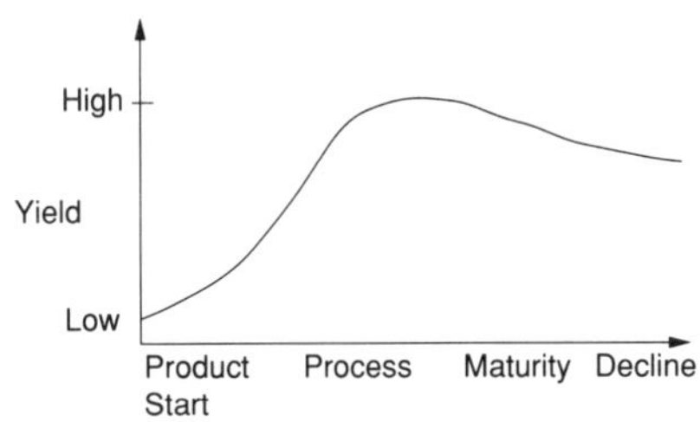

Figure 8.2 Variation of wafer yield with product maturity.

Table 8.1 Wafer fabrication costs.

Factor	Composition (%)
Overhead and utilities	8
Materials	44
Equipment depreciation	25
Labor	
Direct	17
Indirect	6

Yield improvement can be translated to economic improvements. Consider the impact of fabrication yield improvement in the example shown below. Fabrication, die sorting, and assembly yield determine the overall yield.

Process	Change	*Yield*			Overall
		Fabrication	Sort	Assembly	
Base line		0.80	0.50	0.95	0.38 (38%)
	to 0.81	0.81	0.50	0.95	0.384
	to 0.85	0.85	0.50	0.95	0.404 (40.4%)

Consider the above wafer yield improvement. A wafer yield improvement from 80% to 85% increases the overall yield from 38 to 40.4%. If the fabrication area starts at 10 000 wafers per month, there are 350 dies per wafer and the selling price is \$5 per die, the increased revenue is:

$$10\,000 \text{ wafer starts} \times 350 \text{ die per wafer} \times 0.012\ (1.2\%) \times \$5$$

$$= \$210\,000 \text{ per month}$$

The effect translated to actual dollars is significant due to the multiplying effect of the number of dies per wafer.

8.1.3 Basics of lithography

Pattern transfer is the basic lithographic process in the attainment of the final image on each die. The entire process involves the transfer of either a light field or dark field image to a semiconductor surface. Figure 8.3 shows the differences between light and dark field at the mask level. The dark field mask step would result in a mesa-like structure for the gate, or for the source and drain, when

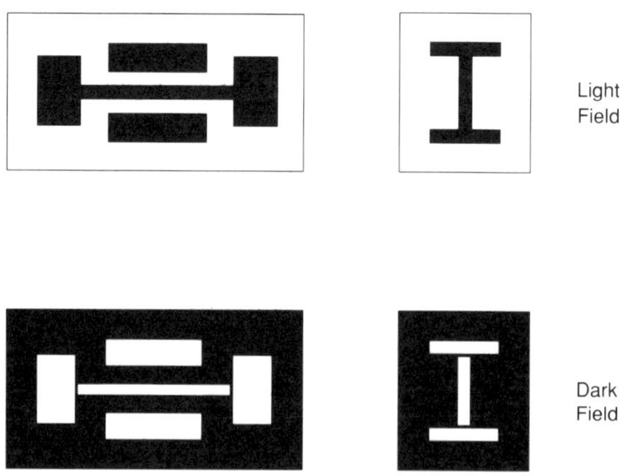

Figure 8.3 Mask-reticle polarities for a field effect transistor showing the gate and contacts.

each structure is developed individually. Based on whether a light field or dark field mask is utilized, the procedure of selecting the photoresist is motivated by the sensitivity required. The entire process reduces to ten basic steps which are outlined in Figure 8.4. It is noted that the surface preparation may also require an etching step which would enhance the adhesion of the photoresist. The photoresist is applied through a spinning process, which in part controls the resist thickness. Also shown in Figure 8.4 are the steps requiring cassette automation, such as multi-wafer cassettes and wafer holders.

The ten process steps include alignment and exposure, which is most critical. Precise alignment pertains to the accuracy of aligning one masking level to a subsequent level. This is achieved through registration test patterns on each mask level. These masks may be simple or complex structures, depending on the degree of registration accuracy required. In most cases, the registration mark can be a simple cross pattern or parallel lines to which each individual masking level is aligned.

The selection of a photoresist depends on the exposure radiation used and the surrounding material sensitivity to this radiation. Ideally, one should select a resist whose sensitivity is in the 10^{-5} to 10^{-7} C/cm^2 range, in order to minimize exposure duration. Typical light sensitive polymers and their sensitivity to exposure by electrons or X-rays is shown in Table 8.2. The common type PMMA is shown to

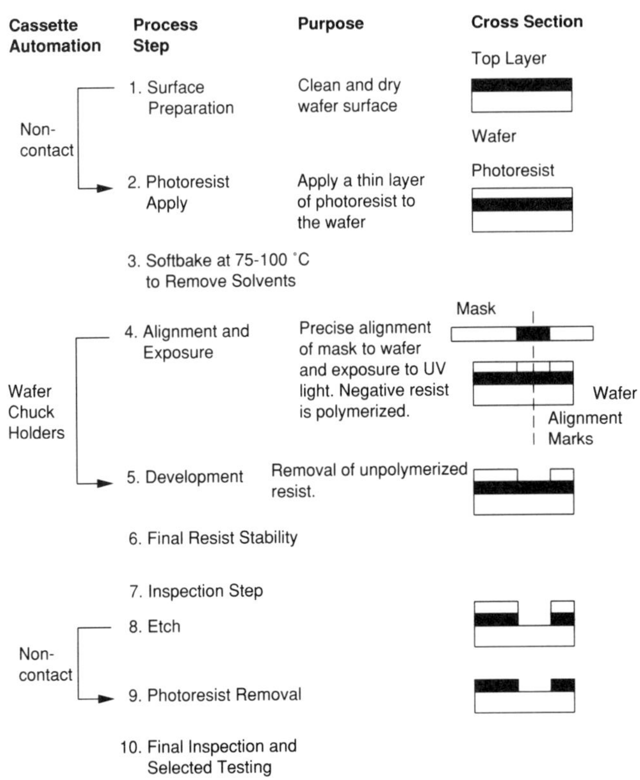

Figure 8.4 The total ten-step process for photolithography resulting in the final wafer inspection [2].

Table 8.2 Sensitivity of typical E-beam and X-ray resists.

Resist	Polymer	Polarity	Sensitivity (C/cm^2)	Exposure radiation
Positive UV	Novolak (m-cresol-formaldehyde)	+	$3–5 \times 10^{-5}$	UV
Negative UV	Poly-isoprene	−	1×10^{-7}	UV
PMMA	Poly-(methyl methacrylate)	+	5×10^{-5}	E-beam
PMIPK	Poly-(methyl isopropenyl ketone)	+	1×10^{-5}	E-beam/ deep UV
PBS	Poly-(butene-1-sulfone)	+	2×10^{-6}	E-beam
TFECK	Poly-(trifluoroethyl chloroacrylate)	+	8×10^{-7}	E-beam
COP	Copolymer-(α-cyano ethyl acrylate-α-amido ethyl acrylate)	−	5×10^{-7}	E-beam X-ray
PMPS	Poly-(2-methyl pentene-1-sulfone)	+	2×10^{-7}	E-beam

be a positive resist which may be exposed by 5×10^{-5} C/cm^2, making it an easily adaptable electron beam resist. Novolak is the positive resist for optical lithography, while poly-isoprene is the negative polymer typically applied to optical lithography.

The differences between negative and positive resists are profound. The positive resist results in a higher aspect ratio and better step coverage, which are the two critical factors in the success of a photolithographic process. The positive resist is also developed by aqueous solutions, while negative resists require organic solvents. In both cases, the resist is spun onto the wafer in a configuration shown in Figure 8.5. In a manufacturing environment, it is dispensed onto a rotating wafer, and after the correct thickness has been attained, the wafer is returned to the cassette where it is transferred to the next processing step. Table 8.3 shows the comparison of negative and positive resists.

The typical experimental apparatus for spinning-on of photoresist in a glove box type enclosure is shown in Figure 8.5. The vacuum chuck configuration holds the wafer in place while it is rotated at a pre-calibrated number of revolutions per minute. The rpm then determines the thickness of the photoresist. As shown in Figure 8.6, the correct selection of the rpm determines the quality of the coating. Elimination of pinholes or the incorporation of dust particles will adversely affect the lithography process and introduce potential reliability problems.

The static spin process is the basic manufacturing step which is of primary importance to the effectiveness of pattern transfer. It is the quality of the resist coating which determines the density of defects transferred to the semiconductor. The surface of the photoresist must be planar, in order to ensure that contact lithography is possible without tilting or distortion. Finally, the application of excessive resist results in edge covering or run-out which again makes wafer handling difficult and reduces wafer manufacturing yield. Excess resist application also results in hillocks and ridges on the resist.

Table 8.3 Comparison of negative and positive resists.

Parameter	Negative	Positive
Aspect ratio (resolution)	Low	Higher
Adhesion	Better	
Exposure speed	Faster	
Pinhole count		Lower
Step coverage		Better
Cost		Higher
Developers	Organic solvents	Aqueous
Strippers		
Oxide steps	Acid	Acid
Metal steps	Chlorinated solvent compounds	Simple solvents

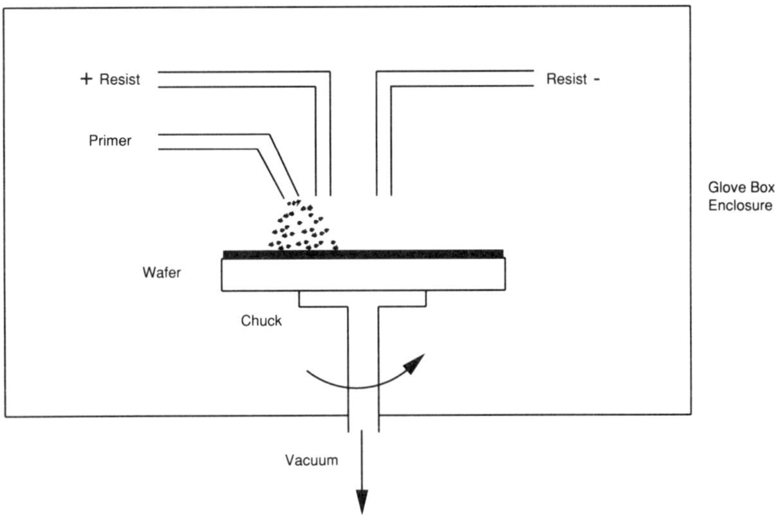

Figure 8.5 Wafer spinning operation for the application of photoresist.

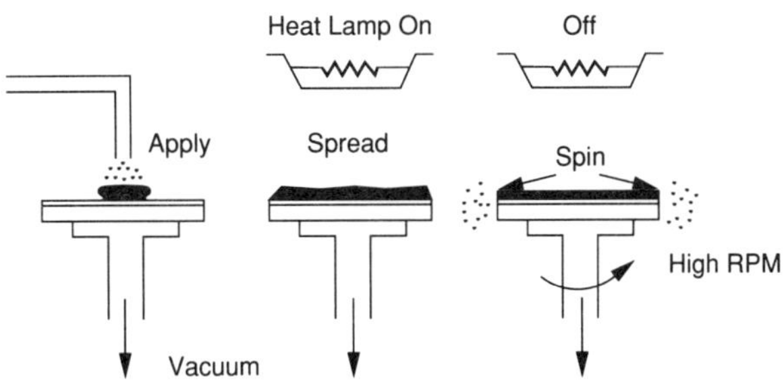

Figure 8.6 Static spin process, showing the three steps towards photoresist application and stabilization.

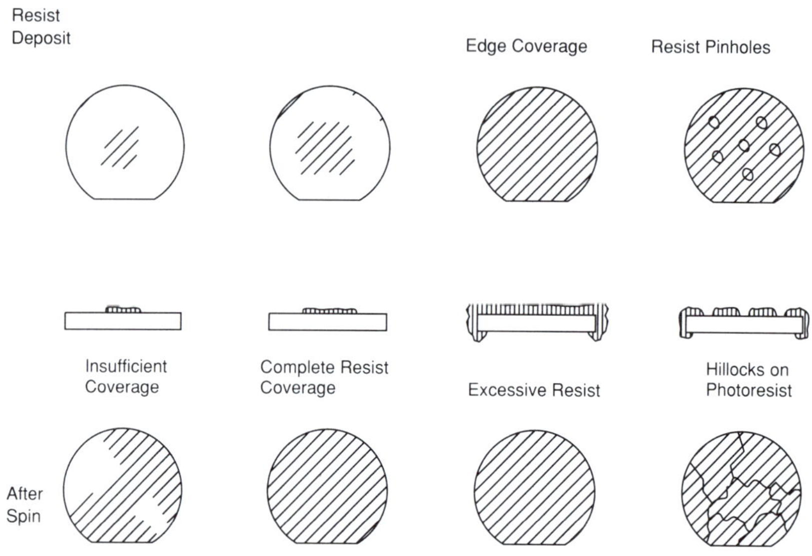

Figure 8.7 The effect of resist deposit on wafer coverage after spin.

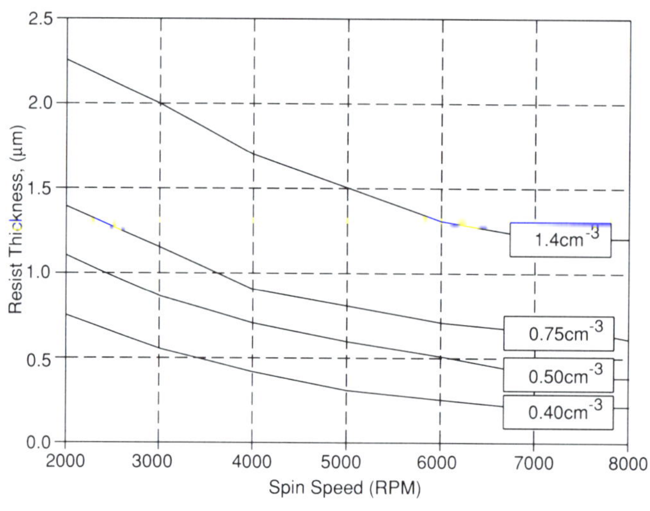

Figure 8.8 Resist thickness versus spin speed for various resist volumes.

Figure 8.7 shows the problems which may be introduced as a result of either excessive or inadequate application of the photoresist. Excess resist coverage may affect the planarity of the wafer. However, insufficient coverage may leave areas without any photoresist, as shown in Figure 8.7. The large uncovered areas then become regions of low yield for the wafer. The amount of resist deposit and whether it results in sufficient coverage is a function of photoresist type and spinning speed, as shown in Figure 8.8. The initial volume of the applied resist must also be known precisely. The resist thickness versus spin speed is shown in Figure 8.8 for the KTI

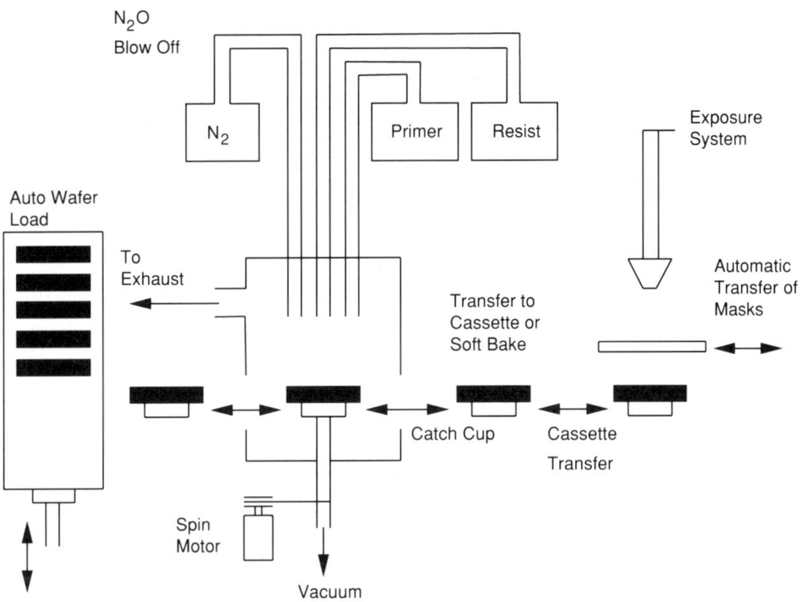

Figure 8.9 An automated resist application process.

resist. Depending on the resist chosen, a number of thicknesses may result, which makes the exact calibration a critical step in establishing the necessary parameters for photolithography. The entire process may be automated, as shown in Figure 8.9, by utilizing a cassette transfer system. Finally the transfer may be made to the exposure system which also includes the mask insertion.

8.2 OPTICAL LITHOGRAPHY

Contact, proximity, and projection printing are three basic methods of optical pattern transfer of photomask images onto the resist-coated wafer. In contact printing, a photomask is pressed against the resist-covered wafer and exposed by light of wavelength near 400 nm [1]. This intimate mask and wafer contact yields the best possible theoretical resolution. However, the contact produces defects in both the mask and the wafer which may lead to the discard of the mask. In addition, the defects make this method of optical pattern transfer unacceptable for VLSI manufacturing. Possible types of defects include pinholes in the chromium film, scratches, intrusions, and star fractures.

There is no contact between the mask and the wafer in proximity printing. Thus, it has the advantage of longer mask life-time. Proximity printing not only reduces the contamination problems associated with contact printing, but also degrades the resolution. Projection printing offers higher resolution than proximity printing together with large separation between the mask and the wafer. It completely avoids the mask and wafer contamination problems associated with contact exposures [2].

8.2.1 Modified illumination technology

The ideal isolated dark line is usually obtained at 180° phase-shifter edge or the chromeless phase-shifting mask. Unfortunately, only closed-loop patterns can be formed with this method. In order to expand its application, the effects of the unnecessary part of the phase-shifter edge must be eliminated. Fortunately, the useless dark line can be easily eliminated by the use of a modified illumination technology [3].

In projection lithography systems with fly-eye elements, a virtual source is created from an array of approximately mutually incoherent point sources. In general, the complex amplitude for an image corresponding to a coherent point source on the Gaussian image plane is given as a function of the point source location. The location of the light source on the x axis is defined as

$$\sigma' = \sin\alpha_c' / \sin\alpha_r \tag{8.1}$$

where α_c' is the angle relative to the optical axis and the constant α_r is the maximum aperture angle at the object plane, as shown in Figure 8.10.

In order to form the fine pattern along only a single direction, we can modify the shape of the source that is arranged along only a single x axis of the light source area. Modified illumination technologies, for example, ring-shaped light source technology and separated light source technology, have been developed to demonstrate resolution and depth-of-focus improvements without a phase-shifting mask. Figure 8.11 shows the optical intensity near the phase-shifter edge for both x and y directions. The solid lines represent the perfect focus and the dashed lines are for the 0.75 μm defocused condition. A deep valley in intensity is obtained at the phase-shifter edge for the x direction.

The above is especially true under the defocused condition where the increase in intensity at the edge of the phase shifter is significant. It also becomes important to consider defocusing effects on under-exposure of resist films. This technique can be effective for delineating the very fine gate patterns required for devices with gates arranged in parallel.

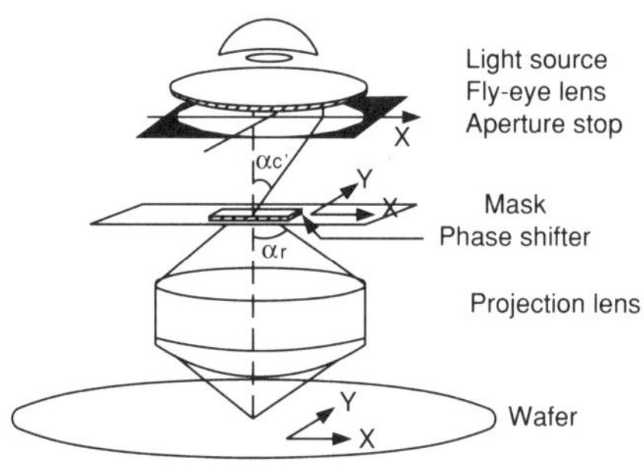

Figure 8.10 Definition of the x and y axes for the light source, mask, and wafer, source location and incidence angle on a plane perpendicular to the optical axis [3].

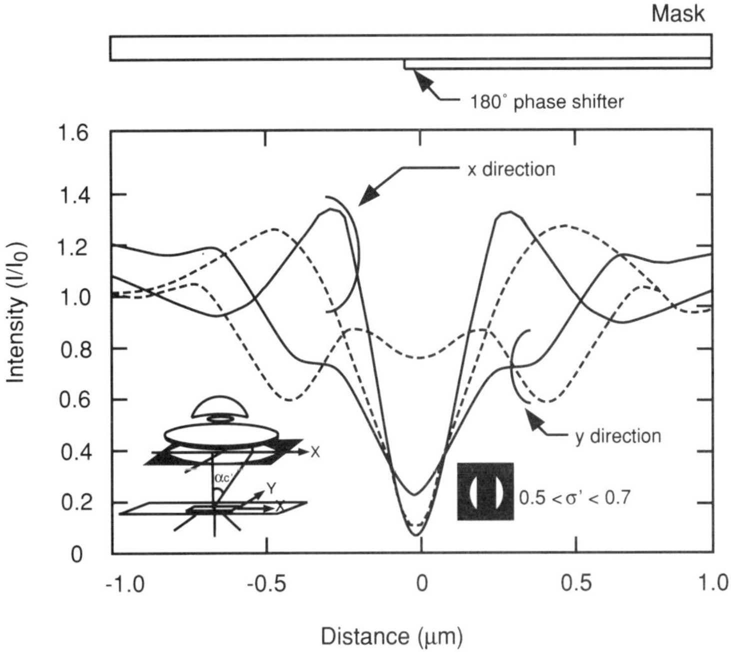

Figure 8.11 Image intensity dependence on the direction of the phase-shifter edge when using the separated light source. Solid lines are for perfect focus and dashed lines are for 0.75 μm defocused conditions at a wavelength of 365 nm and NA of 0.54 [3].

8.2.2 Resolution limit

The basic technique for improvement of the resolution of optical lithography involves the use of higher numerical aperture (NA) optics and shorter exposure wavelengths. Recent progress in lens materials, optics design, lens systems manufacturing technology, as well as the development of a new short-wavelength light source, has advanced optical lithography into the sub-halfmicrometer region. This advance has great importance for the VLSI industry. Furthermore, process engineers are familiar with this technology because it enables process continuity and reasonable throughput [4].

For VLSI processes, however, the lack of depth of focus (DOF) is a very serious problem. The DOF decreases rapidly with increasing resolution. A certain DOF is required to delineate fine patterns on VLSI substrates, because it is difficult to make a projected image plane coincide exactly with the substrate surface. The required DOF is determined by factors such as the topography of the VLSI substrate, wafer flatness, field curvature of the projection lens, focusing accuracy, and others. In particular, large topography features on the VLSI substrate cannot be eliminated because several layers have been stacked in order to pack many functions into a limited area [5, 6].

In projection lithography, it is well known that, from Rayleigh's equations, the theoretical resolution limit R_{th} and the corresponding depth of focus DOF are given by [4]

$$R_{\text{th}} = k_1\lambda/NA \tag{8.2}$$

$$\text{DOF}_{\text{th}} = k_2\lambda/NA^2 \tag{8.3}$$

where k_1 and k_2 are constants that depend on the resist processes. The quantities λ and NA are the exposure wavelength and the numerical aperture of the projection lens, respectively. It can be seen from these equations that the resolution limit can be improved by shortening λ and increasing NA. However, by doing so, this will also decrease DOF rapidly. If the theoretical DOF_{th} becomes smaller than the required focus tolerance, as when sub-halfmicrometer patterns are delineated by optical lithography, the theoretical resolution limit expressed by equation (8.2) will no longer be a good guideline. In this case, for determining the practical resolution limit, the following characteristic of optical images must be considered: the DOF decreases with shrinking pattern size when the same optical system is used. Thus the practical resolution limit is the usable pattern feature size satisfying the actual DOF requirement.

8.2.3 Practical resolution

Two optical factors which lead to deterioration of image quality are investigated here. One is the decrease in pattern feature size W. The other is the increase in focus error D, the deviation from the optimum focal plane. Though image quality is a function of W and D, it also depends on the optical system parameters λ and NA. To eliminate this optical parameter dependence, a normalized pattern feature size factor ρ and a normalized focal error factor δ are introduced

$$\rho = W/(\lambda/NA) \tag{8.4}$$

$$\delta = D/(\lambda/NA^2) \tag{8.5}$$

Theoretical considerations and optical image simulation confirmed that projected optical images having the same values for ρ and δ are completely matched [7, 8]. Thus by examining the optical image as a function of ρ and δ, we can discuss any optical image produced by systems having arbitrary λ and NA.

In general, image contrast γ is considered to be a good indicator of quality for line and space (L/S) patterns. It is defined as [4]

$$\gamma = (I_{\max} - I_{\min})/(I_{\max} + I_{\min}) \times 100(\%) \tag{8.6}$$

Here, $I_{\max}$ is the light intensity at the center of a clear part of the L/S pattern, and $I_{\min}$ is the intensity at the center of an opaque part. Variation of image contrast γ with respect to ρ and δ is calculated by optical image simulators.

8.2.4 Advanced image formation techniques

By introducing a 180° phase difference between the light beams transmitted through neighboring clear parts of the mask patterns, the phase-shifting method creates finer optical images than the conventional method. In particular, the spatial frequency response to L/S patterns can almost be doubled when a phase difference is alternately introduced. It has also been reported that the phase-shifting method can enhance the focus latitude. Thus this technique is anticipated to improve resolution by removing the limitation for increasing NA. The main application of this method is considered to be limited to L/S patterns because of the effectiveness of the method in forming mask images. In general, it is difficult to apply phase-shifting to arbitrary VLSI patterns, making it necessary to establish a CAD method. Thus the discussion here is confined to the application of phase-shifting to L/S patterns.

In applying the phase-shifting method to wiring patterns in actual LSIs, neighboring features must not be shorted. This means that each wiring feature must correspond to a clear part of the mask with a phase of 0° or 180°. This suggests that a negative-tone resist process must be used for the phase-shifting method. If the conventional positive-tone resist process is employed, where the clear mask area is the inverse of the actual wiring pattern, the phase difference is shifted resulting in a dark (exposure) line in the transmitted area and therefore exposing an undesired short between parallel features. Thus high-performance negative resist processes are necessary for practical applications of this method, no matter which wavelength is used.

The FLEX (Focus Latitude Enhancement Exposure) method increases the depth of focus quite effectively by using plural image planes created at different levels along the light axis. This technique is also expected to improve practical resolution by removing the limitation on increasing NA in the same manner as the phase-shifting method. Because FLEX is most effective for isolated transmitted patterns like contact holes, the discussion here is confined to application to hole patterns [4]. In applying FLEX to the hole patterns of actual LSIs, positive tone resist processes must be used, since this method is effective for isolated transmitted patterns. High-contrast positive resist processes are desirable for the wavelengths of excimer lasers.

8.3 ELECTRON BEAM LITHOGRAPHY

Electron beam exposure systems have been used at the leading edge of micro- and nano-fabrication for many years. The superior lithographic resolution capability of electron beam systems has effectively been employed in exploratory research and development work. The first large-scale manufacturing application, however, did not make use of high resolution. Instead, it exploited the ability of electron probe systems to directly generate patterns under computer control. Electron beam pattern generation for mask manufacturing has become the dominant technology throughout the semiconductor industry [9].

The most desirable electron beam lithography process is to write directly on the semiconductor wafer eliminating the expensive and time-consuming mask step entirely. The unique flexibility for maskless pattern generation has an associated disadvantage: the image elements or pixels must be exposed serially. This imposes a severe limitation on exposure speed or the rate at which pattern information is transferred onto the semiconductor wafer. From a mask where the pattern information is stored, optical projection systems transfer up to 10^9 pixels onto the wafer in parallel. In contrast, the conventional electron beam system writes the pattern with a finely focused Gaussian round beam, exposing one pixel at a time. This throughput handicap limited Gaussian round beam systems to research applications.

The development of the shaped beam approach enhanced the exposure efficiency significantly. It also made the use of electron beam direct write systems in semiconductor manufacturing lines economically acceptable. For this direct-write strategy to succeed, the electron beam systems had to keep pace with the increasingly stringent requirements of semiconductor manufacturing. The most stringent of the requirements has been, and still is, throughput. In spite of significant exposure speed enhancements, the throughput of direct-write tools has steadily declined. The reasons are the continually increasing density and complexity of integrated circuits and the expanding wafer size which have led to explosive growth in the number of pixels to be exposed per wafer.

In an attempt to reverse the trend of throughput erosion, further increase of parallelism in pixel exposure, character projection and more recently cell projection techniques have been introduced. These projection schemes provide a trade-off between pattern generation flexibility and pattern writing speed. Furthermore, they are more efficient in the exposure of repetitive patterns like memory arrays.

8.3.1 Technology trends

The introduction of the variable shaped spot technique provided 1–2 orders of magnitude improvement in writing speed over single pixel machines through parallel exposure of pixels. A shaped spot can project square or rectangular pattern elements containing up to several hundred pixels. A further enhancement in writing speed is possible through the projection of repetitive pattern characters or cells which can contain thousands of pixels. Figure 8.12 shows schematics of both

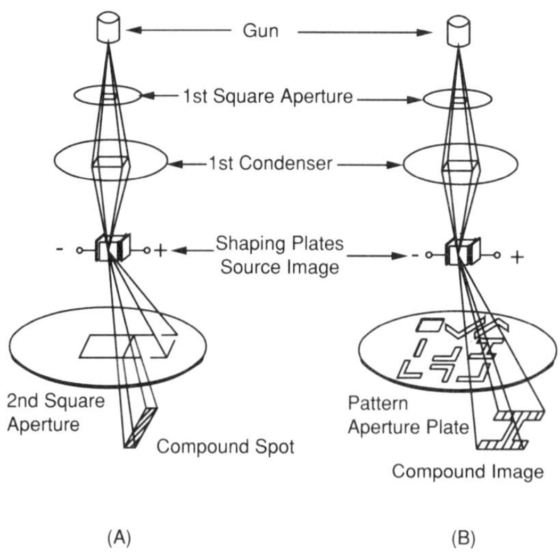

Figure 8.12 Beam shaping techniques; (A) variable shaped spot, (B) character projection [9].

the variable shaped spot and the character projection technique [9]. A character projection prototype system had been built in support of a bubble memory project during the late 1970s. This demonstrated the basic feasibility of the technique with exposures of working bubble memory race tracks on garnet wafers. The highly repetitive nature of memory arrays is conducive to the technique and compatible with the rather limited selection of different characters in the shape aperture. Scanning/projection techniques promise significant increase in pixel exposure without sacrificing pattern generation capability, which has been the most important virtue of electron beam lithography in semiconductor manufacturing.

Highest exposure efficiency has always been the potential of mask-based projection electron beam lithography. This electron analogue of the optical stepper has been pursued since the beginnings of electron beam lithography. Unlike scanning systems, projection electron beam lithography systems have not achieved manufacturing maturity and have not yet been applied in semiconductor production. One key limitation has been the fabrication and use of absorbing stencil masks. All projection electron beam systems, with the exception of photo-cathode systems, use stencil masks in transmission mode.

8.3.2 Mask manufacturing

With the advent of VLSI in the late 1970s, electron beam systems began to replace the slow opto-mechanical pattern generators and photo-repeaters in the mask shops of the semiconductor industry. A 64K DRAM pattern with over 10^5 components could run for 24 hours on an artwork generator, while an electron beam system could complete the job in less than an hour [9]. The trend of DRAMs toward ever higher levels of integration has continued since then with a fourfold density improvement every three years, driving the lithography requirements for minimum features and mask overlay.

The major trend in lithography has been toward $4\times$ and $5\times$ reduction steppers and away from the earlier $1\times$ technology. By reducing demands on pixel resolution, pixel writing speed, and pattern placement accuracy, this approach brought temporary relief for mask making. The return to $1\times$ masks for synchrotron based X-ray lithography and other 1:1 exposure techniques, as well as phase shifting masks for 64M and 256M DRAMs represent formidable challenges. At IBM, the internally developed variable shaped beam system EL-3 has been applied in advanced mask making.

The variable shaped beam approach has proved effective in dealing with throughput demands for both direct write and mask making. Unlike the Gaussian beam, it produces a rectangular beam spot of varying size which decouples resolution from the size. Consequently, up to several hundred pixels can be written in parallel without compromising resolution, providing an effective pixel writing rate of up to 1 GHz. In addition, design grid variations can be accommodated by moving the edge of the beam intensity profile at any required increment without affecting throughput. The design grid resolution of EL-3 is 12.5 nm and the edge acuity is 70 nm. This allows for precise image size control at sub-half micron critical dimensions.

8.4 X-RAY LITHOGRAPHY

In X-ray lithography, an X-ray source illuminates a mask, which casts shadows into a resist-covered wafer. Materials useful for the absorptive and transmissive parts of the mask, the atmosphere in the exposure chamber, and the resist, are in large part determined by the absorption spectra of these materials in the X-ray region [1]. X-ray lithography was first reported in early 1972. Although it has been used very effectively in research and development, it has not yet been employed in high volume integrated circuit manufacturing.

Enhanced resolution is the reason cited most frequently for using X-rays in microlithography. Another reason cited in favor of X-ray lithography is that it is a parallel process, i.e., a large number of pixels can be transferred into resist at the same time. Parallel processes are also well known for electrons and ions and these may even provide exposure times comparable to or even shorter than X-ray lithography. The latter is desirable as an IC manufacturing technology because it is better matched than any other technique to the microlithography system problem [10].

8.4.1 Various forms of X-ray lithography

Several forms of X-ray lithography have been developed or proposed. These are depicted in Figure 8.13. Point-source X-ray lithography (Figure 8.13(a)) is the simplest form and has achieved the highest resolution (17.5 nm lines and spaces). It has been used to fabricate a number of different types of Si and GaAs devices, integrated circuits, and quantum-effect devices. However, it has not yet been applied to the manufacture of commercially high density ICs. The divergence of X-rays produces a slight, uniform magnification but, contrary to a widespread misunderstanding, no pattern distortion. Source divergence causes pattern overlay to depend on precise control of mask-to-substrate gap.

The point source can be collimated by any of three known techniques as depicted in Figure 8.13(b). However, this has not yet been implemented in a practical way. Figure 8.13(c) illustrates synchrotron-based X-ray lithography. The radiation is emitted in the horizontal plane and forms a thin sheet which must be swept back and forth in the vertical direction to cover an entire mask field. The radiation is highly collimated, intense, and broad-band. Figure 8.13(d) describes a newcomer, soft X-ray projection lithography, which has received a great deal of attention recently. Reflection from the several surfaces requires multilayer interference-type mirrors. Reasonable reflection efficiency can be achieved only at wavelengths of 13 nm or greater [10].

8.4.2 Limits of X-ray lithography

Until about 1988, it was widely believed that the range of photoelectrons released when an X-ray photon is absorbed, dictated the resolution limits of X-ray lithography. For this reason, to achieve sub-20 nm features, researchers used 4.5 nm X rays. Recent experiments by Early *et al.*, and numerical simulations by Murata *et al.*, showed that for wavelengths longer than 0.8 nm, resolution limits imposed by

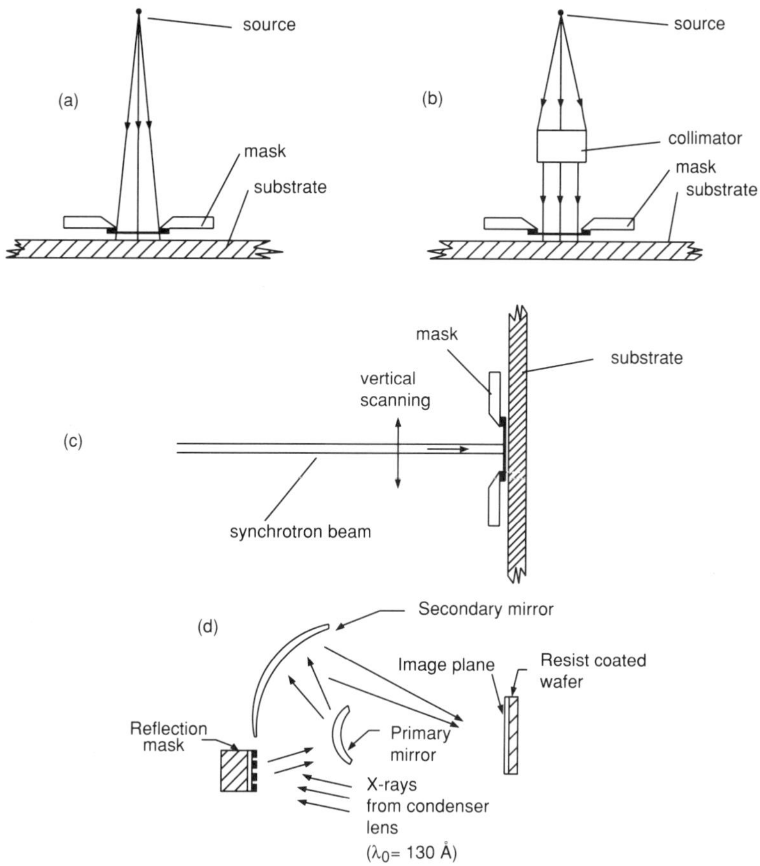

Figure 8.13 The several forms of X-ray lithography. (a) point-source X-ray lithography; (b) point-source X-ray lithography collimation; (c) synchrotron based X-ray lithography; (d) soft X-ray projection lithography [10].

photo and Auger electrons are below 30 nm [11]. So, if a proper X-ray wavelength is chosen (0.8 to 1.6 nm), the resolution limits of X-ray lithography are dictated exclusively by diffraction.

In X-ray lithography, diffraction occurs in the gap between the mask and the resist film. Until very recently, it was assumed that the diffracted field could be calculated by simple analytical techniques using Kirchhoff boundary conditions (KBC). Calculations by B.J. Lin and others indicated considerable ripple in the diffracted image, with an implication of line-width control problems. A more careful analysis has shown that the KBC are inapplicable to X-ray lithography because an X-ray absorber is more akin to a dielectric waveguide than a traditional aperture mask [12]. As a result, the ripple is not present in the image. Moreover, Cerrina *et al.*, and others, have shown that, when the finite X-ray source size and bandwidth are taken into account, no edge ripple is present. Furthermore, all features encountered in IC manufacturing can be printed with appropriate line-width control and large depth-of-focus [13]. This has been confirmed experimentally.

The feature size limits in X-ray lithography are simply a matter of how small the mask-to-substrate gap, G, can be made. This depends on the flatness of the mask, the flatness of the substrate and the presence of dust or other asperities on the substrate. Masks are now made optically flat (less than 0.25 μm), and substrates can be made comparably flat by means of adaptive pin chucks. Thus, gaps of 4 μm should be feasible even in a manufacturing environment. This corresponds to a minimum feature size of 50 nm using a 1 nm wavelength [12]. Dust particles, which were the scourge of contact photolithography with hard chrome-on-glass masks, do not cause damage to X-ray masks, because the membrane is simply pushed back by the dust. This causes local distortion but not mask damage. In modern clean rooms and wafer handlers, dust particles larger than 1 μm are virtually eliminated.

X-ray lithography depends on the mask being free of distortion and the substrate of inhomogeneous distortion or continental drift. Homogeneous distortion, i.e., uniform wafer expansion or contraction, can be corrected with a point source by adjusting the mask-to-substrate gap. Mask distortion due to absorber patterning can be eliminated. That due to handling in an aligner is less certain. Presumably, sufficiently stiff mask frames and kinematic mounting in the aligner can eliminate distortion problems, at least down to levels compatible with 100 or 500 nm features.

8.5 KEY LITHOGRAPHY EQUATIONS

The key equations in lithography will be reviewed in this section, i.e. those which may be used to calculate the limitations to resolution. Projection methods are divided into two types, depending on the distance between the mask and the wafer.

8.5.1 Contact printing

The standard for comparison of different techniques is the lithography parameter:

$$Q = W\sqrt{2/(\lambda g)} \tag{8.7}$$

where W is the minimum line width, and g is the contact distance between the mask and the wafer. The beam divergence prior to illuminating the mask is given as α. Figure 8.14 shows contact lithography geometric conditions.

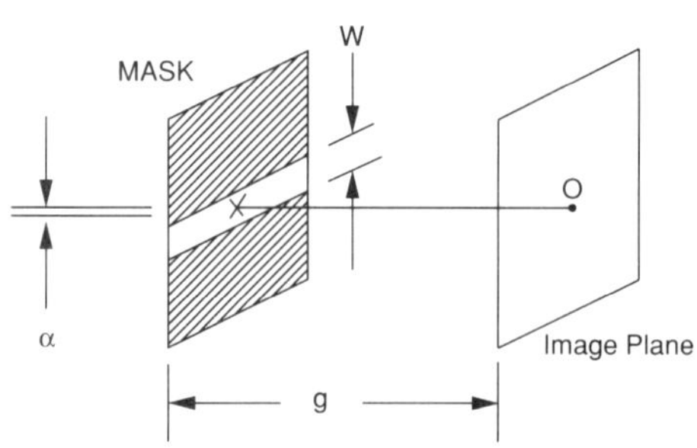

Figure 8.14 Contact lithography geometric conditions.

The major loss is due to diffraction off edges of the mask. The diffraction limit for a source of wavelength λ is attained for $g < W^2/\lambda$, resulting in a lithography parameter of $Q > 2^{1/2}$. The optimized resolution is obtained for $Q = 2.0$. A resolution of 2 μm is possible at a mask to substrate separation of 10 μm.

8.5.2 Projection printing

A higher resolution with larger separation between mask and wafer can be attained by including an optical projection system between the mask (object plane) and the wafer (image plane). The parameters which must be analyzed in projection lithography are resolution, alignment and depth of focus. Figure 8.15 shows the basic projection printing geometry.

For an object at x_0, $y_0 = 0$, the image of this source will have cylindrical symmetry, as determined by the intermediate aperture system. Therefore, the image may be represented by a set of Bessel functions (J_i). The image amplitude is given as:

$$I_i(x_i, y_i) = \frac{\pi}{(2\lambda f)^2}\,[2J_1(s)/(s)]^2 \tag{8.8}$$

$$s = \left(\frac{\pi}{\lambda F}\right)\sqrt{x_i^2 + y_i^2} \tag{8.9}$$

where J_1 = first order Bessel function in terms of position parameter $s(x, y)$. The superposition of Bessel functions will allow one to calculate the system resolution. Table 8.4 shows typical resolutions for projection aligners.

In order to illustrate the calculation of the lithography parameter for a projection aligner, we will carry out the Q calculations for a Perkin-Elmer system. A proximity printer operates with a 10 μm mask to wafer gap and a wavelength of 430 nm. Another printer uses a 40 μm gap with a wavelength of 250 μm. One must compare

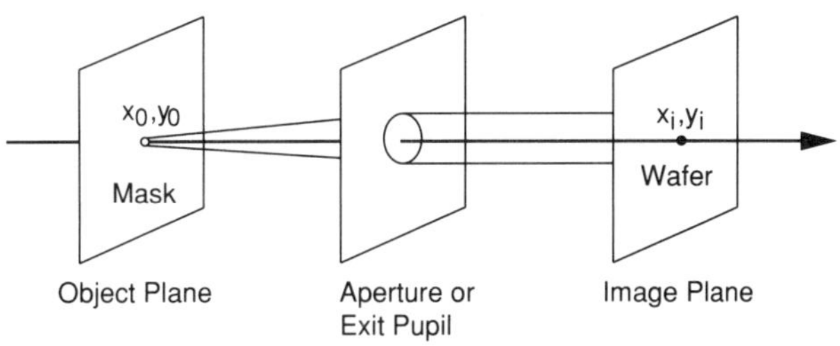

Figure 8.15 Projection lithography imaging system.

Table 8.4 Characteristics of projection printers.

	Resolution (μm)	Alignment (μm)
Perkin-Elmer	1.9	±0.35
GCA	0.8	±0.20
Nikon	0.8	±0.15

both printers, and determine which one would offer the higher resolution:

$$Q = W\sqrt{2/(\lambda g)} \tag{8.10}$$

Case A.	$\lambda = 0.43\ \mu m$	$g = 10\ \mu m$	$Q = 0.68$ W
Case B.	$\lambda = 0.25\ \mu m$	$g = 40\ \mu m$	$Q = 0.45$ W

"A" gives higher resolution since its Q value is 0.68 W in comparison to 0.45 W for case B.

8.5.3 Electron beam lithography

Electron lithography offers higher resolution, because 10–50 keV electrons have a smaller wavelength. The resolution is limited by electron scattering, not by aberrations. The various types of resists available are classified into:

Negative electron-beam resists: electron-beams induce cross-linked polymers making them less soluble in water.

Positive electron-beam resists: electron-beams crack bonds, and exposure leads to lower molecular weight and greater solubility.

The distribution of energy of electrons in resist is:

$$f(r, z) = a_1 \exp\left(-\frac{r^2}{\beta_f^2(z)}\right) + a_2 \exp\left(-\frac{r^2}{\beta_b^2(z)}\right) \tag{8.11}$$

where $\beta_f(z)$ = width of forward scattered beam in the resist,
$\beta_b(z)$ = width of back scattered beam in the resist.

Resist resolution is higher in thinner resists. The resist thickness must be kept to 0.2–0.4 μm in order to attain optimum resolution. The major source of problems with electron-beam lithography is misregistration which leads to manufacturing yield reduction. The effect of misregistration may be calculated in the following way.

If Δ_t = total registration error, and
Δ_i = misregistration of the ith level

then

$$\Delta_t = \left(\sum_i \Delta_i^2\right)^{1/2} \tag{8.12}$$

if the contributions are mutually independent. From equation (8.12) an average total misregistration error may be determined.

8.5.4 Electron optics

One must understand key aspects of electron optics in order to set the resolution limits for lithography. The emission current is given by:

$$J_c = AT^2 \exp\left(-\frac{E_w}{kT}\right) \tag{8.13}$$

for a tungsten filament. The beam brightness is the current density per unit solid angle:

$$\beta = \frac{4I_b}{(\pi d_0 \alpha_0)^2} \tag{8.14}$$

where d_0 is the beam diameter.

The maximum value of the brightness is

$$\beta = \frac{Je\ \text{eV}}{\pi kT} \tag{8.15}$$

The position of the image with respect to the object distance and focal length is

$$\frac{1}{l_0} + \frac{1}{l_i} = \frac{1}{f} \tag{8.16}$$

The image diameter is given as

$$d_i = Md_0 \tag{8.17}$$

where M is the magnification, defined as l_i/l_0.

The size of the final beam is limited by diffraction, spherical aberration and chromatic aberration. The diffraction spot diameter is:

$$d_{\text{diff}} = \frac{1.2\lambda}{\alpha} = 10^{-3}\ \mu\text{m} \tag{8.18}$$

For spherical aberration, we have,

$$d_s = \left(\frac{1}{2}c_s\right)\alpha^3 \tag{8.19}$$

and for chromatic aberration,

$$d_c = c_s\left(\frac{\Delta V}{V}\right)\alpha \tag{8.20}$$

Another source of aberration is due to Coulomb repulsion:

$$d_{ee} = \left[\frac{LI}{\alpha V^{3/2}}\right] \times 10^{-8}\ \mu\text{m} \tag{8.21}$$

where L = length of optical column.

The total beam diameter is then

$$d^2 = d_i^2 + d_s^2 + d_c^2 + d_{df}^2 \tag{8.22}$$

Example: For an electron beam system at 25 keV, the length is 1 m, spot current is 300 nA, and the aperture is 10^{-2} rad. The energy spread is 0.2 eV. Given $C_s = 10$ cm, $C_c = 62.5$ cm what are the resolution limits?

We first calculate the individual aberration effects given by equations (8.19), (8.20), and (8.21):

$$d_{ee} = \left(\frac{LI}{\alpha V^{3/2}}\right) \times 10^8 = 0.08\ \mu\text{m}$$

$$d_s = \frac{1}{2} c_s \alpha^3 = 0.05\ \mu\text{m}$$

$$d_c = c_c \left(\frac{\Delta V}{V}\right) \alpha = 0.05\ \mu\text{m}$$

The total beam diameter is then calculated to be:

$$d = (d_{ee}^2 + d_s^2 + d_c^2)^{1/2} = 0.11\ \mu\text{m}$$

This is the minimum beam diameter possible, neglecting the interactions between the beam and resist.

8.5.5 X-ray lithography sources

In X-ray lithography, the wavelength, λ, is very small, therefore diffraction may be ignored. The same resists are used, but with the incorporation of heavier elements to increase absorption and sensitivity. Cl additives are usually added in order to enhance absorption.

The sources are high energy electrons in a ring pattern, and emit X-rays at wavelength λ_p. The wavelength is related to the electron energy through the equation

$$\lambda_p = 2.35 \frac{R}{E^3}$$

A 0.83 eV machine with radius $R = 2.1\ \mu$m gives X-rays with $\lambda = 8.4$Å, which is the typical wavelength for X-ray lithography. In the following example, the factors which affect yield are examined. This example summarizes the key issues related to pattern transfer.

Example
Given:

- 1.5 μm minimum feature size to be printed.
- Each wafer has 150 chip sites, and each chip area is 0.2 cm^2.
- 5 mask levels are required.
- 2000 finished wafers must be produced each day (20 h/day = 3 shifts).
- Resist defect density = D_0

$$D_0 = 1.4\ T^{-3} \qquad T = \text{ thickness}$$

- Chip yield at each mask level is:

$$Y = (1 + gD_0a)^{-1}$$

  g = fraction of defects which make the chip bad.
  a = active area of the chip.
- 50% of the defects are fatal
- The throughput in wafers/h is $125 - 50\ T$ for $0.5\ \mu\text{m} \le T \le 2.0\ \mu\text{m}$

(a) What resist thickness should be used?

t, μm	D_0	Y	Y^5	No. of good chips/wafer
2.0	0.175	0.982	0.917	137.6
1.5	0.415	0.960	0.816	122.4
1.25	0.717	0.933	0.707	106.1
1.0	1.4	0.877	0.519	77.9
0.75	3.3	0.751	0.239	35.9
0.50	11.2	0.472	0.023	3.5

(1) D_0 is calculated from $D_0 = 1.4(t)^{-3}$
(2) Y is calculated from $Y = (1 + gD_0a)^{-1}$

$$g = 0.5,\ a = 0.2\ \text{cm}^2$$

(3) For 5 masking levels $Y_f = Y^5$
(4) Number of good chips/wafer = 150 Y^5
(5) Number of wafers processed/h/tool = $125 - 50t$
(6) 2000 finished wafers/20 h day requires
10,000 exposures/20 h or
500 exposures/day/tool

No. of tools = 500/TP

TP = throughput

We tabulate the following:

	No. of wafers/h	No. of good chips/h	No. of tools required
	25	3440	20
	50	6120	10
$t = 1.25\ \mu$m $\Rightarrow$	62.5	6631	8
	75	5843	7
	87.5	3141	6
	100	350	5

The resist thickness that allows the maximum number of good chips/h is approximately 1.25 μm.

(b) If each exposure tool costs 350 000, what is the difference in equipment cost for a process using 1 μm and 1.5 μm of resists?

For $t = 1.5\ \mu$m need 10 tools
$t = 1.0\ \mu$m need 7 tools

three additional tools are needed at the 1.5 μm resist thickness at a cost of 1.05 M dollars.

8.6 SUMMARY

The pattern transfer technology and associated reliability and quality problems have been presented. Starting from the basic ten-step lithographic process, the specific processes related to optical, electron, and X-ray lithography were reviewed. In addition the economic factors of the pattern transfer steps were shown to be critical. The factors which affect image quality and depth of field must be overcome in order to achieve a successful lithographic process. The problems of electron beam lithography may more easily be overcome through the assessment of each of the individual aberrations.

REFERENCES

[1] Sze, S.M., *VLSI Technology*, McGraw-Hill, New York, 1988.

[2] Giacomo, J.D., *VLSI Handbook*, McGraw-Hill, New York, 1989.

[3] Asai, S., Hanyu, I., and Takikawa, M., "New effects of modified illumination in optical lithography," *IEEE Electron Device Letters*, **14**(3), March 1993, 110-112.

[4] Fukuda, H., Imai, A., Terasawa, T., and Okazaki, S., "New approach to resolution limit and advanced image formation techniques in optical lithography," *IEEE Transaction on Electron Devices*, **38**(1), January 1991, 67-75.

[5] Koyanagi, M., Sunami, H., Hashimoto, N., and Ashikawa, M., "Noval high density stacked capacitor MOS RAM," *IEDM Tech. Dig.*, 1978, 348-351.

[6] Ema, T., Kawanago, S., and Nishi, T., "3-Dimensional stacked capacitor cell for 16M and 64M DRAMs," *IEDM Tech. Dig.*, 1988, 592-595.

[7] Oldham, W.G., Nandogaonkar, S.N., Neureuther, A.R., and O'Toole, M., "A general simulator for VLSI lithography and etching processes: Part I — Application to projection lithography," *IEEE Trans. Electron Devices*, **ED-26**, 1979, 717-722.

[8] Matsuzawa, T., Ito, T., Tanuma, M., Hasegawa, N., and Sunami, H., "A three dimensional photoresist image simulator: TRIPS-1," *IEEE Electron Device Letters*, **EDL-6**, 1985, 416-418.

[9] Pfeiffer, H.C., "Advanced E-beam systems for manufacturing," *Electron-Beam, X-ray, and ion-beam Submicrometer Lithographies for Manufacturing II*, Martin Peckerar, (Ed.), Proc. SPIE 1671, 1992, pp. 100-110.

[10] Smith, H.I. and Schattenburg, M.L., "Why bother with X-ray lithography?," *Electron-Beam, X-Ray, and Ion-Beam Submicrometer Lithographies for Manufacturing II*, Martin Peckerar, (Ed.), Proc. SPIE 1671, 1992, pp. 282-298.

[11] Early, K., Schattenburg, M.L., and Smith, H.I., "Absence of resolution degradation in X-ray lithography for λ from 4.5 nm to 0.83 nm," *Microelectronic Engineering*, No. 11, 1990, 317-321.

[12] Schattenburg, M.L., Li, K., Shin, R.T., Kong, J.A., and Smith, H.I., "Electromagnetic calculation of soft-X-ray diffraction from 0.1 μm gold structures," *J. Vac. Sci. Technol. B, No. 9*, 1991, 3232-3236.

[13] Guo, J.Z.Y., Chen, G., White, V., Anderson, P., and Cerrina, F., "Aerial image formation in synchrotron-radiation-based X-ray lithography: the whole picture," *J. Vac. Sci. Technol. B, No. 8*, 1990, 1551.

9

METALLIZATIONS FOR DEVICES AND CIRCUITS

9.1 FAILURE MECHANISMS OF VLSI AND GaAs IC METALLIZATIONS

The reliable metallization design starts with understanding failure mechanisms. This chapter is based on the physics of failure of metallizations.

9.1.1 Corrosion of Al interconnects

The development of metallizations for VLSI circuits and for GaAs devices must follow an initial understanding of failure mechanisms. We will first examine corrosion as a metal migration phenomenon and will then review electromigration.

(a) Corrosion occurs when there are two electrodes and an electrolyte forming a classic galvanic cell.

(b) Corrosion occurs at the anode. The corrosion rate is a complex function of several variables, such as electrode material, electrolyte, cell potential, current density, temperature, polarization effects, passivation, and so on. The corrosion rate is also affected by circuit passivation such as SiO_2 or Si_3N_4. The most common metallization is aluminum.

Aluminum is susceptible to corrosion both in acidic and basic solutions. Corrosion of Al film interconnects has been observed to depend on the environment, relative humidity, phosphorous content of the underlying SiO_2, chlorine contamination, passivation layer, pinholes in protective coatings providing easy passage to H_2O molecules, bias condition, packaging, and use conditions.

Corrosion of Al film interconnects is minimized by:

- Reducing and/or eliminating surface contamination (such as Cl_2 molecules). The presence of halogens enhances surface migration and therefore increases leakage currents.
- Employing high integrity pinhole free protective layers such as SiO_2 and Si_3N_4 further eliminates paths for moisture migration.

Table 9.1 Possible Al corrosion reactions.

(a) $Al \longrightarrow Al^{3+} + 3e^-$	E_0 = 1.66 eV
(b) $Al \longrightarrow Al^+ + 1e^{-1}$ (c) $Al^+ + 2H_2O \longrightarrow Al^{3+} + 2OH^- + H_2\ (g)$	(Mechanisms for hydrogen evolution at anode)
(d) $2Al + 6HCl \longrightarrow 2AlCl_3 + 3H_2\ (g)$ (e) $Al + 3Cl^- \longleftrightarrow AlCl_3 + 3e^-$ (f) $AlCl_3 + 3H_2O \longrightarrow Al(OH)_3 + 3HCl$	(Acidic solutions and reactions involving chloride ions)
(g) $Al + NaOH + H_2O \longrightarrow NaAlO_2\ 3/2\ H_2\ (g)$ (h) $Al + 3OH^- \longrightarrow Al(OH)_3 + 3e^-$ (i) $2Al(OH)_3 \longrightarrow Al_2O_3 + 3H_2O$ (j) $2AlO_2^- + 2H^+ \longrightarrow Al_2O_3 + H_2O$	(Basic solutions and reactions involving oxides and hydroxides)

- Using the minimum phosphorous content in SiO_2 adequate to assure device stability, and
- Hermetically sealed packages which effect a further level of moisture reduction.

The above factors must all be considered when designing a corrosion resistant system.

Corrosion resistance of Al–Cu alloys is best in the heat treated and naturally-aged condition when all of the copper is in solution or in the form of GP1 zones. This factor indicates that Al microstructure, to first order, can minimize corrosion.

Copper, when present as $CuAl_2$, does not increase corrosion severely, since the potential difference between $CuAl_2$ and the copper saturated matrix is only a few hundredths of a volt. Thus, there is no substantial change in corrosion resistance in increasing the copper from 5 to 15%. The maximum amount of copper in the aluminum is therefore 5%. Aluminum, however, presents a number of corrosion paths, as shown in Table 9.1.

9.1.2 Metallization reliability issues

The following is a discussion of all possible reliability problems in metallizations. These problems must be solved in order to develop manufacturable metallizations, for circuits both in silicon and GaAs technologies.

Contacts. Leakage problems occur by surface layers, especially when the metal has been incorrectly removed.
Barrier layers. Diffusion barriers can degrade and introduce diffusion paths.
Pinholes in interlevel dielectrics. Dielectrics between first and second level metallizations may fail via pinholes.
Electromigration. This is the key wear-out failure mechanism for metallizations.
Corrosion of leads. This mechanism is driven mainly by stress effects, and is present primarily in plastic packages.
Internal stresses. The presence of high stress points within a package can lead to de-bonding effects, lead separation, and bond pad lifting.
Cracks at chip corners, also related to internal stresses.

Bonds–interface problems. This mechanism leads to bond lifting and failures and will be discussed in the next section. The bonding problems are essentially related to the presence of intermetallics such as Al/Au at the interfaces.

9.1.3 Gold bonding and IC failures

Gold bonding failure mechanisms are usually related to gold–aluminum bonds on the IC chip. In cases where gold–aluminum beams are present, carbon impurities may lead to cracked beams. Such has been the case with power transistors, and with analog circuits due to the elevated temperature environment. In this section, we will examine IC bond failures as an extension of the discussion on metallization failure mechanisms.

The failure of aluminum to gold bonds has been observed and has been reported extensively in the literature [1–4]. A number of conflicting observations have resulted from these investigations stating that (a) impurities inducing Kirkendal voiding are the primary cause of failure, or (b) intermetallics at the interface are responsible for bond failures. The recent work by Philofsky [5], Goldfarb [6], and Horsting [7] indicates that bond failure is connected with impurities in the gold.

In contrast with aluminum to gold bonds, beam lead failures occur as a result of encapsulant expansion and thermal cycling of the beam lead devices. The encapsulant expands with temperature and the beams crack after a minimum of two cycles of $\Delta T = 180–250\,^{\circ}\mathrm{C}$. However, even though contrasting failure modes are present, this investigation shows that both failures are due to carbon contamination in the plated gold.

The work described in the present chapter confirms the viewpoint that the presence of impurities in the plated gold is the predominant cause of failure. A new factor is also introduced: subsurface carbon impurities can make a difference between bonds failing or remaining undergraded. In order to unambiguously identify the impurities, the following techniques were used concurrently: scanning electron microscopy (SEM) electron microprobe analysis, Auger electron spectroscopy (AES) and nuclear deuteron probe analysis (DPA). When used with ion sputter etching [8], the AES techniques can result in a composition depth profile. However, there is a potential problem arising from surface layer removal by sputtering. Nuclear deuteron probe analysis relies on a deuteron to proton reaction [9]. The depth profile is obtained by nuclear resonance reactions [10]. This technique is totally non-destructive, since sputter related problems are not present and is quantitative in both depth determination and total concentration.

9.2 THE GOLD BOND FAILURE MECHANISM

Aluminum-Gold bonds. Power transistors from two different processing lots were observed to have at least one defective bond. The bonding system for these devices consisted of three gold-plated nickel posts (emitter, base, collector), in which bond wires connected the aluminum pads on the transistor to the posts. The bonding wire was 40 micrometers in diameter and made of standard 1% silicon–aluminum alloy material.

Bonding was achieved by thermo-compression, resulting in as-bonded pull strength of the order of 20 grams. The life test failure resulted in a separation of the bond at the post upon subjecting the bond to 1000 hour shelf life at 180 °C. Approximately 50% of the bonds failed due to "bondlifting", without evidence of an applied force such as bending or twisting. In many cases the bond was physically intact, but so weakened that it separated at the initial application of force. In order to further accelerate the process, additional transistors (sample size 15) were exposed at 250 °C for up to 48 hours. These transistors had good bonds prior to the 250 °C life test as determined by a destructive analysis test.

The objective of the investigation was therefore to: (1) measure the physical and chemical differences between "good" and "bad" bonds to the transistor posts, and (2) estimate the effect of these differences on bond strength as a function of time.

Beam lead bonds. Beam lead bonds of large-scale integrated circuits were observed to have cracks or microcracks in the beams after 250 °C high temperature reverse bias tests. Two types of opens were observed on the failed IC. The first type was a combination of silicon to beam interface separation and broken beam on the edge of the silicon chip. The second type was that of a broken beam at the heel or midspan. Electron probe analysis was applied to all cracked areas in order to identify the presence of contaminants. The gold plating of the beams was measured to have 4% elongation. Typical elongation should be in the 20–25% range. The higher elongation value would result in a larger elastic region, and hence greater plastic deformation before cracking.

9.2.1 Failure mechanism related to contamination

9.2.1.1 Scanning electron microscope (SEM)

Typical bad bonds were characterized by an incomplete AuAl intermetallic formation, with gold still present at the interface or at localized spots around the periphery of the bond. It is evident that bond separation had occurred during the life test, since initial visual inspection and pull tests revealed no problems. The bonds, when pulled, also showed the presence of pure gold and the AlAu intermetallic.

Analysis of the exposed bond area at higher magnification indicated the presence of four types of defects. A Type I defect consists of darkened inclusions in the bond area. These can act as stress concentration centers, resulting in a degraded bond strength. Type II defects consisted of a carbon buildup region on the posts, probably in the form of graphite particles which adhered to the gold plating either before or during processing. Type III defects were identified as absorbed carbon surface films, while Type IV defects were gross discontinuities in the gold plating. Any one of these defects will result in weakened bonds, and in particular, defects I and II will lead to a time dependent stress related failure mode.

A series of microsections of good bonds also showed that the gold underneath the aluminum had been completely consumed by the formation of an intermetallic compound. In failed bonds, defects I–IV were always present, in addition to unconsumed gold.

SEM examination of the beam lead failures indicated cracks at grain boundaries of the plated gold. Intergranular failures were observed at the midspan of the beam, while shear type failures were observed at the upper portion of the beam. The shear failures have been related to side movement of the chip during bonding and will not be further discussed in this investigation.

9.2.1.2 Auger electron spectroscopy (AES)

AES analysis was performed on Type I, II, and III defects. The inclusions of Type I were identified as carbon, with a depth of as much as one micron. A typical AES spectrum is shown in Figure 9.1, with a surface carbon peak at 272 eV. The carbon is so prevalent that it obscures the lower energy gold peaks. At a depth of 0.2 μm, the peak had remained at the same intensity, indicating very little variation in concentration. After sputter etching for 3 hours, the carbon peak started to decrease in intensity. These results indicate that the carbon is subsurface, and further exposure to time–temperature and stress may result in cracks at the interface. AES spectra for Type II defects indicated that carbon is in the form of agglomerated surface and near-surface regions, since a decrease in carbon intensity was observed after 45 minutes of sputter etching. Type III defects, however, were shown to be essentially surface defects.

9.2.1.3 Deuteron probe

The same transistor posts which exhibited the Type I–IV defects and were analyzed by AES were also examined by means of nuclear reactions induced by a 1.6 MeV deuteron beam. Measurements were performed on several different spots on each post by rotating the sample. A beam of heavy hydrogen ions from a Van de Graaff accelerator was directed on to the target. Nuclear reactions were induced in the carbon, resulting in the emission of protons. The emerging protons were detected by a solid state detector, which gave electrical pulses proportional in height to the energy of the detected protons. For a thin layer of carbon on the surface, the protons have the same energy, except for a few of lower energy which can be neglected. If the carbon layer is underneath the surface, the protons leaving the metal will have less energy. The change in measured energy between protons coming from a surface layer to those coming from a buried layer can be used to determine the depth of the buried layer.

The results of the data analysis are shown in Figures 9.2 and 9.3. The concentration of carbon is given in atomic % as a function of distance below the surface of the gold layer. All samples showed evidence of surface and subsurface carbon. Since the depth resolution obtained in these measurements was about 0.35 micron near the surface, it is not possible to determine whether the near-surface carbon is exactly on the surface or within 0.35 micron of the surface. Figure 9.2 shows a large near-surface carbon peak greater than 15% and a subsurface peak at 5.4 μm. Separate measurements in the angular regions from 0 to 90° and from 180 to 270° were performed and a variation in near-surface carbon of a factor of 2 was observed.

Subsurface carbon was observed in all of the posts analyzed with concentrations varying from 0.05 at%, to 0.5 at%, as shown in Figure 9.3. At a depth greater than 5 μm, subsurface carbon concentration was observed to increase, probably due to the gold–nickel interface. In conclusion, the DP analysis indicates that both buried and near-surface carbon was present in all of the transistor posts examined. The

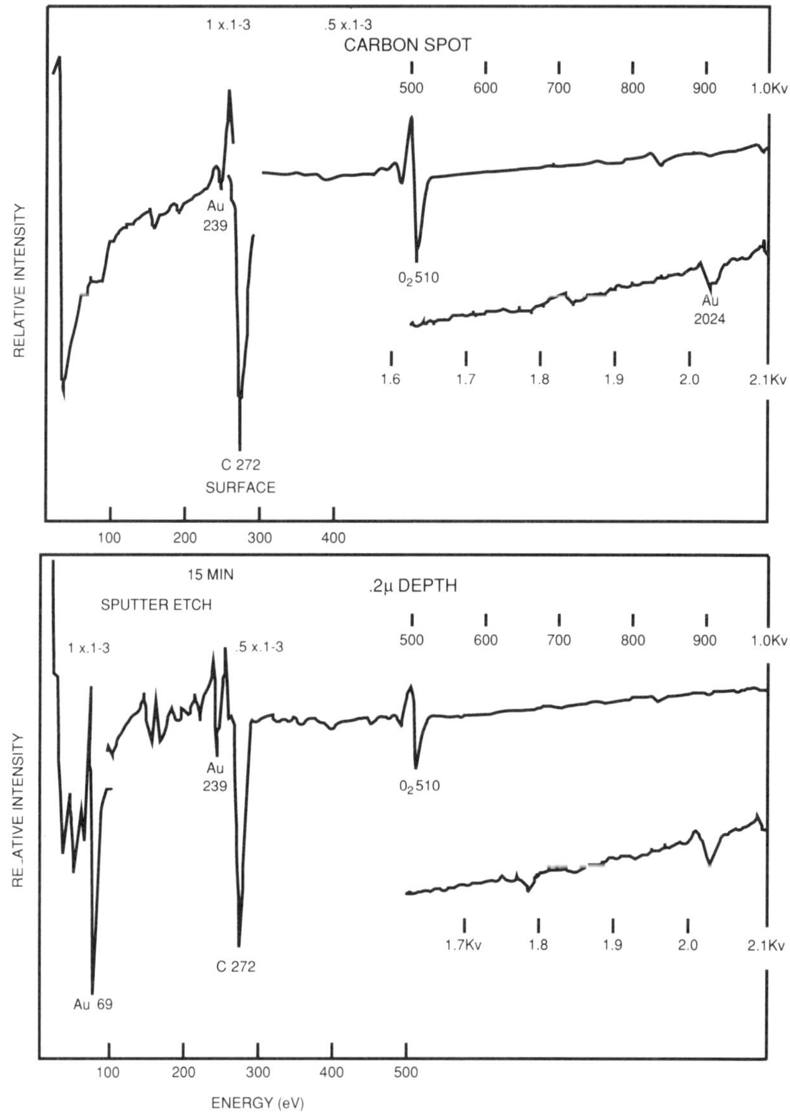

Figure 9.1 AES spectra of a carbon spot showing carbon on the surface and at 0.2 μm depth.

interior peaks may be due to very thin layers containing appreciable amounts of carbon, or due to large quantities of carbon particles concentrated at these levels. DP analysis of good posts did not indicate the presence of subsurface carbon, and the surface, near-surface carbon was always below 0.05 at%.

Deuteron probe analysis of failed beam leads also indicated the presence of subsurface carbon with concentration varying from 0.05 at% to 5.0 at%. A near

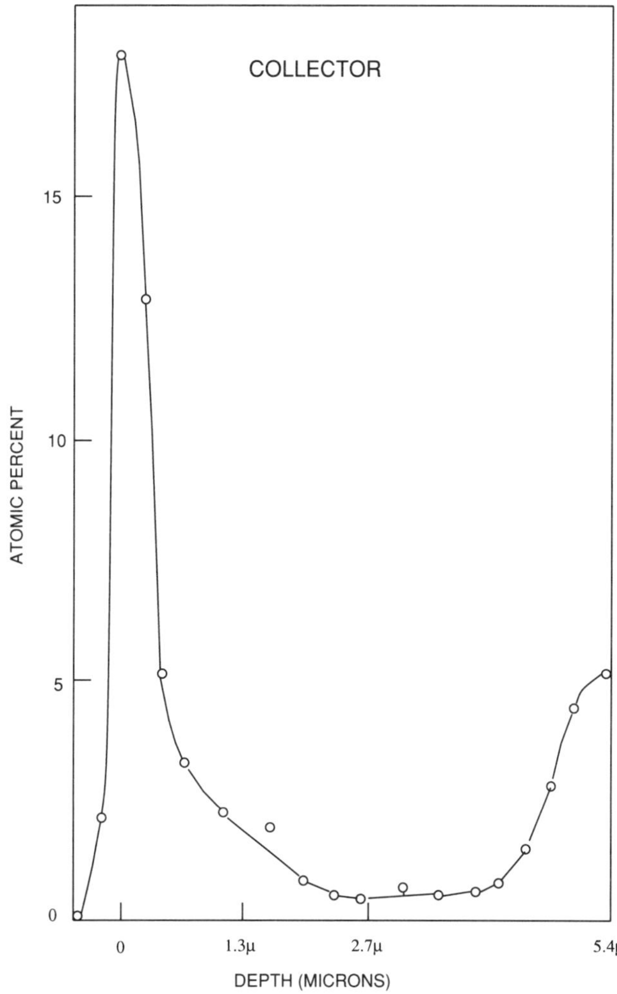

Figure 9.2 DP analysis of a collector post showing subsurface carbon.

surface carbon peak of concentration of 0.05 at% to 1.0 at% was also present. The interior carbon may be due to carbon segregation of grain boundaries, resulting in intergranular failures. Control integrated circuits which did not fail, however, did not show the presence of subsurface carbon.

9.2.1.4 Electron microprobe analysis

The electron microprobe analysis in failed transistors detected three different defective type areas. Types I–III defects were clearly identified, as well as the Type IV defect, namely large discontinuities in the gold. Subsurface carbon inclusions were detected by performing a line scan analysis on cross-sections of the emitter and base posts, as shown in Figure 9.4. The subsurface carbon of this figure was at the 6–10 at% range. Surface carbon was detected at the 18–20 at%

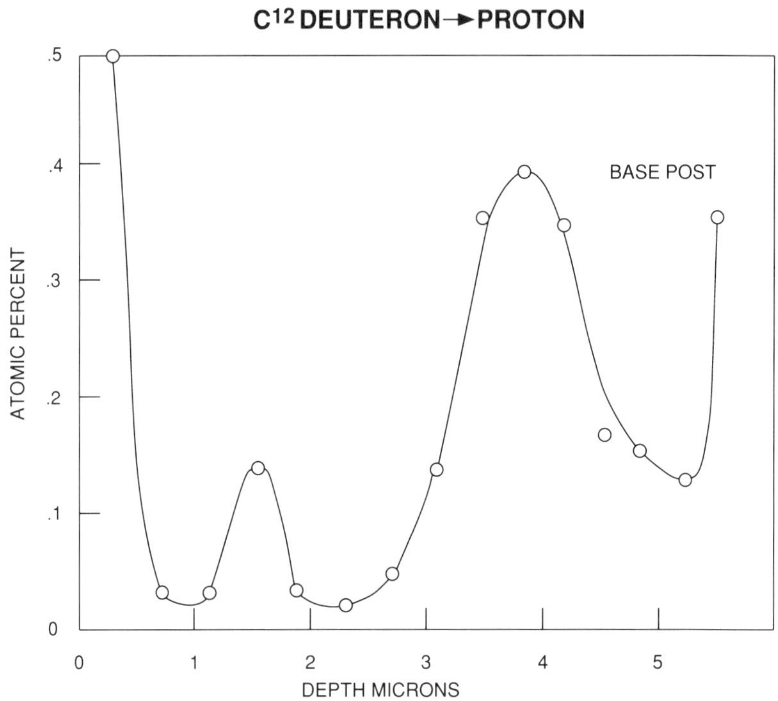

Figure 9.3 DP analysis of base post showing subsurface carbon.

level. Type IV defects, discontinuities in the gold plating, are shown in Figure 9.5, indicating nickel interdispersed within the gold plating. These areas did not indicate the presence of carbon contamination. The results of Figures 9.4 and 9.5 are in good agreement with both the AES and the DPA results.

Electron probe analysis of cracked beams was conducted using the line scan and spot analysis modes. A typical line scan indicated carbon buildup in the area of a cracked beam. In various regions, as shown in Figure 9.6, the buildup of carbon was discontinuous. Spot analysis of regions showing carbon indicated that the electron beam was located on a grain boundary region. This result supports the hypothesis that carbon impurities in the plated gold result in weakened grain boundaries.

Energy dispersive X-ray analysis has indicated the following contaminants present at levels of 0.5 at% to 1.0 at%: copper, iron, manganese, nickel, iron, chlorine titanium, and sulfur. The presence of these elements at the bond interface will result in a time dependent failure mode. The contamination level present is in excess of expected levels in "good" gold-plated layers. In Table 9.2, typical stereographic analysis of the present gold plating from a new semiconductor grade gold bath is displayed.

In addition, energy dispersive X-ray analysis of plating from the new bath did not detect the above impurities, showing that an uncontaminated gold bath has impurity levels below 100 ppm.

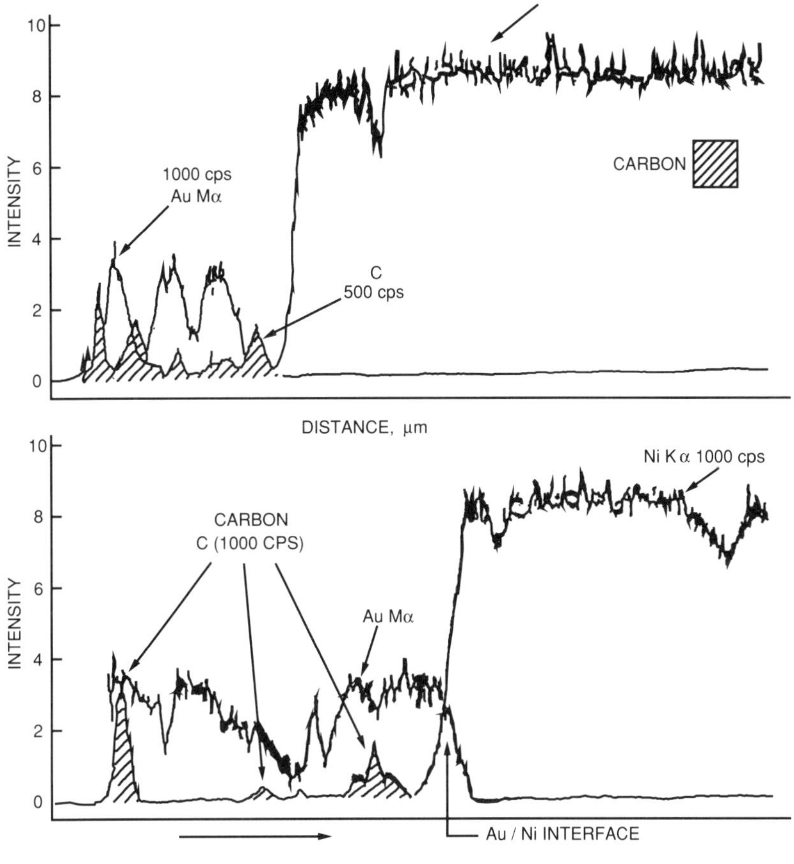

Figure 9.4 Electron microprobe line scan analysis of Au, C, and Ni.

Table 9.2 Typical stereographic analysis of vendors' gold bath.

	Mn	Ni	Co	$Fe_{(ppm)}$	B	Cu	Ag	Na	Cl	S
A	10^4	10^4	10^3	10^2	100	50–100	10–50	10–50	100	100
B	N.D.	1–10	N.D.	N.D.	N.D.	1–10	10–50	1–10	1–10	1–10

Notes: N.D. = Not detected
A = Present gold plating
B = New bath

9.2.2 The failure modes

9.2.2.1 Gold to aluminum bonds

Bond pull data was taken on a set of control transistors subjected to the 250 °C, 48 hour life test and compared to the bond pull data from the 180 °C, 1000 hour life test. Both sets of data indicated the existence of a bimodal distribution of pull

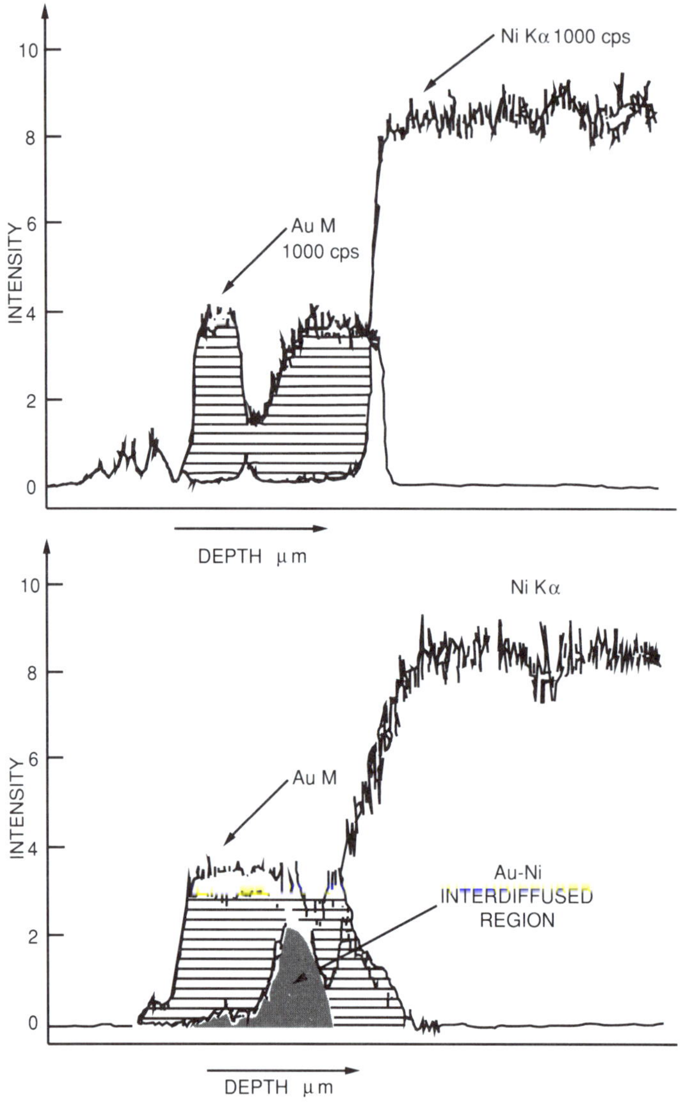

Figure 9.5 Electron microprobe data showing lack of uniformity in the gold plating. Also shown is interdiffusion between gold and nickel.

strengths. Figure 9.7 summarizes the data for the 250 °C, 48 hours life test. The presence of the bimodal distribution along with the (1) carbon subsurface inclusions and (2) high levels of contamination, indicates a contamination related failure mode. In good packages, which were characterized by contamination levels less than 100 ppm, the pull strength distribution remains a Gaussian function even after extensive burn-in. In packages with extensive carbon and gold bath contamination after burn-in, a bimodal distribution results. The carbon inclusions act as stress

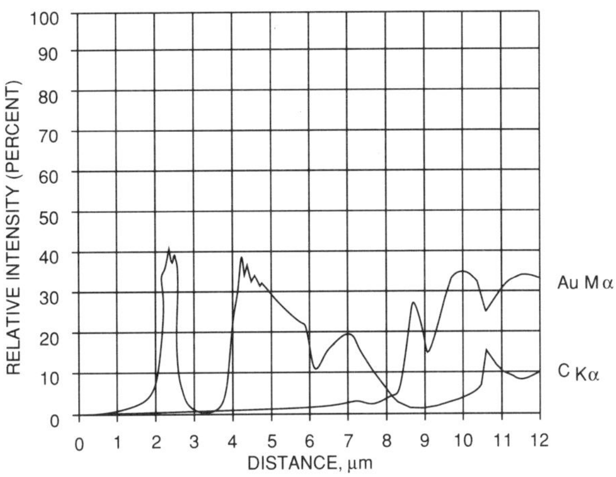

Figure 9.6 Electron probe analysis of cracked beams showing a line scan analysis of an area near a crack.

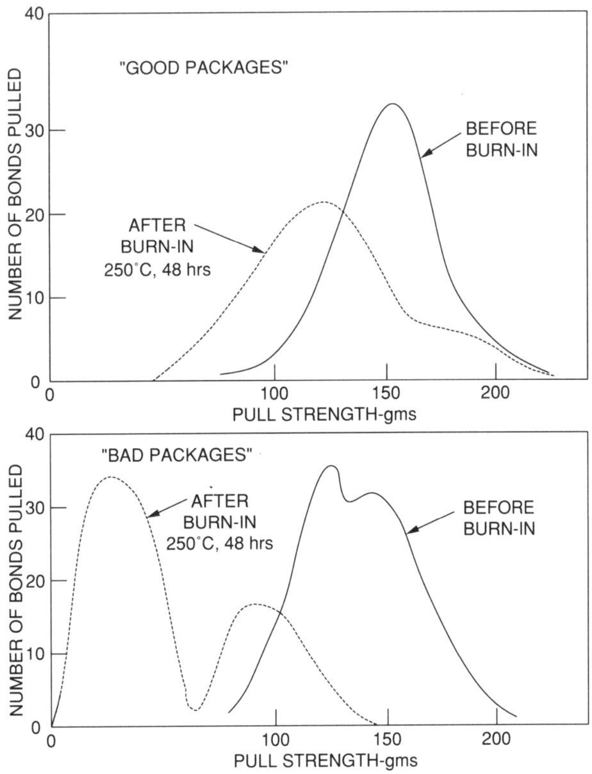

Figure 9.7 Pull strength distributions for good and bad packages showing a bimodal distribution after the burn-in.

concentration regions, resulting in microcracks which grow by a micro-void diffusion process [11], when exposed to temperature–time stress. The remainder of the impurities also induce the classical Kirkendal voiding [6, 8].

During the bonding operation for pure gold, a diffusion front moves unimpeded through the gold layer and stops at the nickel interface. In the case of contaminated gold, the diffusion front starts moving into the gold but stops at carbon inclusions or at small microvoids. Failure usually occurs where the diffusion front stops. Impurities in the gold are expelled by the moving zone and are concentrated directly ahead of the moving diffusion front. When the concentration becomes critical, then precipitation occurs and small voids develop. The voids which form by a time–temperature process are further augmented by microcracks around carbon inclusions, resulting in bond failure. The combination of the two processes results in the perfect bimodal failure distribution shown in Figure 9.7.

9.2.2.2 Gold beam lead devices

The present study indicates that when packaged silicon integrated circuits undergo temperature changes for a variety of reasons, beam fatigue and intergranular cracking occurs as a failure mechanism. Beam fatigue is accelerated due to the presence of carbon contamination in the gold plating. This observation agrees with previous studies [9] of beam fatigue. Experience shows that, usually, devices with subnormal quality of gold fail early in temperature cycling. Devices with good connections and greater ductility in the plated gold (40–60% instead of 4–5%) will eventually fail by beam fatigue, if the magnitude of the temperature cycle is large enough. In order to further test this conclusion, strength analysis of gold plated from a contaminated beam lead electroplating solution was compared with a pure (less than 100 ppm contamination) solution. Table 9.3 shows that for a carbon contaminated gold plating, an elongation of 4–5% was obtained versus 20–25% elongation from the control solution.

The combination of diagnostic techniques, consisting of electron scattering (AES, SEM, electron microprobe) and nuclear scattering (deuteron probe), has unambiguously detected the presence of bond surface and subsurface contamination in both Au–Al bonds and beam lead bonds. The predominant contaminant was carbon although a significant number of other contaminants were present at levels above 100 ppm. These resulted in microcracks in the gold to aluminum bonds and voids which formed by a temperature–time process. In beam lead bonds, carbon contamination resulted in accelerated fatigue failure during temperature cycling. The presence of carbon segregating at gold grain boundaries resulted in intergranular cracks.

Table 9.3 Tensile test of gold plating.

Sample	Thickness	Width	Length	Tensile (psi)	Elongation (%)
C contaminated	4 mm	0.5 mm	0.5 mm	18 000	4
C contaminated	4 mm	0.5 mm	0.5 mm	14 000	5
Pure Au	7 mm	1 mm	.5 mm	19 000	20–25

9.3 ELECTROMIGRATION IN MICROCIRCUIT METALLIZATIONS

In addition to corrosion metal migration and interface degradation, microcircuit metallizations are susceptible to many failure mechanisms, of which the major one is electromigration. This can result in open and short circuit modes of failure. This chapter will explore the electromigration failure mechanism which results from electron transport, and which impacts the design and manufacture of metallizations. Electromigration is characterized by an atomic flux (J_A) related to the ion flux (j_K) through the equation

$$J_A = \frac{ND}{kT} Z^* e\rho j_k$$

where e is the electron charge, ρ is the resistivity, and D is the atomic diffusivity. This equation will be discussed in detail below.

A non-vanishing divergence of atomic flux is a requirement for electromigration to occur. The divergence may be due to a temperature gradient or to microstructural inhomogeneities. Since electromigration is cumulative, it affects the failure rate. Therefore, the median time to failure (MTF) in the presence of electromigration is given by the equation

$$MTF = AJ^{-n} \exp(Q/kT)$$

where A = constant,
J = current density,
n = exponent,
Q = activation energy ($\approx$ 0.6 eV Al, $\approx$ 0.9 eV Au),
T = temperature,
k = Boltzmann's constant

The electromigration exponent ranges from $n = 1$ to $n = 6$. The failure mechanism results in voids, cracks, and hillocks. Two stages of electromigration are shown in Figure 9.8, hillocks near the bond pad with undetected cracks at the opposite interconnect. At the final stage, voids combine, forming a large cracked region. The median life is accompanied by a characteristic σ, a standard deviation. As σ gets larger, so does t_{50}, and the failure rate curve is shown to spread and become diffused.

9.3.1 Electromigration under circuit conditions

This section summarizes the predominant theoretical views of electromigration, as related to actual dynamic circuit conditions. Two major areas have been covered in the theory: steady state current applications, and pulsed current applications. However, before explaining the theory of electromigration, we list the pros and cons of aluminum, which, in both pure and alloy form, has been the subject of most of the electromigration work.

Aluminum is the preferred metallization in the planar process for silicon semiconductors for several reasons [12]:

(1) Ease of deposition — it can be sputtered or evaporated at fairly low temperatures.

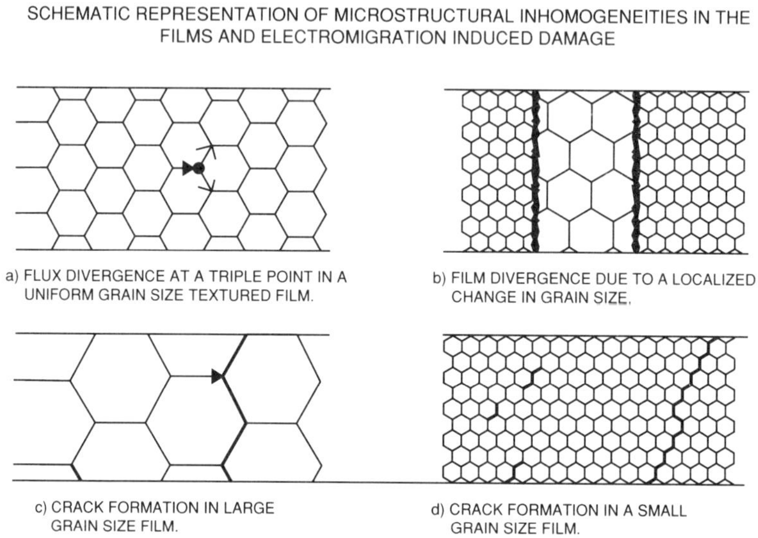

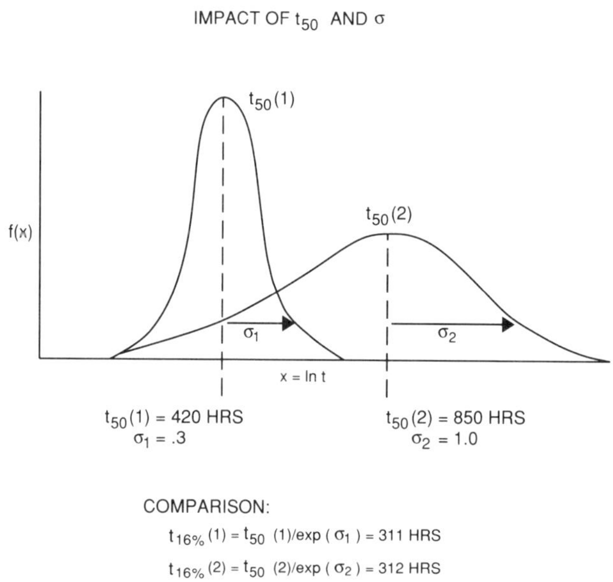

Figure 9.8 Electromigration showing hillocks and then metallization separation.

(2) Ease of patterning — phosphoric acid is usually used as the removal agent.

(3) Aluminum forms ohmic/non-rectifying contact with silicon.

(4) Schottky barrier contacts can be formed through careful process control.

(5) Aluminum sticks to the die because it reacts slightly with SiO_2.

(6) Aluminum will slightly reduce SiO_2 in thin layers and can use up the very thin skin of native oxide which forms contact windows.

(7) Glass deposited on an aluminum surface sticks easily.

(8) Aluminum is a good electrical conductor.

However, there are also several drawbacks to the use of aluminum:

(1) It migrates easily at low current densities and temperatures. Gold is at least 10 times better.

(2) Aluminum can absorb up to approximately 3% silicon. This creates alloying and aluminum spikes.

(3) It corrodes easily in ionic solutions.

(4) Addition of silicon is risky because silicon can precipitate out as nodules in an aluminum film.

Electromigration is the transport of microcircuit current conductor metal atoms due to electron wind effects. If, in an aluminum conductor, the electron current density is sufficiently high, an electron wind effect is created. Since the size and mass of an electron are small compared to the atom, the momentum imparted to an aluminum atom by an electron collision is small. If enough electrons collide with an aluminum atom, then the atom will move gradually. The movement of aluminum atoms will cause a depletion at the negative end of the conductor and will form voids or hillocks along the conductor, depending on the local microstructure. Hillocks will form as aluminum atoms pile up on conductor defects. Figures 9.9 and 9.10 illustrate electromigration and its characteristics. The creation of hillocks and voids is a time-dependent process, and is shown in Figure 9.10.

There are several mechanisms for electromigration. The predominant one is the movement of atoms along the crystal grain boundaries. An activation energy of 0.45 eV has been calculated and used to characterize the temperature dependence. Electromigration is heavily influenced by aluminum grain size, with smaller sizes more easily transported.

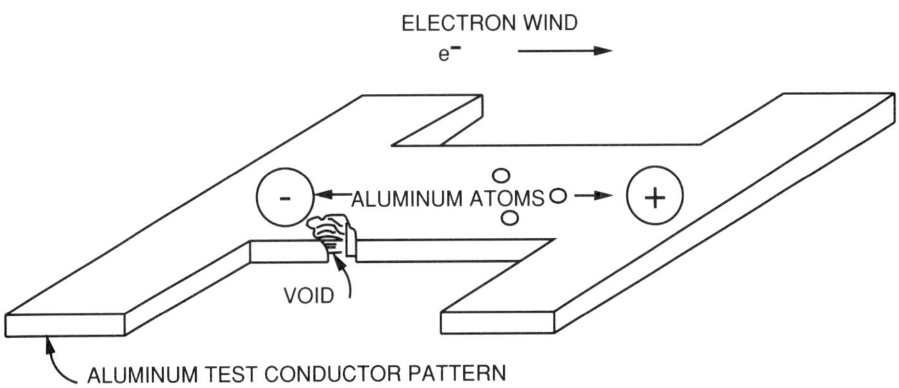

Figure 9.9 Schematic of electromigration in the presence of temperature gradients and interdiffusion.

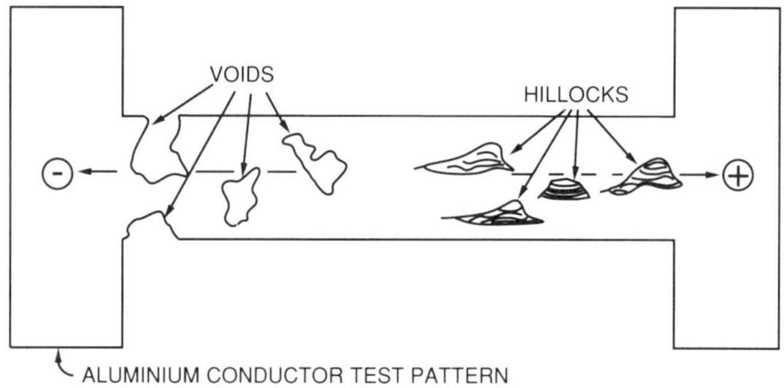

Figure 9.10 SEM micrograph showing void formation after reliability test as a result of electromigration in metallization lines.

Black [13] provides an excellent summary of the physics of electromigration. He notes that a thermally activated metal ion is essentially free of the metal lattice and is acted upon by two forces in an electrically conducting single band metal:

(1) the force created by an electric field applied to the conductor;

(2) the rate of momentum exchange between conducting electrons colliding with the ion.

Shielding effects reduce the electric field effect so that the predominant force is the momentum exchange. The metal ions are driven to the positive end of the conductor while vacancies move toward the negative end. The vacancies condense to form voids. The ions pile up to form hillocks and whiskers at discontinuities, such as a scratch or oxide which reduces conductor width, thinning or microcracking at oxide steps which reduce conductor thickness. The rate of mass transport R is described by:

$$\begin{aligned} R = F &\times \text{(electron momentum)} \\ &\times \text{(number of electrons passing through a unit volume per second)} \\ &\times \text{(effective target cross-section)} \\ &\times \text{(aluminum activated ion density)} \end{aligned}$$

where F is a constant.

The rate of mass transport is related to median time to failure and conductor cross-section area by

$$R = \frac{Fwt}{\text{MTF}}$$

where F = constant,

w = conductor width,

t = conductor thickness,

MTF = median time to failure.

Cross-sectional area is important because it determines the minimum void size to create an open circuit. Momentum considerations and an exponential aluminum ion density yield the following result:

$$\text{MTF} = \frac{wt}{AJ^2} \exp(Ea/kT)$$

where the constant A incorporates the following physical properties:

- Volume resistivity of the metal.
- Mean free time between collisions or the electron free path and average velocity.
- Effective ionic scattering cross-section for electrons.
- Frequency factor for self-diffusion of aluminum in aluminum.
- A factor relating rate of mass transport with median time to failure.

The stress dependence of the activation energy E_a was studied by McPherson [14]. He developed a generalized Eyring model to analyze thermally activated failure mechanisms. The general form of the Eyring model is the following, where TTF is the time-dependent variable given as:

$$\text{TTF} = AT^a \exp(E_a/kt) \exp\left[B + \left(\frac{C}{T}\right)\right] S$$

The model relates the time to failure to two different stresses: thermal and mechanical. It predicts a stress-activated activation energy, provided that two conditions are met: (1) the applied stress must be of the same order of magnitude as the strength of the material; (2) a stress acceleration parameter g (see below) must be a function of temperature. The model is quite general incorporating dielectric breakdown, as well as electromigration. Its application has suggested that the current density exponent n varies with current density from $n = 1$ at $j < 1 \times 10^5$ A/cm^2 to $n > 2$ at $j > 1 \times 10^6$ A/cm^2. The mathematical analysis begins with consideration of free energy of activation and stress functions. The stress function is represented as the difference between the application stress and a "breakdown strength". A time to failure (TTF) equation may be derived:

$$\text{TTF} = A_0[\sinh(gz)]^{-1} \exp(Q/kT)$$

where $Q = \Delta H + az$,
z = generalized applied stress,
ΔH = free enthalpy,
a = temperature dependent constant,
A_0 = a constant,
g = stress acceleration parameter.

Even if the breakdown strengths are normally distributed, the observed TTF under stress z will be lognormally distributed. As is mentioned elsewhere, TTFs do exhibit this behavior in many cases. The basic physics is the same as described by Black, with the following additional consideration provided by McPherson:

> For a vacancy to move, certain localized distortions of the metal lattice (or along grain boundaries) must occur. The energy required to create this distortion is the activation energy of the associated diffusion process. This energy can be obtained from lattice phonons by raising the temperature of

> the metal. At room temperature, vacancies in a metallization can diffuse freely and in an isotropic fashion. However, once a current is introduced, the vacancy diffusion becomes directional.

In this model, the free energy of activation controls the process rate, while the energy depends on both the work contribution and the entropy change associated with the application of an external stress. The role of stress in electromigration is discussed more in the next section.

Partridge and Littlefield [15] note that electromigration models can be consolidated by including a variable power of the current density, which can be expressed as a function of the activation energy. A factor directly proportional to the current density is subtracted from bulk values. In this way, as the current density approaches zero, the activation energy approaches bulk values. They also note that their empirical data agree with a vacancy diffusion mechanism, in which grain boundaries are only one of the contributors to vacancy concentrations. Schafft *et al.* [16] report an empirical result that the current density exponent should be 1.53.

Kwok [17] notes that the atomic flux is due to electromigration inside metal grains and at the grain boundaries. The flux can be expressed by

$$J \propto (1/Kt)Dj\rho eZ^*$$

where D is the diffusivity, j is the current density, ρ is the resistivity, and eZ^* is the effective charge (sum of ionic charge from electrostatic forces and charge from electron wind forces). At device operating temperatures, grain diffusion is the predominant mode of mass transport.

The above information focused on electromigration in the constant current environment. There has also been considerable research conducted with respect to pulsed or time-varying currents which would be present in a microwave circuit. For example, Liew *et al.* [18] have developed a model for predicting interconnect electromigration time-to-failure under arbitrary current waveforms. The model is based on vacancy recombination. Under an arbitrary current condition, electromigration time-to-failure (TTF) is given by the following expression, valid for frequencies above 1 kHz:

$$\mathrm{TTF}_{\mathrm{acdc}} = \frac{A_{\mathrm{dc}}(T)}{|J|^{m-1}J(1 + R_1 R_2)}$$

where J is the average current density, m, A_{dc}, and A_{ac} are experimentally defined parameters and $R_1 = A_{\mathrm{dc}}A_{\mathrm{ac}}$, $R_2 = (|J| - J)/J$. If $m = 2$, and, $A_{\mathrm{ac}} \gg A_{\mathrm{dc}}$, for DC and uni-directional current (pulse DC), we have,

$$\mathrm{TTF}_{\mathrm{dc/pulse\ dc}} = A_{\mathrm{dc}}(T)/|J|^2$$

An expression is also derived for TTF under AC conditions. Liew *et al.* have used this model in a circuit electromigration reliability simulator, which is described in the design section of this chapter.

Suehle and Schafft [19] have noted that the median TTF enhancement (due to pulsed current) depends on current density and joule heating (a decrease in enhancement). They note that metallizations will survive longer under pulsed current than under continuous current. A relaxation phenomenon is considered to be the operative mechanism.

A somewhat simpler formulation of electromigration in pulsed current situations was constructed by Brooke [20]. The proposed model is:

$$\text{MTF} = A_0(J_{\text{dc}}) \exp(E_A^{-n}/kT)$$

where A_0 is a constant and J_{dc} is the time-average current density. The value of n was determined by experiment and found to be dependent on the sample metal and to range between 2.6 and 5.4, close to constant current values for the same metals. This model applies for a rectangular current waveform and for a duty cycle between 20 and 100% and 175 °C.

The formulation by Black and its application to pulsed currents by Brooke are the most commonly invoked theoretical formations of electromigration. As a general form, the above equation provides a simple basis for relating the median time to failure, MTF, of a conductor to current and temperature.

$$\text{MTF} = AJ^{-n} \exp(E_a/kT)$$

Theoretical treatments of grain size and grain boundary orientation have not been developed. Other factors not addressed in theoretical terms are defects, layering of conductors, and conductor steps. All have been explored experimentally.

9.3.2 Empirical electromigration results

The theory associated with electromigration has addressed only some of the governing factors. Fortunately, experimentation in this field is not very complicated, so that gaps in the theory can be filled with empirical evidence. For the most part, an electromigration experiment consists of setting up a sample, running a current through it, and measuring either TTF or void concentration. This, of course, is an oversimplification, since a considerable amount of care goes into the preparation of the test samples. Geometry, topography, and chemical composition must be well controlled. Similarly, data collection must be performed with care. Collection of TTFs is relatively straightforward. However, the analysis of void concentration and distribution (as well as hillock distribution) requires electron microscopy, which necessitates great care in sample preparation. The analysis of TTF data will be discussed at the end of this chapter.

It is not the purpose of this section to discuss experimental technique in detail. The reader can find the details of experiments in the references. In this section, the focus will be on the results of electromigration experimentation, especially in areas not covered well by the theory. The reader will find discussion of grain size and orientation, multilayer technology, topography, line geometry, and voids.

Verification of the theory of electromigration and conductor failure was performed by Black [13]. He explored the effects of gradients in temperature, current density, and the ion diffusion coefficient. As expected, voids formed in the areas where positive gradients existed and hillocks formed where there were negative gradients. It was noted that conductors with a positive gradient in metal ion diffusivity will fail earlier than gradient free conductors. Black also confirmed the existence of etch pits, which occur under the influence of elevated temperature and high current density at aluminum–silicon contracts. It was noted that device failure can occur when an etch pit (filled by aluminum) grows across an underlying junction, resulting in a short. His experiments explored the effects of aluminum structure in three forms: small crystal films, large crystal films, and large crystal films with a glass overcoat (to reduce surface diffusion).

The experiments were based on the premise that small crystal aluminum had a lower activation energy than large crystal aluminum. Results supported this premise, with large crystal activation energy estimated at almost twice that of small crystal activation energy. Activation energy of the glassed large crystal aluminum was even higher.

Baerg *et al.* [21] demonstrated the usefulness of RR, the ratio of electrical resistances at 298 K and at 77 K, for monitoring the quality of Al–Si metal films. They established a correlation between RR, electromigration MTF and median grain radius (MGR). They noted the usefulness of the lognormal distribution in plotting MTF data. The RR technique is intended as a post-deposition in-line process monitor. As other authors have noted, e.g., Black [13] and Amerasekera and Campbell [22], grain boundaries, film surfaces, and impurity sites play important roles in electromigration. In this case, RR is reduced from the bulk single crystal value of 12.6. They found that the RRs of sputtered thin films ranged between 6.0 and 9.5, with the highest values obtained from sputtering under high vacuum and thorough control of deposition temperature. MGR also was measured. Both RR and MGR were found to be correlated with MTF, as shown below. Although RR and MGR can be used as quality checks on metallization, Baerg *et al.* prefer RR because it is simpler. RR is measured using a Van der Pauw four-probe method, while MGR requires the use of a scanning electron microscope.

Towner [23] found that, for aluminum alloy films, lifetimes could be represented by lognormal distributions when grain size is smaller than linewidth and by either lognormal or logarithmic extreme value distributions if grain size exceeds linewidth. These results are attributed to the relative orientation of grain boundaries within a line. Series orientation leads to extreme value distributions. Parallel, as well as series, orientation leads to the lognormal distribution.

Hinode and Homma [24] explored the impact of refractory metal (titanium nitride and tin) on electromigration in layered conductors and found that layering with refractory metals degraded the electromigration immunity of the aluminum layer. They also found that grain growth was suppressed, and that reducing this suppression increased conductor lifetimes by one or more orders of magnitude. Their technique for arriving at these conclusions was the monitoring of conductor resistance in specially constructed samples. They noted that there were three influencing factors: diffusion of the refractory metals into aluminum, crystal orientation affinity, and interface reaction. Figure 9.11 shows the effect of aluminum grain size on conductor lifetime.

Towner [25] studied electromigration induced short circuit failures in multilayered conductors. The layering examined was aluminum on titanium–tungsten. His results are summarized as follows:

(a) The probability of a short circuit failure in conductors increases with the thickness of the titanium–tungsten layer.

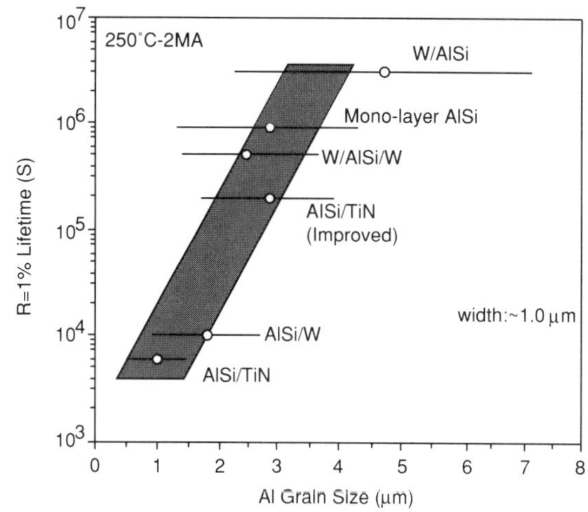

Figure 9.11 Stable and unstable silicide metallization systems based on enthalpy change at elevated temperatures.

(b) Interlayer short circuit failures can be characterized by a lognormal distribution of lifetimes.

(c) Both open and short circuit failures exhibit similar activation energies (0.5–0.6 eV).

(d) In tungsten–titanium–aluminum conductors, short circuit lifetime increases linearly with increasing dielectric thickness.

(e) The current density exponent for short circuit failure in tungsten-titanium-aluminum conductors was between two and four.

Oates [26] noted that electromigration in aluminum conductors is influenced by the underlying topography. He investigated step spacing in two-level metal technology and showed that regions with poor coverage (10–20%) are particularly vulnerable to electromigration failure. At the step, the current density may be increased significantly due to metal thinning. Consequently, it is important that the steps be covered adequately and uniformly. Metal step spacing was found to be a critical factor. Oates notes that for critical step spacings, poor step coverage results, with a consequent vulnerability to electromigration. However, if spacing is adjusted to improve coverage, the benefit will not be as large as expected, because for a certain range of spacings, a new failure mode is introduced. The exact mechanism for the new mode was not determined. The step spacings for the new failure mode were found to be from 3–4.375 μm.

The issue of step coverage was also addressed by Kim *et al.* [27]. They studied electromigration of DC bias-sputtered aluminum and found step coverage to be excellent. They found electromigration resistance to be inferior to aluminum sputtered without bias. Failure was caused by voids along grain boundaries and an extremely small grain size. A wide distribution of grain sizes was also reported. Strausser *et al.* [28] noted that the most significant materials parameter that changes at steps is grain size. They reinforced the view that, in terms of topographical factors affecting electromigration, frequency and severity of steps are the primary variables. There is a high likelihood of the formation of a grain boundary at a step. As the grain grows at the corner, different orientations occur on either side and a boundary is formed. The grain boundaries are points for the entrapment of voids.

Kwok [17] performed an excellent study of the effect that conductor line geometry has on lifetime, due to electromigration in aluminum–copper submicron lines. He found the following:

(1) Lifetime decreases slightly with increasing line length and then levels off after a critical value.

(2) As line width decreases, lifetime decreases, goes through a minimum, and then increases for less than a critical width.

(3) Lifetime increases with decreasing film thickness.

(4) Critical length and width decrease with decreasing linewidth and film thickness respectively.

(5) Grain morphology is an important parameter in the control of electromigration lifetime.

LaCombe and Parks [29] evaluated conductor failure distributions as functions of varying length and width. They were found to follow the lognormal distribution down to the 0.3% point. All failures were open circuits resulting from voids, which formed at the edge of a line and grew inward. Hillocks did not occur anywhere except within 10 μm of a large void. The researchers concluded that the length dependence of early failures was consistent with an independent segment model. Later failures were found not to be consistent with the segment model and may have been influenced by more global parameters, such as current, average temperature, or average void thickness. Earlier work by LaCombe and Parks [30] noted an overall increase in conductor resistance, primarily due to the formation and growth of voids. They did notice that the resistance will increase and decrease several times before final failure. This was attributed to some self-healing. They cautioned against using early resistance data to assess conductor lifelength because failure mechanisms other that just void formation may become significant.

Several resistance change techniques for measuring electromigration effects have been developed [31, 32]. These are based on void formation models, according to which voids form and enlarge, increasing current density and local temperature gradients. The change in conductor resistance is used as an indicator.

Sadana *et al.* [33] used transmission electron microscopy to study the location and shape of voids in aluminum and aluminum alloys. They found both to be dependent on the type of impurity that was added. They concluded that precipitation of added impurities at grain boundaries did not always ensure electromigration resistance. They explained void formation and shape by considering flux divergence at triple points.

Note that void nucleation occurs upstream of a grain boundary. Once voids nucleate, erosion takes place from the aluminum surface surrounding them and results in irregularly shaped voids. If the grain boundaries are perfectly hardened, and the grains contain fine impurity precipitates, then the voids are located on the downstream side of the grain boundaries.

A method for minimizing void formation was discovered by Koyama *et al.* [34]. They noted that void formation could be suppressed by mercury light irradiation of plasma enhanced chemical vapor deposition silicon nitride film coatings. The light beam induces stress relaxation of the nitride film, which is responsible for the relaxation. The voids can be created through annealing at relatively low temperatures, especially when plasma enhanced coatings are used. Koyama *et al.* tested the voiding in samples that had been annealed at 350, 400 and 450 °C. One mechanism they proposed for the formation of voids was the presence of hydrogen in the lattice. This creates microvoids that aggregate into aluminum voids under the stress of the nitride film. The hydrogen lowers the surface tension of the aluminum and allows microvoids smaller than 1 μm to remain. Such small voids are not stable in pure aluminum. Hydrogen inside the microvoids prevents collapse of the voids. The microvoids aggregate and grow more stable after the applied stress of passivation. A particular finding of theirs was that heat treatments at 200–400 °C alone do not reduce aluminum voiding after the final aluminum sinter. Light beam irradiation (200–2000 nm) on to the nitride films does suppress the voids after final annealing at 450 °C for 30 minutes.

Rosenmeyer *et al.* [35] explored the effects of bending stresses on electromigration and found that void formation was greatest in the region of highest stress. They noted that stresses and stress gradients have only negligible effects on vacancy diffusion.

9.3.3 Reducing electromigration data

Experimentally, the parameters available for reduction are either the current density or the temperature. The median life is given as:

$$t_{50} = AJ^{-n}\exp(Q/kT)$$

For a fixed current, we can then extrapolate t_{50} for a desired temperature:

$$\frac{t_{50}(T_1)}{t_{50}(T_2)} = \exp\left[\frac{Q}{k}\left(\frac{1}{T_1} - \frac{1}{T_2}\right)\right]$$

As a second experiment, we can fix the temperature and vary the current density. The ratio of t_{50} is then given as:

$$\frac{t_{50}(J_1)}{t_{50}(J_2)} = \left(\frac{J_1}{J_2}\right)^{-n} \qquad n = 1.5$$

Table 9.4 shows the estimated MTFs for refractory based metallizations

One may also calculate failure rates due to electromigration, since t_f obeys a log-normal distribution

$$f(t) = \frac{1}{\sqrt{2\pi}} \times \frac{1}{\sigma} \times \frac{1}{t} \times \exp\left[-0.5\left(\frac{\ln t - \ln t_{50}}{\sigma}\right)^2\right]$$

Table 9.4 Estimated MTFs of Ti:W/Al and Ti:W/Al + Cu.

Film conductors at 85 °C			
Current density	Exponent	Ti: W/Al (hours)	Ti: W/Al + Cu (hours)
1×10^6 A/cm^2		17 882	53 230
5×10^5 A/cm^2	$m = 1.5$	50 592	150 557
	$n = 2.0$	71 548	212 920
2×10^5 A/cm^2	$n = 1.5$	199 982	595 130
	$n = 2.0$	437 175	1 330 250

where $f(t)\mathrm{d}t$ = probability of observing a failure in time interval dt. The failure rate is

$$\lambda(t) = \frac{f(t)}{1 - F(t)}$$

and $F(t)$ is the cumulative failure density function. Finally, the stability of interconnects must also be included especially in terms of elevated temperature reliability. The trend is towards silicide gates. The enthalpy change at elevated temperatures is shown in Figure 9.12. The key systems for reliability are shown to be the refractory silicides.

9.3.4 Design and manufacturing guidelines

Both theoretical and experimental work in electromigration have led to changes in design and manufacturing to mitigate its effects. For example, manufacturers have taken the following steps.

(1) Control the shape of the oxide steps.

(2) Control the metal deposition process carefully. The resistance ratio technique of Baerg *et al.* should be considered as a quality control device.

(3) Add 3–4% copper to the aluminum to retard metal migration. Copper lines the grain boundaries and prevents their movement through the metal. Alloys of aluminum–copper, aluminum–silicon, and aluminum–copper–silicon have shown no failures due to electromigration at temperatures between 150 °C and 215 °C.

(4) Deposit a glass layer on to the metal to partially cool it and pin down metal atoms.

(5) Factor the current density/metal migration mechanism into designs. A current density of 10^6 A cm^2 should not be exceeded in the metallization, although some manufacturers claim this can be done by a factor of five in some devices with no adverse effects.

Common failure modes associated with electromigration in discrete and integrated devices are the following:

(1) High resistance or open circuits due to the formation of voids.

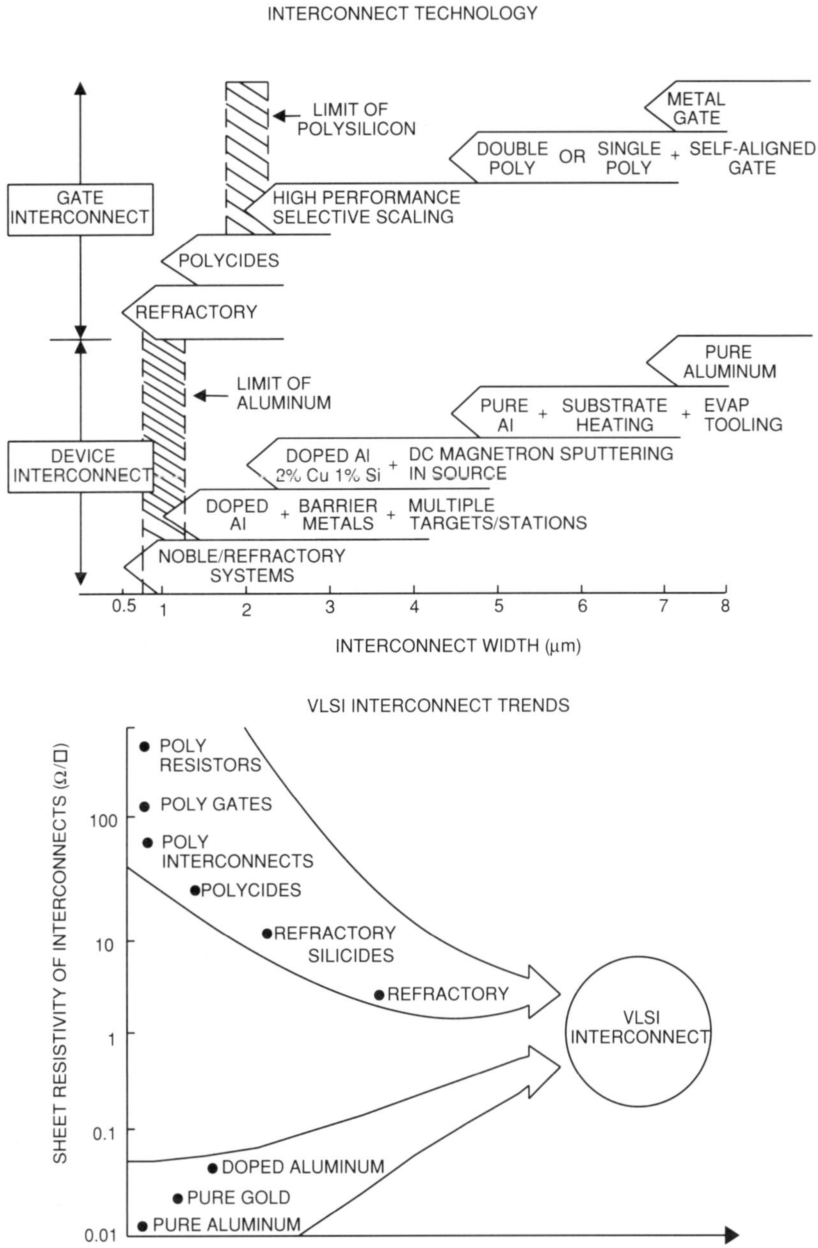

Figure 9.12 Interconnect technology.

(2) Enhanced alloying and shorting due to penetration of aluminum into silicon surfaces.

(3) Increased contact resistance and non-uniform current flow due to hillock formation, which results from high current densities and thermal cycling.

Bond lifts also have been observed. Failures may result from defective metallization layers caused by flaws in the pattern mask or errors in the manufacturing process, and the inclusion of contaminants at metallization interfaces. The basic result of electromigration research is the development of new design and manufacturing guidelines for metallizations.

REFERENCES

[1] Cunningham, J.A., "Expanded contacts and interconnections to monolithic silicon integrated circuits", *Solid State Electronics*, **8**,1965, 735-745.

[2] Selikson, B. and Longo, T.A., "A study of purple plague and its role in integrated circuits", *Proc. Inst. Elec. Electronic Engrs.*, **52**, 1964, 1638.

[3] Blech, I.A. and Sello, H., "Some new aspects of gold-aluminum bonds", *J. El. Chem. Soc.*, **113**, 1966, 1052-1054.

[4] Philofsky, E., "Purple plague revisited", *8th Annual Proceedings, Reliability Physics*, 1970, 177-185.

[5] Philofsky, E., "Design limits when using gold-aluminum bonds", *9th Annual Proceedings, Reliability Physics*, 1971, 114-119.

[6] Goldfarb, S., "Wire bonds on thick film conductors", *Proc. 21st Ecc. IEEE Conf.*, 1971, 295-302.

[7] Horsting, C.W., "Purple plague and gold purity", *10th Annual Proceedings, Reliability Physics*, 1972, 155-158.

[8] Christou, A., Jarvis. L., Weisenberger, W., and Hirvonen, J.K., ""SEM, auger spectroscopy and ion backscattering techniques applied to analysis of Au/refractory metallizations", *J. Elec. Mat.*, **4**, 1975, 329-336.

[9] Daus, J.L. and Howland, F.L., "Fatigue failure of encapsulated gold-beam lead and TAB devices", *IEEE Trans. on Comp., Hybrids and MT*, **CHMT-1**, (2), 1978, 158-166.

[10] Dunning, K.L., "Depth profiles of aluminum and sodium near surfaces — Nuclear Resonance Method", Thin Solid Films, **19**, 1973, 145-154.

[11] Berg, H. and Hall, E.L., "Dissolution rates and reliability effects of Au, Ag, Ni and Cu in lead base solders", *11th Annual Proceedings, Reliability Physics*, 1973, 10-20.

[12] Devaney, J.R., Hill, G.R., and Seippel, R.G., *Failure Mechanisms, Techniques, & Photo Atlas, 1983*; Failure Recognition & Training Services, Inc., 10-24.

[13] Black, J. R., "Electromigration failure modes in aluminum metallization for semiconductor devices", *Proc. IEEE*, **57**, (9), September 1969, 1587-1593.

[14] McPherson, J., "Stress dependent activation energy", *Proc. IEEE IRPS*, April 1986, 12-18.

[15] Partridge, J. and Littlefield, G., "Aluminum electromigration parameters", *Proc. IEEE IRPS*, April 1985, 118-125.

[16] Schafft, H., Grant, T., Saxena, A., and Kao, C., "Electromigration and the current density dependence", *Proc. IEEE IPRS*, April 1985, 93-99.

[17] Kwok, T., "Effect of metal line geometry on electromigration lifetime in Al-Cu submicron interconnects", *Proc. IEEE IRPS*, April 1988, 185-188.

[18] Liew, B.K., Fang, P., Cheung, N.W. and Hu, C., "Reliability simulator for interconnect and intermetallic contact electromigration", *Proc. IEEE IRPS*, March 1990, 111-118.

[19] Suehle, J. and Schafft, H., "Current-density dependence of electromigration t_{50} enhancement due to pulsed operation", *Proc. IEEE IRPS*, March 1990, 106-110.

[20] Brooke, L., "Pulsed current electromigration parameters", *Proc. IEEE IRPS*, April 1987, 136-139.

[21] Baerg, W., Wu, K., Davies, P., Dao, G. and Fraser, D., "The electrical resistance ratio (RR) as a thin film metal monitor", *Proc. IEEE IRPS*, March 1990, 119-123.

[22] Amerasekera, E.A. and Campbell, D.S., *Failure Mechanisms in Semiconductor Devices*, 1987; J. Wiley & Sons, 38-39.

[23] Towner, J., "Are electromigration failures lognormally distributed?", *Proc. IEEE IRPS*, March 1990, 100-105.

[24] Hinode, K. and Homma, Y., "Improvement of electromigration resistance of layered aluminum conductors", *Proc. IEEE IRPS*, March 1990, 25-30.

[25] Towner, J., "Electromigration — induced short circuit failure", *Proc. IEEE IRPS*, April 1985, 81-86.

[26] Oates, A., "Step spacing effects on electromigration", *Proc. IEEE IRPS*, March 1990, 20-24.

[27] Kim, M., Skelly, D. and Brown, D., "Electromigration of bias — sputtered Al and comparison with others", *Proc. IEEE IRPS*, April 1987, 126-129.

[28] Strausser, Y., Euzent, B., Smith, R., Tracy, B. and Wu, K., "The effect of metal film topography and lithography on grain size distributions and on electromigration performance", *Proc. IEEE IRPS*, April 1987, 140-144.

[29] LaCombe, D. and Parks, E., "The distribution of electromigration failures", *Proc IEEE IRPS*, April 1986, pp. 1- 6.

[30] LaCombe, D. and Parks, E., "A study of resistance variations during electromigrations", *Proc. IEEE IRPS*, April 1985, 74-80.

[31] Hoang, H., Ondrusek, J., Mishimura, A., Sigiura, T., Blumenthal, R., Kitagawa, H. and McPherson, J.W., "Effective kinetics variations with stress duration for multilayered metallization", *Proc. IEEE IRPS*, April 1988, 179-184.

[32] Maiz, J. and Segura, I., "A resistance change methodology for the study of electromigration in Al-Si interconnects", *Proc. IEEE IRPS*, April 1988, 209-215.

[33] Sadana, D., Towner, J., Norcott, M. and Ellwanger, R., "Some TEM observations on electromigrated Al and Al Alloy interconnects", *Proc. IEEE IRPS*, April 1986, 38-43.

[34] Koyama, H., Mashiko, K. and Nishioka, T., "Electromigration in Aluminum and Tantalum Silicide Contacts", *Proc. IEEE IRPS*, April 1986, 24-29.

[35] Rosenmeyer, C., Brotzen, F., McPherson, J., and Dunn, C., "Effect of stress on electromigration", *Proc. IEEE IRPS*, April 1991, 52-56.

10

DEVICE PACKAGING AND MICROCIRCUIT CHARACTERIZATION

10.1 PACKAGING AND RELIABILITY

Packaging of semiconductor devices has always had a significant impact on the reliability of integrated circuits. This fact surely became prevalent after thermoset plastic compounds had been established as the dominant packaging materials. Hand in hand with material-related aspects, the assembly processes and their controls started to add a second important influence on device reliability. Finally, with the advent of very-large-scale integration and the accompanying complexity of chip structure and shrinking feature sizes, the design of packages emerged as the third strong factor determining the reliability of the finished product. This review will highlight significant features of all three contributors to plastic package reliability: package design, choice of materials, and assembly/packaging processes. The first nine chapters have covered chip level reliability and manufacturing issues. Packaging and assembly issues will now be addressed.

10.1.1 Market trend towards surface mount packages

For many years, dual-in-line plastic packages with metallic pins for insertion assembly dominated integrated circuit package design. While these packages will be available for many more years, strong trends are now projected towards new kinds of plastic packages, based on surface mount and tape-automated assembly. The projections differ in the absolute rate of increase of market share, but there is agreement that future state-of-the-art semiconductor circuits will be housed in surface mount configurations. The newer package types are plastic leaded chip carriers (PLCC), small-outline packages with gull-wing shaped leads, and small-outline packages with J-shaped leads.

The advantages and limitations of these lead configurations are summarized in Figure 10.1. As an example, Figure 10.2 shows the J-lead small-outline package for the MOS 1-Megabit DRAM, manufactured by Texas Instruments. For this product, the chip size is almost as large as the current 256K PLCC. Thus, the 1-Megabit DRAM package will be driven by silicon area rather than pin count. The design of this package offers advantages in routeability and can be extended in pin count and package width to facilitate other large memory products.

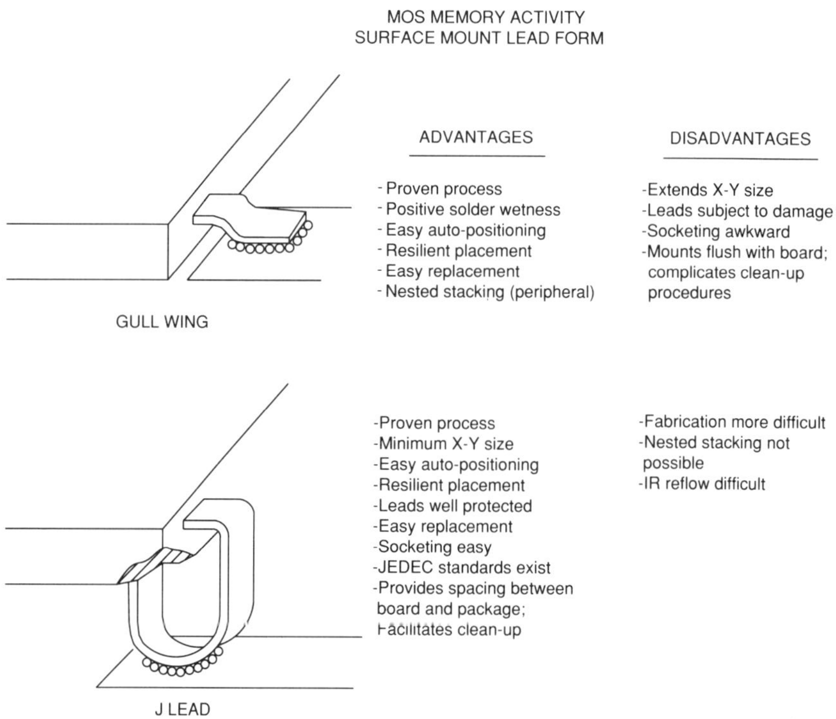

Figure 10.1 Comparison of advantages and disadvantages of the surface mount process.

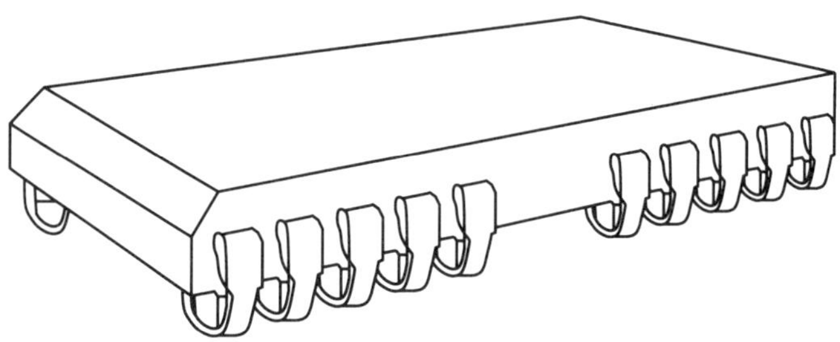

Figure 10.2 J-lead small outline package.

10.1.2 Computer modeling of stress in silicon and molding compound

Plastic packages for microelectronic devices have to reconcile seemingly conflicting requirements. While the IC market favors constant out package dimensions, ever increasing memory (64K, 256K, 1MB) and logic functions dictate larger silicon chip sizes inside the package. Consequently, the discrepancy between the thermal expansion coefficients of silicon on the one hand, the lead-frame metal and the epoxy molding compound on the other, will become crucial. Severe stresses, therefore, will develop in the plastic and silicon chip to the point that the molded plastic starts cracking when exposed to thermal shocks. Finally, the IC market demands ever tighter reliability requirements for temperature stability, as well as moisture resistance. The need of a package design tool based on computer modeling has never been more urgent.

The dilemma of ensuring package integrity can best be handled by defining the stress distribution in the package and then minimizing it through various geometries, materials and processes. The numerical technique called finite element analysis can allow the packaging engineer to characterize the stresses and deformations for various types of loads (thermal, mechanical, etc.). This method is widely used in the mechanical engineering field, but is equally applicable and appropriate for stress analysis of plastic encapsulated IC packages.

In stress analysis, the structure is divided into a large number of small elements, interconnected at a discrete number of nodal points describing the geometry. The nodal displacements from an induced load determine the strain and, given the material properties, the stress for a selected element. When a chip is encapsulated in a plastic molding compound, it sees a temperature of about 150–175 °C. At this mold cure temperature, the entire package can be considered stress free. As the device is cooled, a progressive stress condition sets in due to dissimilar expansion coefficients.

A finite element model of a large chip package has been developed to locate the relatively high stress concentrations at various cross-sections of the device. The stress contours which follow represent regions of equal stress within the package. There is an extremely high concentration of shear stress at the bottom corner of the chip pad. This has been the site for many fatigue-initiated cracks during actual temperature cycling. In Figure 10.3, the package was molded at about 150 °C, and used a copper-based leadframe with a chemically and mechanically treated surface to ensure a high degree of adhesion. The overall package stresses are much lower than the stresses shown in Figure 10.4. The shear stress distribution is located immediately above the bar. This intense concentration of stress explains such failures as, passivation cracks, metal line shifts, and sheared ball bonds. The model has also been used to investigate various offsets between leadframe and bar pad, package widths, and other innovative processing techniques. Once the designer defines the material properties and package geometry, the overall stress distribution of the package can be evaluated. With this knowledge, the limits of package form factors, assembly processes, and materials can be specified.

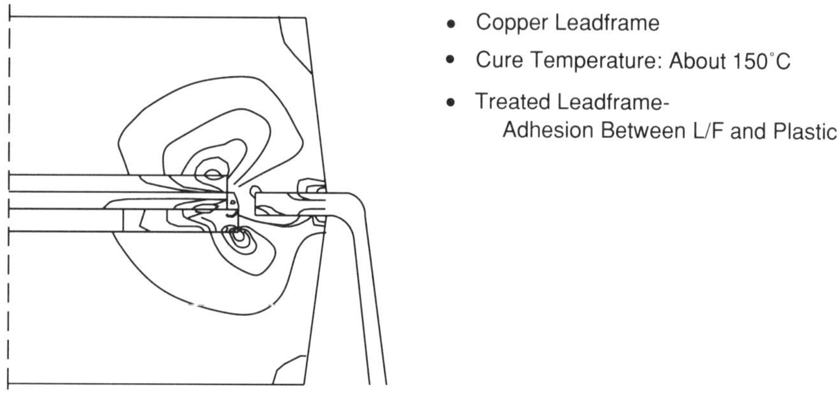

Figure 10.3 Lateral-cross section contour plot.

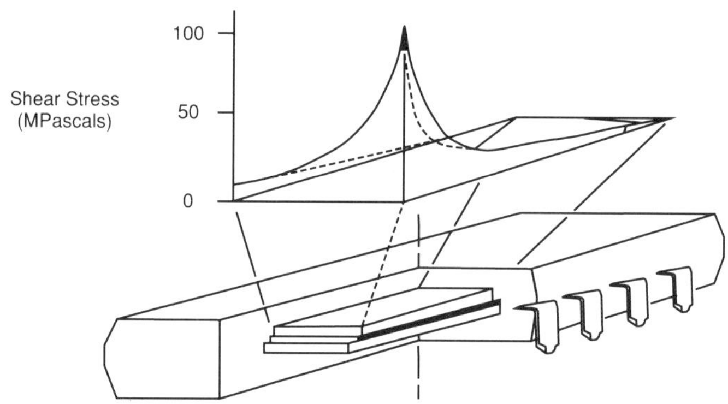

Figure 10.4 Shear stress above bar.

10.1.3 Test structures for package and product reliability

10.1.3.1 Strain gauge for stress measurements

It has been well established that predictions of theoretical device failure mechanism models can be verified by measurements using specific simplified test structures. The same is true for verifying the prediction of computerized package performance models. For the stress modeling described above, integrated strain gauges have been proven to supply quantitative stress data by mapping large Si chips. This data was found to verify precisely the predictions of finite element models.

10.1.3.2 Electrical structure for moisture leakage

Interdigitated Al lines of fine geometries have been used to monitor surface leakage down to 10^{-15} amps. The resistance measurement of Al lines is used to monitor moisture ingress into plastic packages as a function of molding compound, plastic curing time and temperature, protective overcoat over chip, potential degradation caused by prior temperature cycling, etc.

10.1.3.3 *Resistance structure for metal corrosion*

Moisture in combination with contamination (e.g., phosphorus, chlorides, sulfuric acids, amines, human sweat/spit) may lead to Al corrosion. Packages can be evaluated for corrosion resistance by Al line test structures which compare protected and unprotected Al snakes and biased combs. The Al lines may be positioned flat on the insulator, or contoured over insulator steps. The insulator may contain various amounts of phosphorus or other corrosive agents. The test structures are monitored for resistance increases and electrical opens as a function of prolonged testing in autoclave, 85 °C/85% RH, etc. environments.

10.1.3.4 *Resistance structure for metal shear*

Metal shear is caused by high shear stress in plastics, package design parameters, soft metal under brittle protective overcoats, or assembly process conditions such as curing. Shear can result in several defects: cracked protective overcoat, smeared metal lines, or enhanced leakage.

Test structures have been designed to measure quantitatively the effects of shear stress on metal performance, passivation and accelerated testing as a function of chip geometry, package design, and material composition. Al continuity snakes measure resistance shifts and opens, and, when second level metallization is available, leakage currents through passivation cracks. If transistors are included in the test structure, parametric shifts caused by resistance change or leakage current can also be monitored.

10.1.3.5 *Bump structure for controlled collapse*

For evaluating the controlled collapse characteristics of bumps, such as shown in Figure 10.5, a four-point probe type layout has been proposed in literature. Figure 10.6 depicts this test structure. The technique employed is non-destructive to the bumps until failure mechanism information has been obtained; it involves the removal of the supporting silicon structure by RF sputtering. Upon completion of the silicon removal all metallization, including the bump and the chip's runners, is

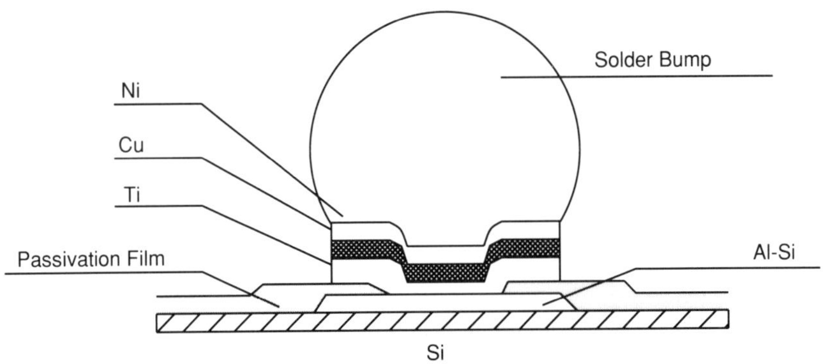

Figure 10.5 Schematic presentation of controlled collapse bump (after R.A. Young, *et al.*).

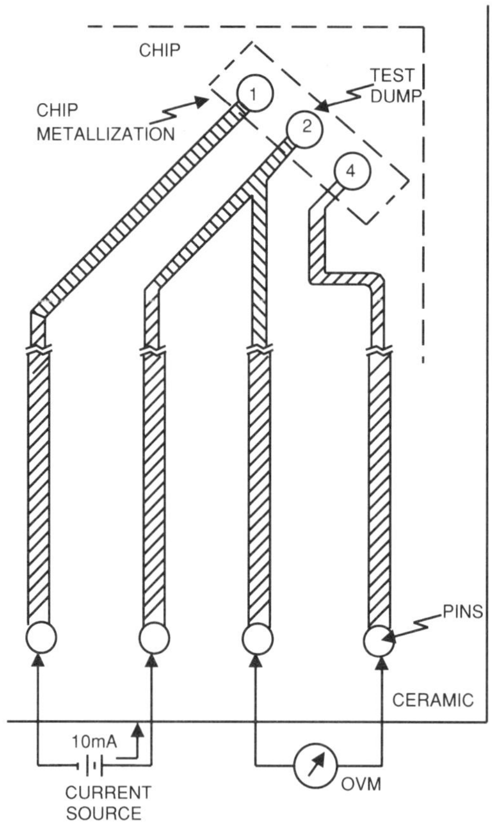

Figure 10.6 Physical layout of test bumps (after E. Winkler [11]).

left intact. The approach was validated by accurate resistance measurements of the bump/metallization paths before and after chip removal. In addition, destructive analysis was accomplished on the same samples by replotting and resectioning through the bumps.

10.2 CHIP PREPARATION FOR PLASTIC PACKAGE RELIABILITY

Two process improvements have been investigated, which will increase plastic package reliability. These are post-bond clean-up, and post-bond protective flux.

In 1981, M. Iannuzzi published *Development and Evaluation of a Pre-Encapsulation Cleaning Process to Improve Reliability of HICs With Aluminum Metallized Chips*. The reactivity of cleaning processes toward aluminum was determined. It was found that H_2O_2, alkali cleaners, and hot water are unacceptable, and that freon is inadequate. However, freon followed by O_2 plasma, followed by cold H_2O was the most effective cleanup. It resulted in 13 000 h on 85C/85% RH without failures, while no overcoat covered the samples.

Also in 1981, K.M. Striny and A.W. Schelling published *Reliability Evaluation of Aluminum-Metallized MOST Dynamic RAMs in Plastic Packages in High Humidity and Temperature Environments.* 4K DRAMs were used as test vehicles with Si_3N_4 passivation. After freon and cold H_2O cleanup, 2 mil conformal coating of RTV rubber was put on a chip/lead frame. This process resulted in 9312 h on 85C/85% RH with only 4% failures.

10.2.1 Flip-chip technology and thermal fatigue

A substantial portion of the total semiconductor market in 1990–95 will be served by flip-chip assembled circuits. In order to accomplish this, the solder bump technology depicted in Figure 10.5 has to be fully developed and characterized in terms of product reliability.

Several non-corrosive metal combinations have been reported in the literature (Cr–Ni–Au, Cr–Cu–Au, Ti:W–Ni–Au, etc.). Solder bumps of various Pb+Sn compositions and geometrical dimensions have been investigated. In addition, hard bumps such as Cu have been tried, which do not collapse and reflow. After bumped chips are assembled by the flip-chip technique, thermal displacement and eventual metal fatigue is observed in temperature cycle investigations.

10.2.2 Molding compounds and thermal characterization of packages

Thermal characterization of packaged integrated circuits generally takes one of three forms:

(1) Computer simulations (thermal network analyzer).
(2) Indirect measurements (infrared detection).
(3) Direct measurements (electrical readings).

Package design work can utilize thermal computer modeling and actual measurements from previous device types. The results from such thermal models can be used as benchmarks for measured results and stored into databases for future design and testing.

Examples of thermal modeling are shown in Figure 10.7 for an IC chip over a printed circuit board, and in Figure 10.8 for a more detailed look at the heat flow through a pin grid array. The thermal flow in a plastic device is analogous to the ceramic example depicted.

As in the case discussed above (stress modeling), the computer thermal analysis programs are finite difference solvers which break up the structure into many small elements (called nodes), connected to each other by conduction "resistors". These programs utilize resistors by solving a thermal energy equation to account for all the energy. Knowing the heat dissipation, an electronic packaging engineer can design the package for better thermal management.

Thermal models of IC packages aid in optimizing the following design parameters:

Form factors (lead-frame, package, etc.).

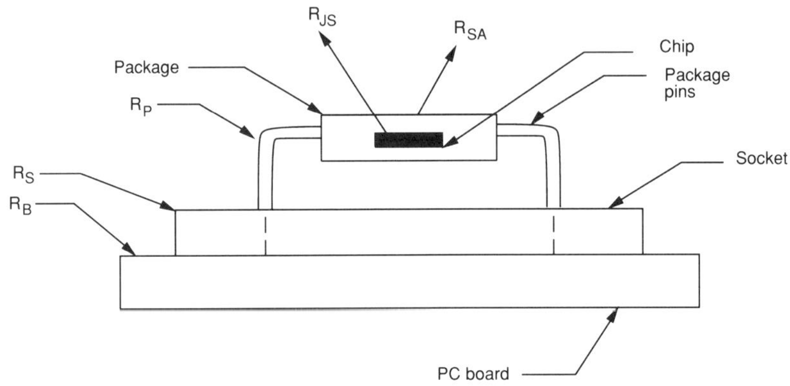

Figure 10.7 Electrical analog of heat flow from IC chip to printed circuit board.

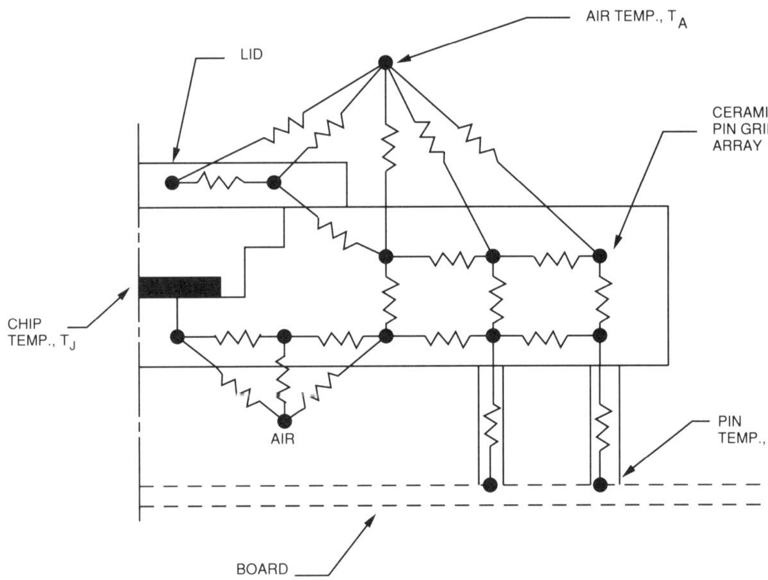

Figure 10.8 Ceramic pin grid array thermal model.

Material selection (chip attach, molding compounds, ceramic substrates, leadframes, etc.).

Device-to-system interfacing.

Thermal modeling has been validated by thermal evaluation tests of chips throughout the semiconductor industry. Thermal modeling does not assume that the thermal properties of materials involved are temperature-independent. It also accounts for non-uniformities in heat distribution within the package, as opposed to direct and indirect measurement methods.

10.2.3 Surface mount technology

The recommended high volume assembly process is defined as follows: screen print and solder paste on PWB; place components; oven dry (optional); batch/in-line vapor phase reflow solder; vapor deflux; inspect/test. The effectiveness of the assembly process can be characterized by the number of unacceptable solder joints formed during the process. Unacceptable joints are defined by their electrical (open circuit) and mechanical (strength and reliability) characteristics.

Open circuits are the first defects detected at electrical test after soldering. Several factors contribute to open solder joints. Lead tip planarity is the most important factor in obtaining acceptable process yields. Lead position, lead finish, solder paste composition and volume affect process yields as well. Another interesting result showed that silver in the process, either as a lead finish or in the solder paste, improves yields significantly. One explanation may have to do with the dynamics of the solder during the reflow process, as they are affected by the different surface forces acting in the silver and non-silver processes.

The most important mechanical characteristic of the solder joint is strength under varying stress conditions. Although not a problem with small devices, there is a concern for the integrity of the solder joint for large packages. Testing is underway by an IEEE task force to determine the solder joint reliability under mechanical and thermal stress. Results to date do not indicate a need for concern with this problem. Possible failure mechanisms for weak joints include voids in solder (Figure 10.9), wicking of solder, and insufficient solder (Figure 10.10).

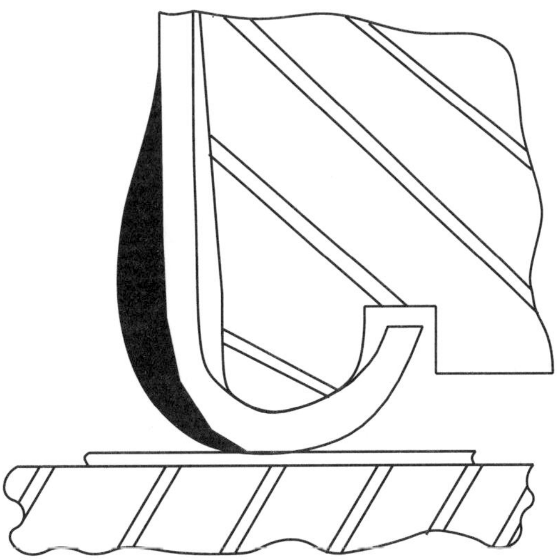

Figure 10.9 Stenciled pad solder wicks up the lead as contaminated board pad metallization does not wet stenciled solder paste.

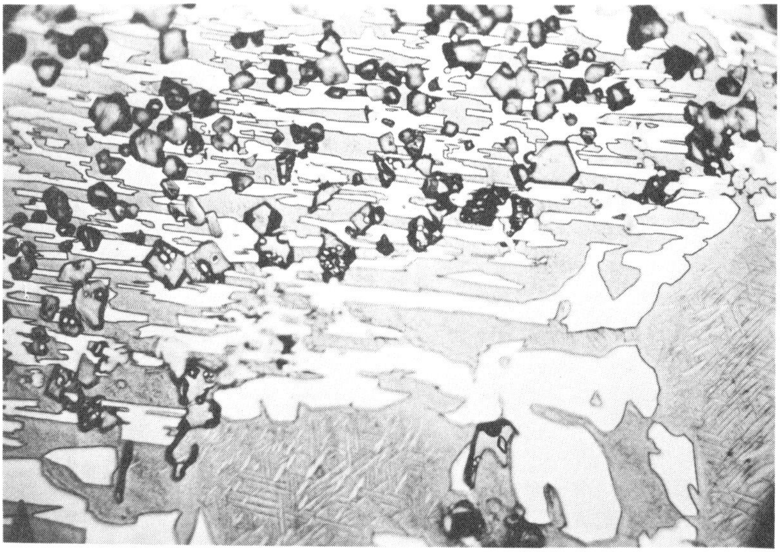

Figure 10.10 Weak solder joint due to board pad contamination or lack of fluxing, at a magnification of 100 times.

As for reliability testing of surface mounted assemblies, attention has to be paid to temperature cycling (−65 to +150 °C). The mismatch of thermal expansion coefficients between plastic package, metallic solder and substrate materials can be minimized by choosing suitable substrates, such as, aluminum nitride, silicon carbide, porcelain-like enamel steel, or copper-clad invar.

Furthermore, the heat dissipation from the operating device to the environment has to be optimized. This can be done by having low thermal resistance from chip to package. To achieve this, low thermal resistance substrates, copper leadframes, integral heat sinks and high conductivity plastics are recommended.

Testing of surface-mounted assemblies remains a challenge. Visual solder joint inspection is difficult for J-leads. The Vanzetti system offers auto-inspection by laser heating and thermal signature analysis. Careful node layout-access is needed for fault isolation testing.

10.2.4 High pin count and tape automated bonding

High pin counts up to 300 can be accommodated by the twin-row plastic chip carrier. This concept adds a second row of surface mount leads that are indented into the package via a "castellated" package perimeter.

- Leads are on 50 mil centers in any one row
- Individual rows are also separated by 50 mils

The twin row chip carriers also have double the lead count of their equivalent JEDEC counter parts.

Std PLCC	Twin row PLCC	Package size (in. sq.)
28	48	0.450
44	80	0.650
52	96	0.750
68	128	0.950
84	160	1.150

It is an additional advantage that the same package dimensions allow standard and twin row PLCCs to share common handling equipment. Twin row PLCCs are capable of meeting logic VLSI I/O needs; they may also replace pin grid array products.

A more commonly accepted approach to high input/output needs is tape-automated bonding (TAB). For high density, high lead count devices, TAB has definite advantages over wire bonding. Several approaches are described in the literature. In one study by T.A. Scharr at Motorola, TAB of a 200 lead, 300 mil square test chip has been investigated. This chip was successfully inner lead bonded with Au–Sn eutectic bonding. Outer lead bonds were made with vapor phase solder reflow techniques. The identified favorable features of TAB included:

- The ability of leads to be bonded in a higher density pattern than wire bonds. Smaller center-to-center spacing has been achieved with TAB than with wire bonds. Al wire with 1.25 mils (31.8 μm) diameter can be bonded on 5.5 mil (140 μm) centers. Stateof-the-art TA bonds have been made to 0.002″ (51 μm) pads on 0.004″ (102 μm) centers.
- The electrical and thermal characteristics of TAB leads are superior to wire bonds. Their rectangular cross-section results in lower electrical and thermal resistance and lower inductance than bond wires.
- TAB devices can be burned-in and pre-tested on tape, before attachment to a package or circuit board.
- Bumping a chip for TAB bonding may effectively seal off the usually open bond pad area against the ambient environment.
- Shorter total bond time, higher reliability and adaptability to automation are also projected with TAB.

Plastic packaging of semiconductor circuits is presently facing three major challenges:

- Find innovative ways to safely accommodate increasingly larger and thicker Si chips in plastic packages of same outline. This section describes how performance modeling of new packages can serve as a guide to selecting the optimum design, material choice and process definition.
- Find solutions to ultra-reliable surface mounting technologies for ever more densely assembled boards.
- Find methods to provide high input/output connections.

In the following section, the analytical techniques for microcircuit evaluation will be presented. The manufacturing aspects of plastic circuits will be presented in Chapter 11.

10.3 INTRODUCTION TO CHARACTERIZATION OF MANUFACTURING AND RELIABILITY

A substantial percentage of all device failures are related to surface, near-surface or interface chemistry, making a knowledge of surface composition a key factor in assuring device reliability. For most microelectronics applications, the main goal in surface analysis is to identify the structural and compositional characteristics of a solid material surface. To accomplish this task, there are over a dozen surface analysis techniques available to the materials engineer today. In determining the appropriate analytical technique for a particular application, the capabilities and limitations of each method should be weighed carefully. For example, sampling depth, lateral resolution, induced damage and sensitivity of the technique should be reviewed to prevent possible misconceptions about data yield.

10.3.1 Analytical techniques

The analytical methods which will be reviewed are scattering or emission techniques in which a primary beam of electrons, ions or photons, focused on the sample surface, causes the emission or scattering of electrons, ions and photons with characteristic energy or mass values. The detection and measurement of these emissions provides surface structure, elemental and compound information. A summary of the excitation and emission modes for these techniques is shown in Table 10.1.

The following is a comparative review of each of these techniques, outlining the physical phenomena of the method as well as capabilities and limitations.

10.3.1.1 Surface structure analysis

Scanning electron microscopy (SEM). The Scanning Electron Microscope has become commonplace in the microelectronics industry and is literally the

Table 10.1 A summary of the principle features of the various surface analysis techniques.

	AES	EDX	WDX	LEIS	HEIS	SIMS	ESCA	LMSS
Elemental detection	$Z>2$	$Z>11$	$Z>2$	$Z>2$	$Z>2$	$Z\geq 1$	$Z>1$	$Z\geq 1$
Detection limits	0.1%	0.1%	0.1%	0.1%	0.1%	10^{-4}%	0.1%	10^{-4}%
Compound information	Some	No 0	No	No	Some	Yes	Yes	Yes
Spatial resolution	500Å	1 μm	1 μm	<100 μm	<10 μm	<0.1 μm	150 μm	0.5 μm
Depth Resolution	10Å	1 μm	1 μm	10Å	>0.5 μm	10Å	10Å	1000Å
destructive	Yes	Yes	Yes	No	No	Yes	No	Yes
Sample charging	Yes	Yes	Yes	No	No	Yes	No	No
Quantitative	No	Yes	Yes	Yes	Yes	No	No	No

starting point for all materials related analyses. The SEM yields photomicrographs illustrating high resolution surface topography at magnifications up to 200 000×.

For secondary electron generation, the SEM employs a focused electron beam which is rastered over the sample surface. The release of secondary electrons by the incident beam is directly related to the brightness generated on an adjoining CRT screen, providing a contrast scale for imaging.

The spatial or lateral resolution of the SEM is typically ~ 100Å; however, a resolving power of ~ 30Å is attainable with the use of a LaB6 source rather than the commonly used tungsten source. This high resolution imaging capability makes the SEM extremely useful in identifying failure sites as well as in preventative quality inspection programs (Figure 10.11).

Merits

- High resolution imaging of surface structure and topography.
- Sample preparation typically easy.

Limitations

- Sample charging due to incident electron beam can result in loss of resolution and imaging capability

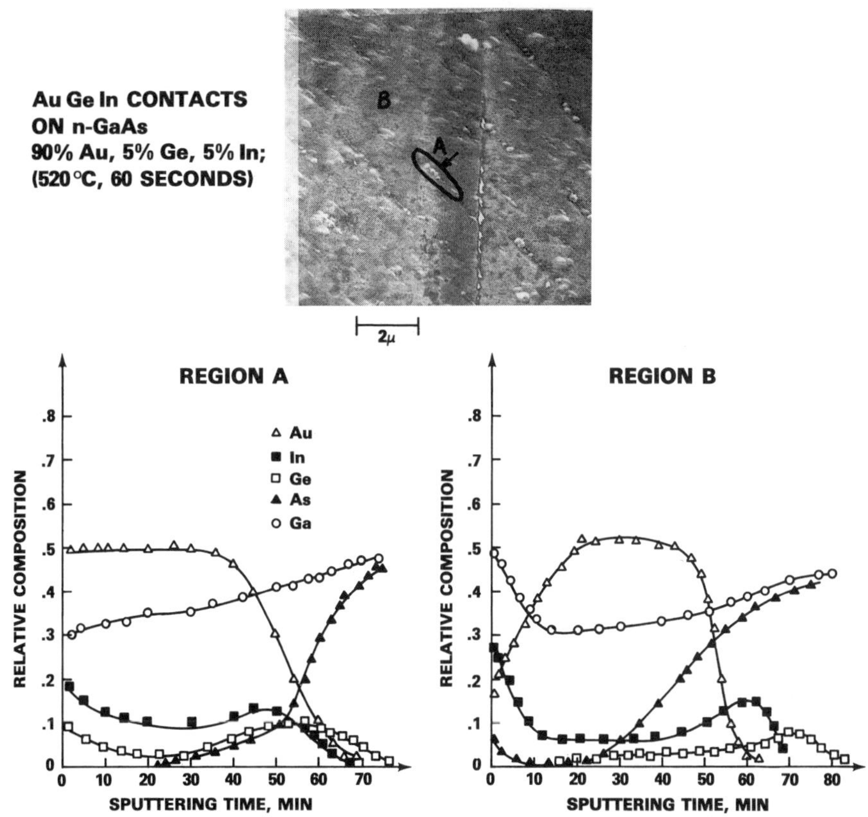

Figure 10.11 Defects in AuGeIn contacts shown by SEM and AES analysis.

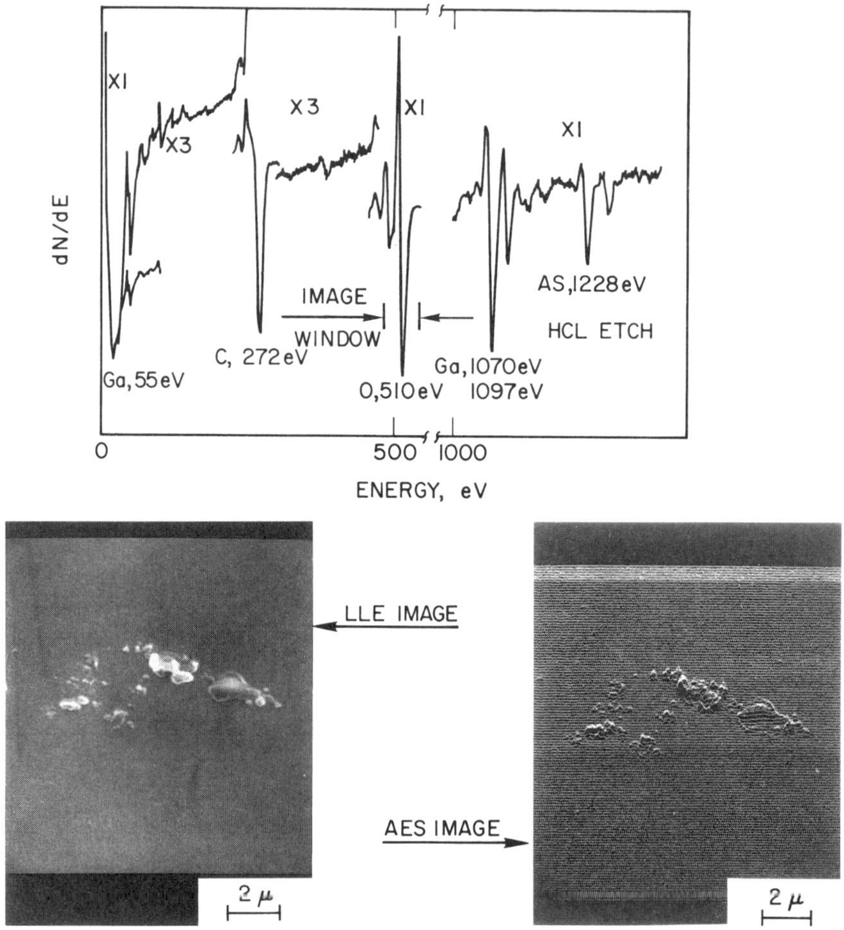

Figure 10.12 AES imaging and low energy backscattered electron imaging of GaAs surfaces.

Backscattered electron imaging. Subtle topographical features which can sometimes be overlooked in a SEM micrograph are often more evident in a backscattered electron image. Like the SEM, this imaging technique uses a primary electron beam incident on the sample surface. However, primary electrons from this excitation beam can be "backscattered", rebounding from the sample surface with little energy loss. In this case, electrons emitted from a sample through electron beam excitation can be grouped into two categories: (1) those which have an energy level of less than 50 eV and emerge from a shallow depth (secondary electrons), and (2) those which have an energy level of greater than 50 eV and emerge from deeper within the sample (backscattered electrons). Backscattered electrons are detected using a scintillator photomultiplier detector or a system of solid state diodes which are often mounted within a SEM chamber to provide multiple imaging capability.

Contrast in a backscattered image is related to the mean atomic number of the material which is being excited by the incident electron beam. The electron

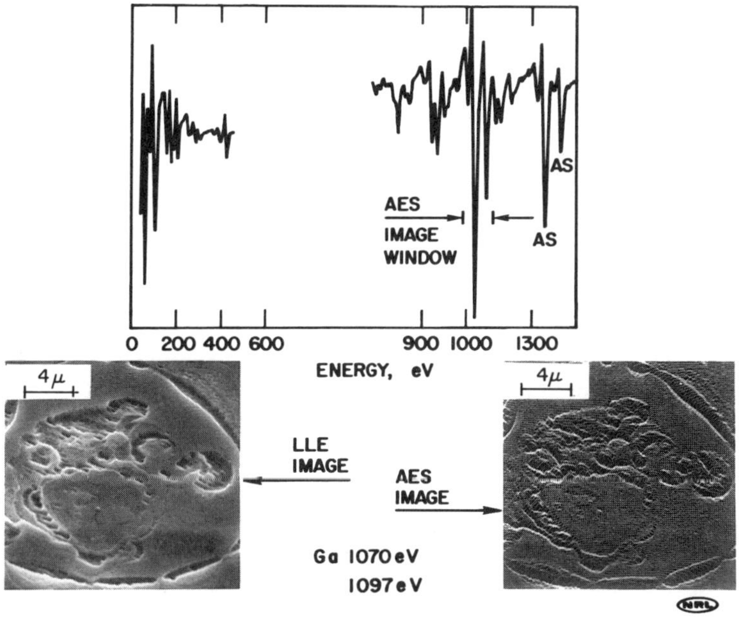

Figure 10.13 AES surface composition image.

backscattering coefficient decreases with decreasing atomic number, represented on the backscattered image as a darkening of an area. This feature makes backscattered imaging or low loss imaging a useful inspection tool by providing information on elemental segregation (Figure 10.12).

The depth from which backscattered electrons escape from the sample surface ($\sim 0.5\ \mu$m) should make the analyst wary of the interpretation of contrast variation. Variations in the thickness of multilayered surface features will cause a change in contrast which can be interpreted as a difference in composition. However, this variation could simply be an "averaging" of backscattered electron emission from two interfacing constituents (Figure 10.13). When used in combination with the SEM, backscattered electron imaging proves to be a complementary technique and a useful precursor to compositional analysis, as shown in Figure 10.13 for the case of surface contaminants.

Merits

- Fewer sample charging effects than with SEM.
- Indicates variations in concentration of elemental constituents.

Limitations

- Large escape depth of backscattered electrons can cause misinterpretation of elemental segregation in multilayered samples.

10.3.1.2 Elemental analysis

Auger electron spectroscopy (AES). One of the most surface sensitive techniques, AES, is used to characterize the elemental composition of surfaces and interfaces at a penetration depth of 10Å or less. All elements in the periodic table, with the exception of hydrogen and helium, can be detected in concentrations as low as 0.1%.

AES involves the energy analysis of low energy electrons emitted by the sample when excited by an incident electron beam. In the Auger process, an incident electron ejects an inner shell electron, with an outer shell electron subsequently filling the vacancy. The energy released by this transition causes the ejection of an outer shell or "Auger" electron. The energy of the excited Auger electron is measured via an electron spectrometer to yield an Auger spectrum, such as shown in Figure 10.12. A series of transitions are possible, resulting in characteristic Auger peaks for each element detected. Although considered an elemental identification technique, energy shifts in Auger peaks for some elements (e.g., Al, Si, Ti) can also be used to determine the oxidized state.

Scanning Auger microscopy (SAM) is based on the Auger spectroscopy technique and employs a set of deflection plates to raster the incident electron beam. The ability to scan the electron beam across the sample surface gives rise to a number of data collection methods. Auger spectra may be collected in the point mode with a focused electron beam, as well as in the scanning mode for analysis of larger areas. Elemental mapping may be performed to determine spatial distribution of surface constituents (Figure 10.14). Auger line scans can also give an indication of elemental distribution by measuring the amplitude of a selected Auger signal as a function of position along a given line across the sample. EDX employs an Li–Si energy dispersive detector to measure the emitted X-ray energy level. Energy dispersive analyzers are often mounted within SEM chambers, utilizing the electron beam for imaging and elemental analysis. EDX is capable of detecting elements above Na (AMU 11) in the periodic table. However,

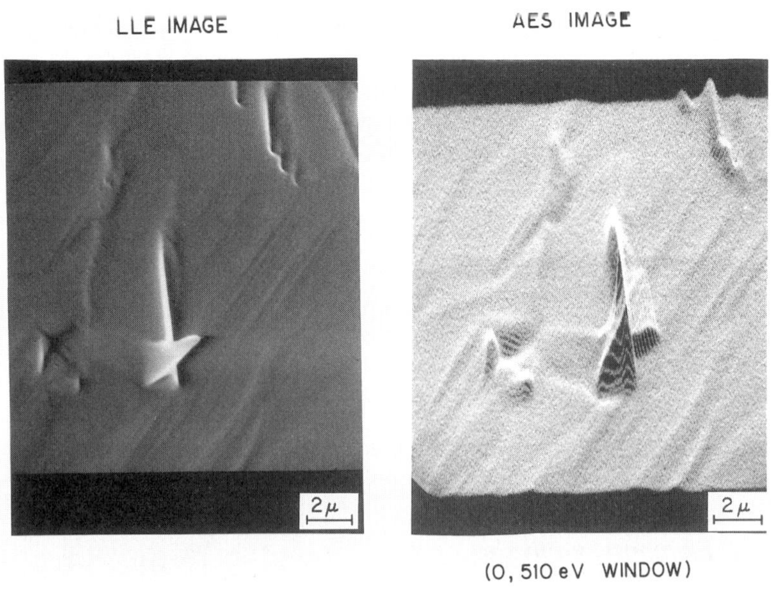

Figure 10.14 Surface distribution map of oxygen.

"windowless" detectors are available for EDX analyzers which increase detection capabilities to include elements as light as carbon.

Since EDX analyzers utilize the SEM scanning electron beam as an excitation source, the versatile data collection techniques employed in SEM are also possible in EDX analysis. Analysis may be performed in the point mode or scanning mode. Elemental maps and line scans are also possible.

In comparison to AES/SAM, the spatial resolution and depth of analysis in EDX is larger, due to the fact that X-rays travel farther in matter than electrons. The escape depth and lateral resolution in EDX is typically $\sim 1\ \mu$m, but will vary with incident beam voltage, angle of incidence, beam diameter and sample matrix. As a result of the larger escape depth, a surface feature which is $\sim 1\ \mu$m in diameter but only $\sim$ 200Å thick may be visible in the image. However, since most of the X-ray information is emerging from areas well below this artifact, the X-ray spectra generated will not be representative of the imaged feature. Also, in relatively thin multilayered samples, a "combined" spectrum, representing X-ray data from the material specific to each layer, will be generated.

EDX is sensitive to the heavier elements with detection limits of $\sim$ 0.1%. When used with matrix standards, this technique can provide quantitative information. SEM/EDX is a unique combination of analytical methods and is extremely useful in areas such as identification of corrosion sites, determining electromigration patterns and studying thick and thin film depositions.

Merits

- Sample charging effects not as severe as in AES.
- Quantitative.
- Rapid form of elemental analysis.
- High sensitivity to metals.

When combined with ion sputter etching (Figure 10.15), AES/SAM is also an effective technique in determining the composition as a function of depth (Figure 10.15), for the analysis of ohmic contacts.

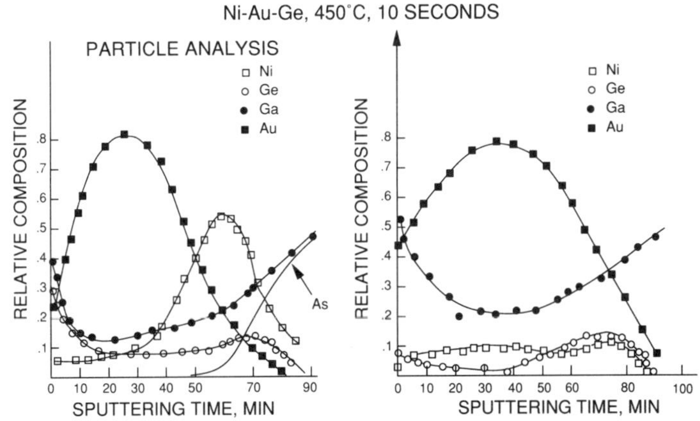

Figure 10.15 AES sputter profile of GaAs ohmic contacts.

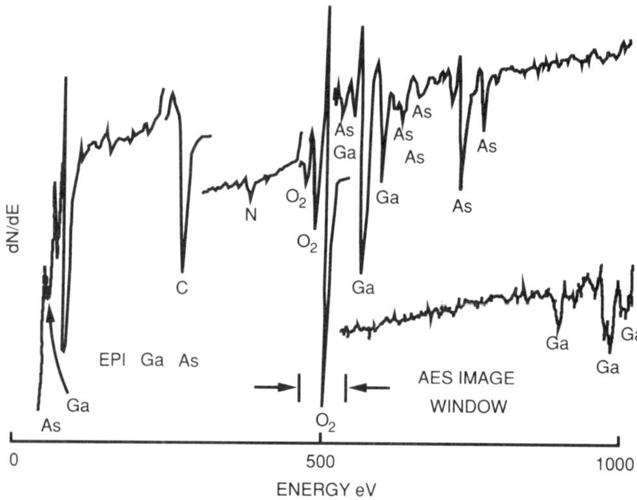

Figure 10.16 AES spectrum of epitaxial GaAs.

AES is highly sensitive to the lighter elements, which are often characteristic of process residuals. Combined with its excellent spatial resolution (~ 500 angstroms), these features make AES a useful tool in analyzing thin oxide layers, halide contamination and diffusion patterns at interfaces. A typical Auger spectrum of epitaxial GaAs is shown in Figure 10.16.

Merits

- Very sensitive to surface structure.
- High sensitivity to light elements.
- Good spatial resolution.
- Depth profiling capability.
- Some information on oxidation state from energy peak shifts.

Limitations

- Sample charging can reduce data yield.
- High vacuum compatibility required.

X-ray electron spectroscopy (XES). XES includes those surface analysis techniques which employ an incident electron beam to induce emission of X-rays from the sample surface. Like the Auger process, an inner shell electron is ejected and an outer shell electron fills the vacancy to return the atom to its lowest energy state. As a result, an X-ray photon is released with an energy equal to the difference between the energy of the electron in its initial and final states. Variations in radiation levels give rise to characteristic spectral peaks for each element.

XES may be divided into two categories based on data collection techniques:

(1) Energy Dispersive X-Ray Analysis (EDX), and
(2) Wavelength Dispersive X-Ray Analysis (WDX).

1. Energy dispersive X-ray analysis (EDX)

Merits

- Highly quantitative.
- Detection of all elements except hydrogen and helium.
- High sensitivity.
- Higher resolution in differentiating various elements, less peak overlapping.

Limitations

- Time consuming method of analysis on an unknown sample.
- Loss of some spatial resolution by broadening area of analysis to acquire a good data base for quantitative analysis.

2. Wavelength dispersive X-ray analysis (WDX)

The older of the two methods, WDX employs a diffracting crystal spectrometer to disperse the emitted X-rays, allowing only a chosen wavelength to be diffracted to the detector and collected as a function of wavelength. WDX is typically the method of analysis used in an electron microprobe, which is the forerunner of the SEM, and employs a stationary electron beam to generate X-rays. With the advent of the SEM and its versatility, WDX detectors are commonly added to SEM systems alone or in combination with an EDX detector.

The detectable element range for WDX includes all elements with the exception of hydrogen and helium. As with EDX, lateral resolution and escape depth are $\sim 1\ \mu$m. In a system with a single WDX spectrometer, data collection for a multielement sample is sequential, allowing the analysis of any one element at any one time. X-ray analysis via EDX is simultaneous, collecting information for a full spectrum of elements during a single analysis. Due to the inherent data collection method employed in WDX analysis, elemental detection limits are lower ($< 0.1\%$). Higher count rates on individual elements provide for a very quantitative method. However, EDX allows a more rapid qualitative analysis with less chance of overlooking an unexpected element. In general, WDX is the preferred method when the detection of light or trace elements and an accurate quantitative analysis is required. For the rapid qualitative analysis of an unknown material, EDX is the accepted method.

10.3.2 Compound analysis techniques

10.3.2.1 Electron spectroscopy for chemical analysis (ESCA)

ESCA is a surface analysis technique which is used to obtain both elemental and compound information from the top ten angstroms of a sample. The ESCA process, which is also known as X-ray photoelectron spectroscopy (XPS), employs an incident X-ray (photon) beam to eject inner or outer shell electrons from the sample surface. The measured kinetic energy of these ejected electrons is related to their binding energy as follows:

$$E_B = h\nu - E_K$$

where E_B = binding energy of ejected electrons,

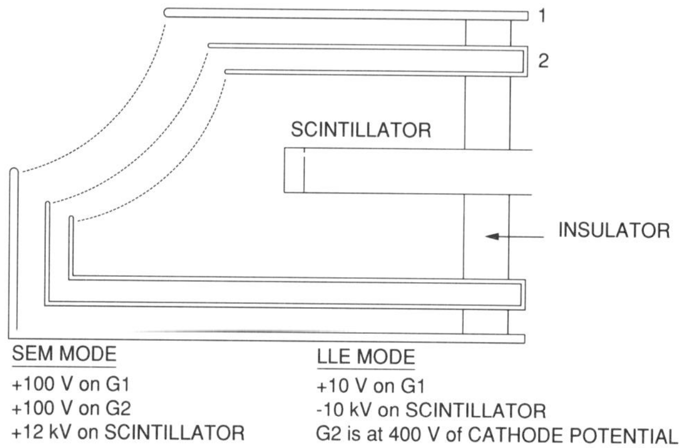

Figure 10.17 Low energy electron detector for the SEM.

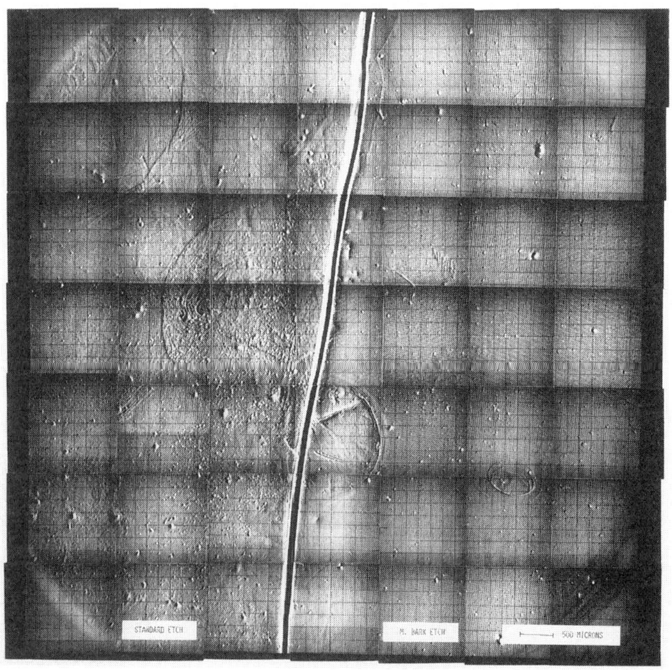

Figure 10.18 Low energy SEM image of surface defects on GaAs.

E_K = kinetic energy of ejected electrons,
$h\nu$ = photon energy of incident beam.

The shifts in binding energy produced by different oxidation and valence states of the same atomic species will appear as shifts in peak positions in an ESCA spectrum. The location of ESCA peaks, therefore, provides not only elemental information but also information about molecular structure.

Detection with ESCA is possible for all elements in the periodic table except hydrogen at detection levels of 0.1%. Because oxidation and valence states can be determined, ESCA is useful in identifying organic as well as inorganic materials. This technique is extremely helpful in determining the primary constituents in corrosion and contamination mechanisms. As with AES, ESCA can be combined with ion sputter etching to yield information about composition as a function of depth. Added features are the low rate of beam induced damage and minor sample charging effects.

One of the major drawbacks to ESCA is its poor spatial resolution, which is typically several millimeters in diameter. However, most instrument vendors are now offering an upgraded optical system which can reduce the analysis area to $\sim$ 150 μm in diameter. Another drawback is the degree of peak overlap encountered with an extremely rich spectrum. Elaborate software programs with curve fitting capabilities are helpful in overcoming this problem but, in general, ESCA is not the ideal method for analyzing complex unknowns.

Limitations

- Destructive analysis of the sample
- Ionization efficiency, and therefore sensitivity to certain elements varies, based on laser power density
- Extremely variable penetration depth

10.3.2.2 *Secondary ion mass spectrometry (SIMS)*

SIMS utilizes a focused beam of energetic ions which strike the sample surface, liberating particles in the form of positive and negative ions. These are extracted by a collector and analyzed according to their mass to charge ratio by a mass spectrometer. Identification of the ejected secondary ions, which consist of a mixture of single atoms and clusters of atoms, yields elemental, molecular, and isotopic information.

SIMS is typically performed in either the "static" or "dynamic" mode. In the dynamic mode, a substantial sputter rate is maintained as data is being collected, resulting in a depth profile. In static SIMS, the sputter rate is maintained at a lower level, so that it is smaller than the data collection rate. This results in a SIMS survey which is used to identify surface level constituents at a penetration depth of $\sim$ 10Å.

A feature which distinguishes SIMS from most of the other surface analysis techniques is its range of elemental detection which includes all elements in the period table. In addition, SIMS has extremely low detection limits, typically in the ppm range and under optimum conditions in the ppb range. Nevertheless, quantification is extremely difficult. This is due to variations in secondary ion yield, which is a strong function of chemical state and can vary over several orders of magnitude. The problem can sometimes be overcome by maintaining the surface in a constant chemical state. This is achieved by using an oxygen ion gun, or including an oxygen jet in the sputter gun system, to oxidize the sample surface. However, there still remains much to be learned about variations in secondary yield before SIMS can be considered a dependable quantitative technique.

Ion microprobe analysis is quite similar to SIMS and the two techniques are often considered to be one and the same. By definition, the ion microprobe method typically involves a localized analysis using a microfocused ion beam, whereas SIMS employs a broader beam for larger area analysis. Depending on the type of sputter gun used, spatial resolution in SIMS or ion microprobe analysis can vary from less than a micron to several hundred microns. In systems employing a scanning microfocused ion beam, it is possible to generate maps illustrating the location of various molecular and isotopic species.

SIMS is extremely useful in analyzing semiconductors and insulators, due to charging effects, which are often difficult to study with electron beam techniques. In addition, its high sensitivity to trace level constituents provides an excellent data base for low concentration profiles.

Merits

- Detection of all elements.
- Provides molecular and isotopic information.
- High sensitivity (ppm range).
- Depth profiling capability.

Limitations

- Destructive analysis of the sample.
- Overlapping peaks make identification of unknowns difficult.
- Matrix effects can cause substantial variations in secondary ion yield.

10.3.2.3 Ion scattering spectroscopy (ISS)

Unlike most of the other surface analysis techniques, ISS is based on energy loss in scattered primary ions rather than the energy or mass of ejected secondary particles. With ISS, the sample surface is bombarded with noble gas ions at a fixed angle of incidence. As these primary ions approach the sample surface, many will be neutralized by electrons from the solid material. However, a small fraction will be backscattered from the sample through binary collisions. After measurement of the energy of the backscattered ions, the amount of energy loss induced by collisions is related to the mass of the atom acting as the collision partner. Therefore, each elemental constituent will generate an ISS peak at a specific energy level. ISS is particularly sensitive to the high mass elements, since the differential scattering cross-section increases with atomic number. Based on the primary ion energy, ISS may be divided into two categories: (1) Low energy ion scattering (LEIS) and (2) High energy ion scattering (HEIS).

10.3.2.4 Low energy ion scattering (LEIS)

With LEIS the top ten angstroms of a sample surface are analyzed through the binary elastic collisions of low energy (1–3 keV) He+ or other noble gas ions with sample surface atoms. The mass of the surface atom can be calculated from the measured energy of the scattered ions, and the known primary ion mass, energy, and scattering angle.

Limitations

- Without windowless detector, elements below Na cannot be detected
- Not effective in analyzing relatively thin surface features
- Overlapping peaks sometimes a problem in data interpretation

10.3.2.5 Laser microprobe mass spectrometry (LMMS)

The newest of the surface analysis techniques, LMMS is still in the developmental stages but has already been utilized successfully in a wide range of applications. The basic technique employs a focused laser beam which causes the ionization of a microvolume of the sample. The generated ions are mass analyzed by a time-of-flight method, and are representative of the atomic or molecular constituents of the sample surface. As with SIMS, elemental, compound and isotopic information may be obtained.

10.4 SELECTION OF CHARACTERIZATION METHOD

It is apparent that for a given application, each technique has its own advantages and disadvantages. In most cases, any single method will not answer all of the questions and a combination of methods may be required. In choosing the appropriate surface analysis technique, there are several key parameters which should be considered. Table 10.1 compares the various techniques.

10.4.1 Nature and range of data yield

For the specific application one must answer the question; is elemental information sufficient or is compound and isotopic information required? Elemental detection limits should also be an initial consideration, since only a few of the techniques provide full periodic table coverage (SMS, LMMS).

10.4.2 Beam induced damage

In many instances, maintenance of the integrity of the sample surface is required for additional analyses (e.g., electrical, thermal, physical) or simply because it is fragile and prone to desorption or decomposition. Therefore, damage induced by the surface analysis probe may be critical in acquiring an accurate analysis. Of the three excitation sources reviewed, photons are probably the least destructive. Ions can cause sputtering and intrinsic damage, while electrons do not affect the sample lattice but may remove adsorbed gas layers or cause carbon deposition from residual hydrocarbons in poor vacuum systems. ESCA is considered one of the most non-destructive surface analysis methods. Its low level of beam induced damage makes this technique useful in analyzing fragile organics and inorganic oxides.

10.4.3 Detection limits

A low detection threshold is essential in analyzing race level constituents in areas such as ion implantation and other dopant diffusion processes. Both LMMS and SIMS have excellent detection limits (ppm) and can be extremely useful in low concentration studies. The inherent depth profiling capability of SIMS makes this technique ideal for analyzing low concentration profiles.

10.4.4 Sampling depth and spatial resolution

The degree to which a technique is specific to the surface region is often critical in the analysis of samples exhibiting process residuals, adhesion failure or incomplete etching procedures. Undoubtedly, AES, ESA and SIMS are the three most surface sensitive techniques with a penetration depth of 10 angstroms or less. However, with the increases achieved through patterning resolution in the microelectronics industry, spatial resolution has become a limiting factor for some of the surface sensitive methods. AES instrumentation is capable of $\sim$ 500Å spatial resolution which makes it the most effective in analyses involving stringent depth and spatial requirements.

A comparative study of all the surface analysis techniques shows that a well planned combination of two or more of them will provide the ability to analyze the surface of almost any sample. The usefulness of employing several methods is reflected in the number of multitechnique instruments on the market today.

An excellent method of obtaining surface information with the SEM is to employ a series of filters so that only surface secondaries are collected. Such a detector is

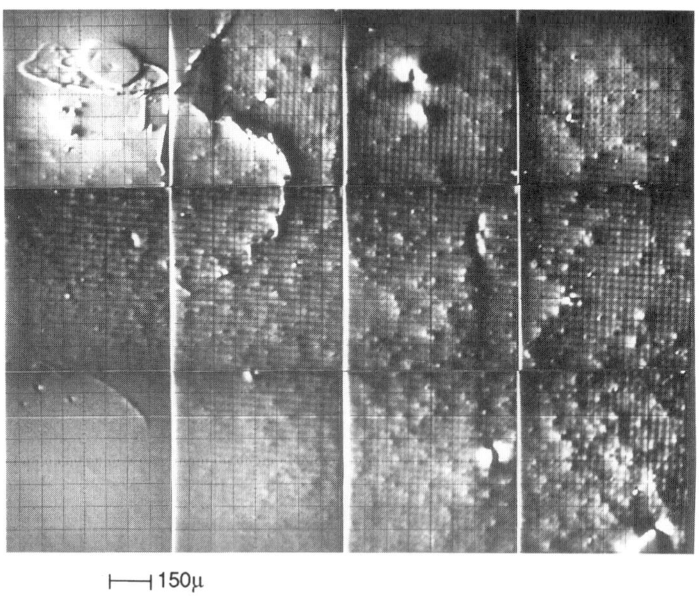

Figure 10.19 Low loss electron image of dislocation loops in GaAs.

shown in Figure 10.17. Typical data of subsurface damage marks on a GaAs wafer are shown in Figure 10.18. The surface inhomogeneities from a typical GaAs etch are shown in Figure 10.19.

REFERENCES

[1] Barr, T.L. and Davis, L.E. (eds.), *Applied Surface Analysis*, ASTM699, American Society for Testing and Materials, Philadelphia (1989).

[2] Carbonara, R.S. and Cuthill, J.R. (eds.), *Surface Analysis Technique for Metallurgical Applications*, ASTM596, American Society for Testing and Materials, Philadelphia (1976).

[3] Carlson, T.A., *Photoelectron and Auger Spectroscopy*, Plenum Press, New York (1976).

[4] Casper, L.A. and Powell, C.J. (eds.), *Industrial Applications of Surface Analysis*, ACS199, American Chemical Society, Washington, DC (1982).

[5] Czanderan, A.W. (ed.), *Methods of Surface Analysis*, Elsevier, New York (1975).

[6] Devaney, J.R., Hill, G.L., and Seippel, R.G., *Failure Analysis, Mechanisms, Techniques, and Photo Atlas*, Failure Recognition and Training Services, Inc., Monrovia, CA (1983).

[7] Heinrich, K.F.J., *Electron Beam X-Ray Microanalysis*, Van Nostrand Reinhold, New York (1981).

[8] McIntyre, N.S. (ed.), *Quantitative Surface Analysis of Materials*, ASTM643, American Society for Testing and Materials, Philadelphia (1978).

[9] Murr, L., *Electron and Ion Microscopy and Analysis*, Marcel Dekker, Inc. (1982).

[10] Woldseth, R., *X-Ray Energy Spectrometry*, Kevex Corp., Burlingame, CA (1973).

[11] Winkler, E., "Integrated circuit packaging trends", *Solid State Technology*, June 1982.

11

MANUFACTURING OF PLASTIC INTEGRATED CIRCUITS

11.1 INTRODUCTION

Plastic packages account for approximately 80% of the worldwide package share, a percentage which has been increasing in recent years [1]. Plastic packages utilize a metal leadframe or metallized circuit pattern to mount the silicon device and to provide the fanout pattern of leads to the pins of the package. Molded plastic is then used to protect the chip and leadframe from physical damage and contamination. Plastic packages, therefore, offer the advantages of cost, ease of handling, low weight, availability, and high reliability. The IC price in a ceramic package may be from 2 to 10 times that of the plastic counterpart. The mean value currently observed is 3.5 times the cost of the plastic package. A cost reduction of 50% may be achieved when all the components of the board are available in plastic rather than ceramic format. The use of MIL-M-38510/MIL-STD-883 for integrated circuits encapsulated in ceramic is not only costly, but limits the product availability. Ceramic packages are also very fragile and, therefore, hard to handle in automated assembly machines [2]. Due to the recent increases in volume, the level of quality assurance of plastic packages has improved. Plastic packaging has shown significant improvements in reliability and has reached the point where performance rivals that of the more costly ceramics and CERDIPs. In fact, plastic packages of the mid-1980s were as reliable as hermetic packages of the mid-1970s.

11.2 PRIMARY PACKAGE CONCERNS

Research in plastic technology is aimed at addressing applications from low lead count to those in excess of 300 I/O. The reliability of these packages depends upon a number of factors, including die size and moisture sensitivity, leadframe and package design, molding compound, material properties, and processing methodology. Unfortunately, larger die size and higher lead count result in difficulties with respect to the high reliability requirements [3].

Shrinking the package size, while increasing the silicon die area, will result in a thinner plastic profile protecting the device. Hence, plastic packages are much more likely to crack and permit rapid moisture and ionic contamination ingression. Even in a good package, the moisture path is shorter than with a conventional DIP because of the thinner walls. As the pitch between the leads gets tighter, the lead thickness is reduced, introducing significant manufacturing problems.

Conventional thermoset molding materials have a thermal expansion coefficient significantly larger than silicon or GaAs and establish a high stress level in the device. Any lack of cleanliness in the manufacturing process, or the presence of impurities in the molding material, will result in corrosion, accompanied by the ingress of moisture into the plastic [3]. The critical reliability problems introduced in Chapter 9 will be further analyzed in this chapter and extended to manufacturing.

11.2.1 Device susceptibility

The most significant failure mechanism associated with plastic packages is corrosion. Microcracks in the plastic from stress relief result in exposure of the underlying metallization to moisture and mobile ions (Na^{2+}, K^{2+}, Cl^- , or Ca^{2+}). Electrochemical cells are formed during operation, which can destroy the metallization and functionality of the device. The trend today is towards producing a "bulletproof" structure which is resistant to moisture and corrosion. Passivation technology, using plasma nitride and other approaches, has improved the reliability of devices through increased crack resistance and elimination of pinholes. More important, however, is the use of a solder bump technology to protect the bonding areas. One way to do this is to use gold bumps with a barrier layer between the gold and the underlying pad metallization. Gold is much less susceptible to corrosion than aluminum and, through use of 1 mil bumps, package survivability is enhanced. TAB technology therefore provides volume improvement and passivation protection for plastics. The intended result in combining a state-of-the-art passivation technique with bumping technology is the creation of a "hermetic" device, which is not susceptible to corrosion, whether packaged or not. This would relieve much of the requirement that the package protect the device against adverse environments. It would then provide only mechanical protection.

Traditionally, plastic packages are known to be susceptible to moisture induced failures. This is particularly worrisome with some SMD, such as SOICs, when subjected to severe temperature shock during board soldering [4]. The key moisture paths, in terms of severity, were found to be trim/form; leadframe design, leadframe material, plating material, molding compound, lead finish operations, flux, and deflashing. Contaminants in a molded plastic package were traced to three major sources: flux residual, molding compound, and die attach. With modern low impurity materials, it is now possible to produce high reliability molded plastic packages. The moisture path in a plastic package is shown in Figure 11.1.

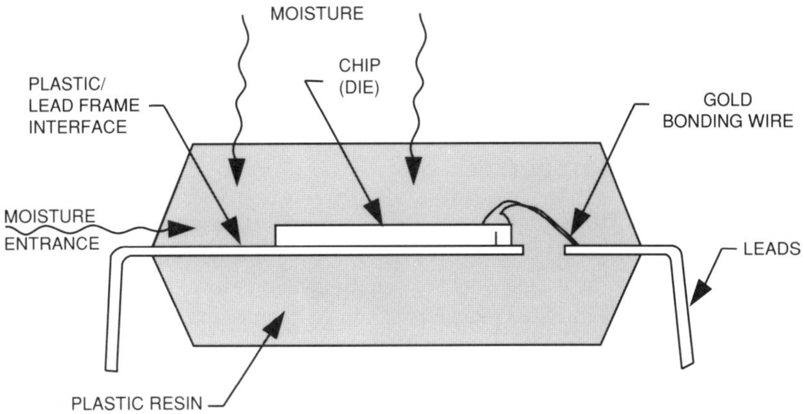

Figure 11.1 Cross-sectional view of plastic package showing moisture path.

11.2.2 Package design

The package design becomes more significant as the die gets larger while the package becomes smaller. This results in a tighter lead pitch. Intelligent design decisions must be made to keep the dimensions within reason. Equally important is the selection of the appropriate molding materials and manufacturing process. Plastic material shrinkage on cooling can distort lead geometries, which is particularly crucial for the smaller leads found in high lead count devices. With more fragile leads coming out of the package, a scheme for maintaining their integrity is mandatory.

11.2.3 Material control

Plastic materials with appropriate properties must be developed and maintained. A thorough knowledge of the parameters which influence reliability is necessary, so that a dependable product qualification procedure can be developed. Since it is impossible to monitor "reliability" during manufacture, lot to lot variations must be controlled within defined limits through incoming inspection testing and supplier process control. Of particular concern is the amount of ionic impurities inherent in the molding material, and its variation from lot by lot.

11.2.4 Process control

Process control is an important factor in maintaining high reliability in plastic packages. There are no 100% screening tests for plastic parts, no non-destructive tests which can be used to weed out the potentially defective parts. Plastic parts can be destructively evaluated on a sample basis only, through typically time consuming tests. There is no room for uncontrolled manufacturing processes until reliability predictions are made on packages. Reliability must be built in, not tested or inspected in [3]. Valuable non-destructive qualification techniques, such as scanning acoustic microscopy and scanning laser acoustic microscope are briefly discussed at the end of the chapter.

11.3 MATERIALS FOR MOLDED PLASTIC PACKAGES

11.3.1 Molding compounds

The molding compound is the most important polymer material used in plastic packaging of IC devices. It provides both mechanical and chemical protection for the device at a cost which is significantly below that of metal, ceramic or glass and even premolded plastic packaging. Its principal advantages are very high yield, high productivity and good reliability.

11.3.1.1 Epoxy molding compounds

Epoxy molding compounds are the most widely used for plastic packaging. They are thermoset plastics with low initial viscosity and high modulus and heat distortion temperature. Cresolnovolac epoxies are the most common resins used for packaging, due to their desirable high-temperature properties, superior chemical resistance and better adhesion-such as silica and alumina are added to the epoxy in concentrations up to 75% in order to improve the properties of the molded package [5]. The fillers reduce curing shrinkage, increase the strength of the epoxy, lower the thermal expansivity, and in the case of alumina, increase the thermal conductivity. Epoxies are generally supplied as colored or black powders, which are loaded into the hot chamber of the transfer molding press, where they liquify. A liquid epoxy compound is sometimes supplied, which can be directly injected into the parting line of the mold, from where it passes through the runners and gates into the mold cavities. Apart from epoxy resin, the molding compound is made up of several different components which are described briefly below:

Curing agents (hardeners). Amines, acid anhydrides and phenols can all be used to crosslink epoxy resins. The reactions proceed through different chemistries. Ring opening is the predominant mechanism for amine curatives, whereas crosslinking occurs through pendant hydroxyls for the anhydrides and phenols [1]. Phenol novolac and cresol novolac hardeners have become the dominant hardeners for microelectronics packaging because of their excellent moldability, electrical properties, and heat and humidity resistance.

Accelerators. Accelerators are used in most epoxy molding compound formulations to reduce the in-mold cure time and thus improve productivity. The most common agents are amines, imidazoles, organophosphines, ureas, and Lewis acids and their organic salts. The critical properties of an accelerating agent are its reactivity enhancement and its latency. Faster cure times should not be at the expense of shorter pot life. Amines provide excellent reactivity but provide poor pot life, relatively poor electrical properties and humidity resistance. Organophosphines provide excellent acceleration, excellent electrical and humidity resistance properties, but poor pot life. Lewis base salts provide moderate acceleration, excellent prior to processing life and excellent electrical properties and humidity resistance.

Fillers and coupling agents. Fillers are used to lower the coefficient of thermal expansion and enhance the thermal conductivity. They also improve mechanical properties, such as the modulus, and prevent excessive resin bleed at the parting line of the molding tool. Other criteria of an inorganic filler include good electrical and mechanical properties, chemical stability, heat and moisture resistance, and low alpha particle activity. Yet another important consideration is low abrasiveness. Ground, fused silica is the most widely used filler in epoxy molding compounds because it possesses the optimum combination of properties.

Flame retardants. Epoxy materials are not inherently flame retardant, so additives must be incorporated for this purpose. Brominated epoxies, like brominated DGEBA epoxy, and antimony trioxide additives are generally used for this purpose.

Stress-relief additives. Epoxy materials are inherently brittle due to their high cross-link density, which decreases polymer chain mobility and relaxation response. Flexibilizing or stress relief agents are added to epoxy molding compounds to inhibit crack propagation, as well as to lower the thermomechanical shrinkage stresses that can initiate cracks in the molding compound or in the passivation layer of the device. Silicones, acrylonitrile-butadiene rubbers, and polybutyl acrylate (PBA) are the major stress relief agents used in epoxy molding compounds.

Coloring agents. Most plastic packages for silicon integrated circuits are black. The filled molding compound is naturally a translucent yellow, so coloring agents like carbon black are added to color the packages black. The level of carbon black in a semiconductor grade molding compound is typically below 0.5%.

Mold release agents. Epoxy resins are excellent adhesives and this makes the release from the compounding and molding equipment difficult. Mold release agents are hence added to the resin to take care of this problem, most commonly silicones, hydrocarbon waxes, metallic salts of organic acids, and fluorocarbons. Of these, the waxes are the most frequently used.

Encapsulants for multichip packaging (MCP). MCP is becoming an essential approach of device packaging, not only for meeting high operation speed requirements, but also for system integration [6]. This is true especially when package volume and weight are critical driving factors. As MCP grows to larger sizes to accommodate the ever increasing demand of system integration, hermetic sealing becomes very expensive and technically difficult, if not impossible, to achieve. This has prompted the evaluation of various encapsulants for nonhermetic sealing. Encapsulation provides "reliability without hermiticity" on advanced MCPs, such as monolithic and wafer scale integration. In addition, it provides cost, weight, and volume savings with a reliability equivalent to hermetic packages. A few of the encapsulants used in MCPs are discussed in Table 11.1.

11.3.1.2 Silicones

Silicones, unlike the organic polymers, are characterized by Si–O–Si linkages instead of the C–C linkages. They are available in a wide variety of forms, such as room temperature vulcanizing (RTV), clear or colored resins or gels, rigid resin,

Table 11.1 Encapsulant candidates for non-hermetic MCP applications.

Encapsulants	Types	Advantages	Disadvantages
Union carbide parylene	Polyparaxylylene	Conformal coating proven military specification	Slow deposition rate
AT&T modified silicone encapsulant (MSE)	Modified silicone	Proven moisture protection	Low mechanical strength
M&T rely-imide 1326	Silicone-polyimide	Good adhesion–high thermal stability	Marginal solvent resistance
Occidental 2010M	Fluorinated silicone-polyimide	High thermal stability	Weak adhesion
Hitachi PIQ8410	Polyimide	Good solvent resistance–high thermal stability	High water absorption
Dow chemical XU13005.021	Benzocyclobutene	Good adhesion–low water absorption	New material

and as transfer molding compounds. All silicones have the advantage over organic resins with their superior thermal stability and excellent electrical properties. They can withstand continuous operation at temperatures up to 250 °C and short time exposures up to 375 °C. Silicones, especially those which may be cured to flexible mass, are also advantageous in that they are virtually immune to thermal shock and exert very little stress on the device or on the delicate wire bonds. One principal disadvantage of silicones is their poor adhesion to most substrates.

Silicone materials are ideally suited for use in demanding automotive environment situations. They can be soft and flexible for vibration damping; they are thermally stable from −50 °C to 200 °C; and are also excellent electrical insulators. The four different applications of silicones for a single hybrid module are discussed below [7]:

- High strength, flexible electronic grade adhesives for the assembly of hybrid modules.
- Thermally conductive flexible electronic grade adhesives to attach the hybrid substrate to the heat sink.
- Soft, totally stress free gel encapsulants to protect the entire assembly from humidity and other pollution.
- High purity coatings or gels for localized and selective chip protection.

High strength adhesive. Whenever a metallic part is to be bonded to a plastic molding, considerable stress will be created during thermal cycling because of differences in thermal expansion characteristics. Any adhesive used at such an interface should allow the bonded surfaces to move slightly relative to each other without cracking. It must also guarantee an impervious and homogeneous seal. This flexibility must be maintained over the whole operating range of temperature and humidity.

Thermally conductive adhesive. Interconnection of substrates having a mismatch of thermal coefficients of expansion can result in failure upon thermal cycling, due to the concentration of stress. One simple technique to transfer heat effectively from the heat source (component) to the heat sink is the use of thermally conductive adhesive.

Silicon gel encapsulants. Flexible primerless silicone encapsulants are available, ranging from an elastomer to a soft dielectric gel. Silicone dielectric gels, SILGARD 527, are polymers based on poly-dimethyl siloxanes with a low level of cross-linking and exhibit good pressure sensitive adhesion to most substrates. Therefore, they are extremely good in protecting electronic circuits against corrosion and metal migration. Silicone gels form a cushioning gel-like mass which combines the stress relief and self-healing properties of a liquid, thereby preventing stress damage to delicate components, even when subjected to heat aging. The unique silicone dielectric gels retain their softness, adhesion and electrical properties after more than 1000 cycles from −50 °C to 200 °C.

11.3.1.3 Polyimides

Polyimide polymers are excellent due to their heat resistance, performance at cryogenic temperatures, and performance in a radiation environment. Their electrical properties and low outgassing make them favorites for use in extreme environments of space, and high temperatures. Polyimides are sensitive to alkalis, organic acids, and moisture. The polymers are often copolymerized with other molecules and are considered either thermosetting or thermoplastic. However, by rigorous definition, most lack cross-linking and are truly thermoplastics with a melting point close to the decomposition point [8]. Typical copolymers are amideimide, imide phenolics, bismaleimide, epoxy imides, and polyester imides.

11.3.1.4 Phenolics

Phenolics are often selected due to their low cost, high-temperature capability, and high mechanical strength. They have poor arc resistance. Phenolic resins are the reaction product of phenol and formaldehyde. Phenolic materials can be molded by compression, transfer, or injection. They are however attacked by strong acids and alkalis and are available in dark colors only.

11.3.1.5 Urethanes

Urethanes, also sometimes called isocyanates or polyurethanes, are thermosetting resins, which can yield rigid or rubber-like solids or foams. When cured, they are tough and possess outstanding abrasion and thermal shock resistance. They also possess favorable electrical properties and good adhesion to most surfaces. They can be used from cryogenic temperatures to 130 °C and can also endure short exposures to about 150 °C. Some polyurethanes are solvent-removable and allow

Table 11.2 Common polymer encapsulants.

Polymer material	Advantages	Shortcomings
Epoxies	Good chemical and machanical protection Low moisture absorption Excellent wetting characteristics Ability to cure at atmospheric pressure and temperature	High stresses
Silicones	Low stresses Excellent electrical properties Good thermal stability Chemically inert	Poor mechanical protection; cost
Polyimides	High thermal stability Good mechanical properties Solvent resistant	Requires high temperature cooling
Phenolics	Good electrical and mechanical properties Good moldability and dimensional stability Excellent thermal resistance Low cost	Attacked by strong acids and alkalis; dark colors only
Urethanes	Tough Outstanding abrasion and thermal shock resistance Favorable electrical properties Some polyurethanes are solvent removable and can be reworked	Sensitive to chlorinated solvents, ketones, strong acids, bases; cannot be used above 250 °F

device rework [5]. Table 11.2 summarizes the advantages and disadvantages of the common polymer encapsulants.

11.3.2 Die attach adhesives

Die attach bonding is used to mechanically attach the die to the leadframe. Solder bonding alloys, organic adhesives, and silver–glass composites are employed. The die attach solders are usually applied as preforms less than 0.05 mm thick. Both the back of the die and the leadframe are metallized to promote solder wetting. Some of the common die attach preform solders are given in Table 11.3. Conductive polymer adhesives are displacing solders for die attach in plastic packages because they are less expensive than high-gold-content solders and their application is more easily automated. Solventless conductive epoxies are the materials of greatest commercial importance, accounting for approximately 80% of the overall die attach market. Most formulations are based on silver particles to provide conductivity. Typical materials contain between 50% and 72% by weight of silver flakes.

Table 11.3 Die attach preform solders.

Composition	Melting range	
	Liquidus temp. (°C)	Solidus temp. (°C)
80% Au/20% Sn	280	280
92.5% Pb/2.5% Ag/5% In	300	—
97.5% Pb/1.5% Ag/1% Sn	309	309
95% Pb/5% Sn	314	310
88% Au/12% Ge	356	356
98% Au/2% Si	800	370
100% Au	1063	1063

11.3.3 Leadframe materials

The choice of leadframe material has important implications for the degree of adhesion of the molding compound to the frame, the thermomechanical shrinkage stresses, the trim and form characteristics, the buckling of the leadframe after molding, and the heat dissipation capacity of the package. Several different metallic alloys share the market for leadframe used in plastic packaging. Alloys 194, 155, and 151 are predominantly copper, and therefore have a relatively high CTE. Table 11.4 gives the compositions of the leadframe alloys used in plastic packaging. The thermomechanical shrinkage stress will be higher with these alloys than with others because of the larger disparity between the CTEs of the silicon die and leadframe. In addition, the adhesion of the molding compound to copper alloys is not as good as to other alloys. The principal advantage of the copper rich alloys is a much higher thermal conductivity. Heat conduction along the leadframe is an important heat dissipation mechanism in molded plastic packages, and the better thermal conductivity of the copper-rich alloys makes them a good choice for high-power devices.

Alloy 42 is another common leadframe material that is popular because of its low coefficient of thermal expansion. This alloy contains iron and nickel with just a trace amount of other materials. It has a significantly lower CTE than the copper-rich alloys, providing corresponding lower thermal shrinkage stresses. It has better adhesion to epoxy molding compounds, probably because it does not form a passivating oxide film as easily as the copper alloys. Its greater rigidity confers better resistance to damage during assembly and molding operations, and less buckling under the compressive loads applied by the molded body. Its principal drawback is lower thermal conductivity. Packages using Alloy 42 leadframes have less power dissipation capacity than similar package designs using copper alloy leadframes. For this reason, Alloy 42 is being replaced by copper alloys as higher heat dissipation becomes a more serious consideration. Newer leadframe materials such as KLF 125 are gaining acceptance because of their high strength advantages in fine-pitch applications where a thinner leadframe is used. Through-hole packages such as DIPs also require high-strength leadframe materials to resist damage upon insertion through the circuit board.

Table 11.4 Compositions of leadframe alloys in plastic packaging.

Alloy	Percentage of composition
Alloy 194	
Copper	97.5
Iron	2.35
Phosphorous	0.003
Zinc	0.12
Alloy 151	
Copper	99.9
Zirconium	0.1
CDA 155	
Copper	97.8
Silver	0.034
Phosphorous	0.058
Magnesium	0.11
Alloy 42	
Iron	58
Nickel	42
KLF 125	
Copper	99.55
Nickel	3.2
Tin	1.25
Silicon	0.7

The copper alloys may form a thick, brittle oxide film on the surface. As a result, heating is facilitated during packaging, thus reducing moisture resistance. But it has been reported that surface finishing of the copper with Sn–Ni alloy plating has improved the oxidation resistance and the adhesive property between mold resin and leadframe [9]. The Sn–Ni plated copper leadframe has an advantage over the alloy-42 leadframe in moisture resistance of the packages. The moisture penetration in encapsulated packages is accelerated by silver plating of the leadframe, because of the low adhesion of silver to resin. To improve the moisture resistance of packages, Al corrosion is reduced with a minimized area of silver plating (ring plating) for IC leadframes.

11.4 MANUFACTURING PROCESS AND PROCESS FLOW

Plastic packages are divided into surface mount and through hole types. The common families of surface-mounted plastic packages are the small-outline package (SOP), the plastic leaded chip carrier (PLCC), and the plastic quad flatpack (PQFP). The common families of through-hole mounted plastic packages are the plastic dual-in-line (PDIP), the single-in-line (SIP), and the plastic pin grid array (PPGA). The various plastic package configurations are given in Figure 11.2 [10].

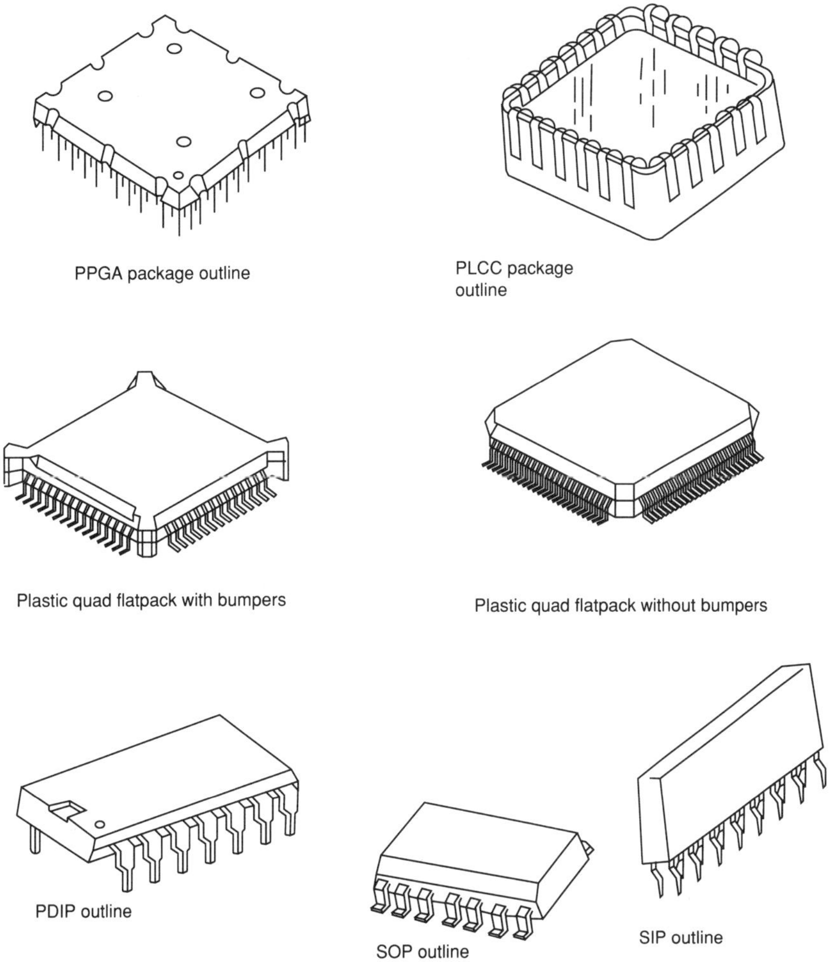

Figure 11.2 Plastic package configuration.

In general, a packaged integrated circuit consists of the silicon chip, metal support or frame, small wires to attach the circuit to the external pins. In addition, a plastic epoxy or silicone encapsulating material protects the chip, wire connections and defines the shape and size of the package. The frame is made of copper, kovar, steel, Alloy 42, Alloy 52, or nickel and is plated with gold and silver either completely or in selected areas. The silicon devices are mounted to the frame by solder (hard or soft), gold eutectic bonding, or an organic conductive formulation of epoxy or polyimide. Gold wires are bonded to the aluminum bonding pads on the device and to the fingers of the frame (Figure 11.3). The assembly is then transfer molded in silicone or epoxy. Following the molding operations, the external pins are often plated with a lead tin alloy, cut away from the strip, and formed [7].

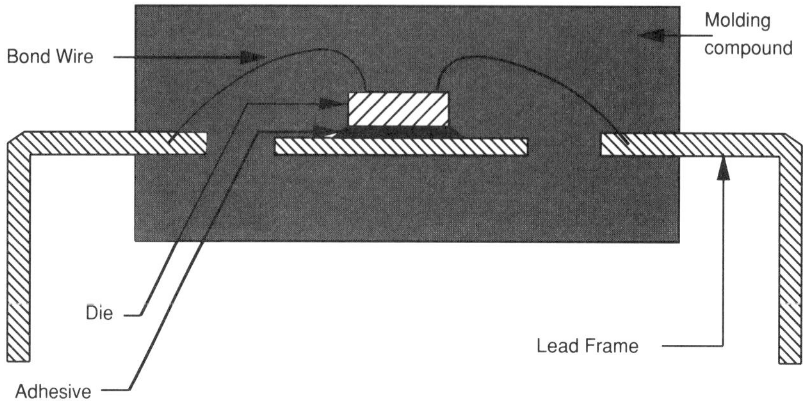

Figure 11.3 Plastic package construction.

Plastic packages are either of the premolded or the post-molded type. In the former, a plastic base is first molded and the chip then placed on it and connected to the fanout pattern through wire bonds or through a direct chip attach. Premolded packages are most often used for high-pin-count devices or pin grid arrays that are not amenable to flat leadframes and simple fanout patterns. Premolded packages are more expensive than post-molding because of the greater number of parts and assembly steps. In the post-molded type, the leadframe upon which the die is attached is loaded into a multicavity molding tool and encapsulated in thermoset or a thermal plastic molding compound using a transfer molding process. About 80% of plastic packages are made using post-molding methods.

There are several molding techniques available for the manufacture of plastic cases. The most widely used in manufacturing is the "transfer molding process" which is discussed in detail below. Others used include the "injection molding process" and the "reaction-injection molding process". A comparison of molding processes is shown in Table 11.5.

Table 11.5 Comparison of molding processes.

Molding type	Advantages	Shortcomings
Transfer molding	Multiple cavities — high yield Relative material savings Short cycle time Low tool maintenance costs	High molding pressure High viscosity Restriction to the packaging of lead frames
Injection molding	Good surface finish Good dimensional control	Material availability
Reaction injection molding	Energy efficiency Low mold pressure Good wetting of chip surface Adaptability to TAB	Few resin systems available for electronic packaging Need for good mixing

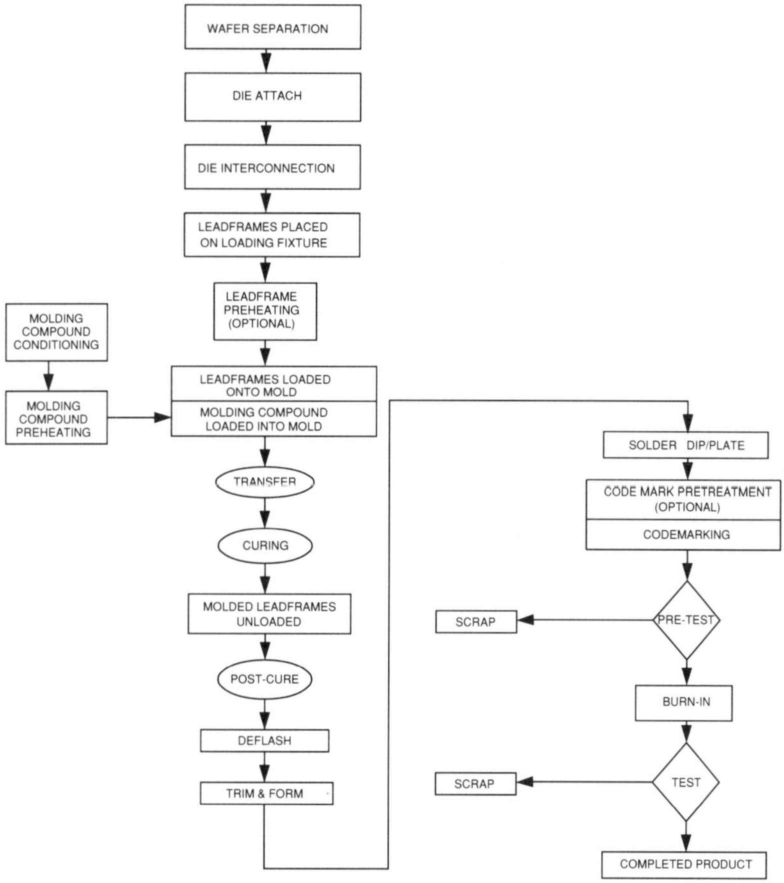

Figure 11.4 Flow chart: transfer molding process.

11.4.1 Transfer molding process

A typical flow chart of the manufacturing of plastic packages by the transfer molding process is shown in Figure 11.4 [1]. The process steps conducted prior to encapsulation in the plastic molding compound are known as the assembly or front-end operations; steps after molding are called back end. Clean rooms of the Class 100 to 1000 scale are used for most front-end assembly operations. Some molding operations are carried out in clean rooms as well, although the machinery hydraulics and dust from the preforms make it very difficult to achieve anything better than Class 10 000. In general, the increasing complexity and miniaturization of silicon devices will drive more of the assembly and molding operations to dust controlled environments. A few of the important process steps are discussed below.

11.4.1.1 Die interconnections

Wire bonding and tape automated bonding (TAB) are the two principal ways of connecting the silicon device to the leadframe in molded plastic packaging. Direct chip attachment using solder bumps is not common, but is more feasible with premolded packaging where the fanout pattern is supported on the premolded base.

Wire Bonding. Within the semiconductor industry, wirebonding is a well defined technique for first level interconnects. This mature technology is used worldwide for semiconductor type interconnects, including both Al ultrasonic and Au thermosonic methods. Au is favored with plastic packages, due to its superior performance under stress encountered in temperature cycling. Au has well known favorable properties in a corrosive environment. The major advantage of Al is cost [11]. Typical wire diameters are 0.0025 to 0.0032 cm (1.00 to 1.25 mil) for gold and up to 0.005 cm (2.0 mils) for aluminum, because of its lower electrical conductivity.

Flip chip. Flip chip assembly eliminates wirebonds between a chip and a substrate. It is also compatible with surface mount assembly. In flip-chip bonding the die is turned upside down and connected to the substrate by some connecting medium. The problem with flip chip is that the device requires special metallurgy and processing. Additionally, a rigid substrate with fine track geometry is needed [11]. This construction technique has seen wide acceptance on underhood electronic assemblies manufactured by GM, Ford, Hitachi, and others.

Tape automated bonding (TAB). Tape automated bonding, is an interconnection technology that uses finely etched metal leads, often supported on a polymeric film, to provide signal and power distribution from active elements (die) to the substrate or the package leads. TAB interconnects the bonding pads of the device through thermocompression bonding or eutectic methods. The pads on the device often have solder bumps to aid the thermocompression bonding. Multiple semiconductor interconnects are created with one machine operation. High production rates are possible with high quality [11]. Table 11.6 shows the comparison of TAB and wirebond.

11.4.1.2 Molds for plastic packaging

There are basically two types of molds used for plastic packaging — cavity chase molds and aperture molds. A cavity chase mold is based on the conventional split mold design where the cavities are machined out of steel blocks and plates and then assembled to produce a large multi-cavity mold (Figure 11.5). The mold is a split cavity design where one half of the cavity is in one side of the mold and the other half is formed in the other side. The runner system typically resides only in

Table 11.6 TAB versus wire bonds (values based on 100 mil bond lengths).

	Al 0.001″ dia.	Au 0.001″ dia.	TAB 0.003″
Bond pull strength	6–1 g	6–10 g	50 g
Thermal conductivity (W/°C s)	8.69×10^{-5}	7.44×10^{-4}	1.01×10^{-3}
Electrical resistance (ohms)	0.1289	0.1385	3.75×10^{-3}

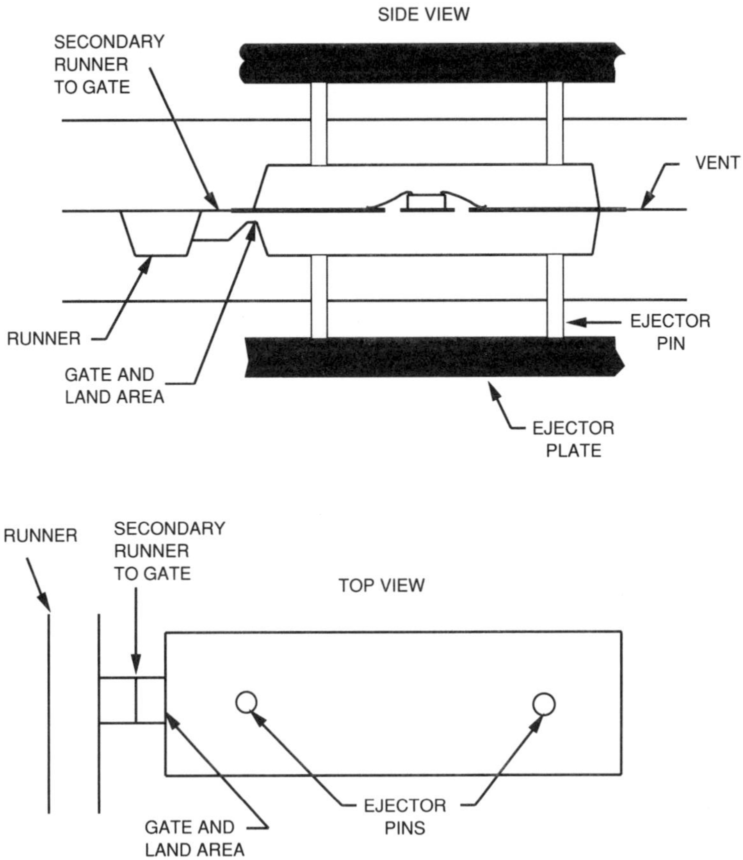

Figure 11.5 Side and top cross-sectional views of the cavity and runner for a cavity chase mold [1].

one half. Ejector pins are required to push the molded part out of the cavity. The typical mold has runners, gate and land areas, vents, ejector plates, a secondary runner to gate, and ejector pins.

Aperture plate molds are developed exclusively for plastic packaging of microelectronic devices. They are assembled from a series of stacked plates (Figure 11.6). The sides of the package body are formed with cutouts in the two apertures plates. The leadframes are loaded between these two plates. The top and bottom of the body are formed by separate plates. The runner system is in the plane above the aperture plates. The surface finish plate forms the top surface of the molded body. In a standard press without microprocessor control, the transfer pressure is controlled by throttling the higher packing pressure through a speed control valve during transfer. Once the mold fills, the transfer pressure builds up to the packing pressure, since the flow through the throttling valve slows to a stop. For presses with microprocessor control systems, the transfer ram switches from transfer to packing pressure either by setting a transfer time interval, a travel distance, or through a feedback sensor that indicates when the mold is filled. The packing pressure is

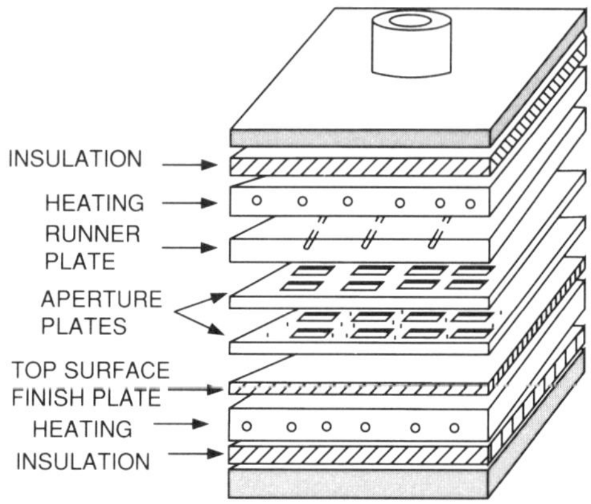

Figure 11.6 A schematic diagram of the construction of the plate mold [1].

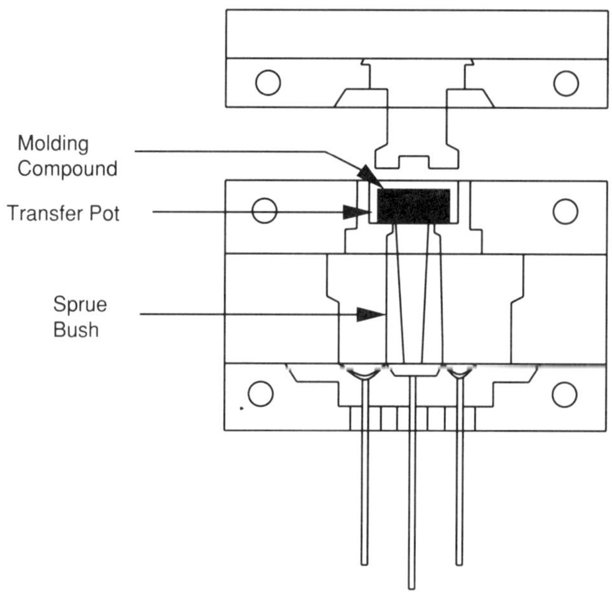

Figure 11.7 Pot type transfer mold.

applied throughout the cure time, which may be as long as three minutes. In a press with a process controller, the packing pressure may be reduced during the cure cycle to minimize resin bleed. The press opens after the cure interval timer expires, at which point the operator removes the molded lead frame from the press with the loading fixture [1]. A typical Pot type transfer mold is shown in Figure 11.7.

11.4.1.3 Post-cure

Post-cure is required for essentially all epoxy molding compounds used in molded plastic packaging. The need for post-cure is derived from the chemical kinematics of the polymerization reaction. The time the molding compound remains in the molding tool is insufficient for it to undergo complete conversion. The reaction rate slows significantly as it approaches complete conversion, making it counter-productive to expend capital and human resources to realize complete conversion in the molding tool. Therefore, the reaction is allowed to continue in the mold till the device has achieved an ejectable degree of conversion, about 60–90% depending on the compound used. Large batches are then placed in convective ovens for 4–8 hours at 170–180 °C for the process of post-curing. Generally, post-curing is performed immediately after molding, but in some cases it is done after the code marking to eliminate an added heatcure cycle for the marking ink.

11.4.1.4 Deflashing and dejunking

During the molding process, some of the molding compound inadvertently may flow on to the leadframe due to equipment, process or design inadequacies. This defect is called a flash. The flash is generally less than 0.001 cm and is thin and brittle. If allowed to remain, the flash would cause problems in the downstream operations of trim and form and solder dipping. It is therefore removed with the help of special equipment like media deflashers, which use high pressure pneumatic streams containing plastic beads, cracked walnut shells or ground apricot pits. The leadframes pass under these pneumatic streams and the flash is removed. Another type of deflash equipment is one which exposes the entire leadframe strip to a hot solvent that reduces the adhesion of the flash to the leadframe, and may swell and soften it as well. The advantages of solvent deflashers are lower deformation to the leadframe and high effectiveness in removing stubborn flash. An important drawback is, however, the undesirable environmental effects caused by their use. Water deflashers, using very high-pressure water streams, directed at the leadframe, are also used.

Media deflashing removes small amounts of molding compound outside the molded body but behind the dam bar. The dam bar has to be displaced from the edge of the molded body to allow it to be cut out cleanly without chipping the package at the parting line. Hence, molding compound will flow out until it encounters the dam bar. In the case where media deflashing is not used, removal of the material behind the dam bar is still necessary, because it could damage the trim-and-form tool. This operation is called dejunking. The options for dejunking include removal in a dedicated apparatus, or as an additional step in the trim and form operation.

11.4.1.5 Trimming and forming

During this process, the lead fingers are electrically isolated and the leads formed (bent into shape), using a punch press with a trim-and-form tool. This generally is comprised of several stages, which sequentially step through the cutting and forming of the lead fingers. Most tools and presses operate automatically. Cassettes of molded strip are loaded at one end and completed units are collected in tubes at the other end. Any discrepancies or inaccuracies in the dimensions of the leadframe or the molded body can cause trim-and-form problems. These could arise from buckling of the leadframe due to excessive shrinkage of the molding compound, or a shift in location of the molded body on the leadframe caused by damage or wear of the molding tool. Excessive and irremovable flash of molding compound on the leads can cause jamming and premature wear of the trim tool. Dewebbing, that is, cutting out the dam bars and shorting bars which connect lead tips together, is first completed. The lead is then formed using the form tool. It is not bent against the plastic molded body, but against a thin anvil to avoid damage to the package body. The machinery for trim and form is essentially a reciprocating punch press. Typical punch rates are about 80 cycles per minute. Cam operated presses are also used for this process.

11.4.1.6 Solder dipping and plating

As most of the IC devices are finally soldered to a circuit board, solder coating of the leads is required to facilitate solder attachment. A thin coat of solder, typically less than 0.0005 inch is applied to the leads. The leads can be solder-plated after molding or can be dipped in molten solder. Solder-plating of the leads prior to molding is possible in cases like TAB. In cases of wire bonding, solder dipping is done after the molding operation. Solder-plating is usually performed before trim-and-form, but solder dipping is done after singulation.

11.4.1.7 Pretest

Pretest inspection is usually conducted after trim-and-form, and code marking. This is also a semi-automated operation in that the packaged devices are moving on a conveyor, possibly the same line that brought them through the code marking station. During pretest, inspection of the position and forming of the leads, of defects in the molded body and of code marks is done. Leads can often be formed improperly or bent in the trim-and-form die or in the subsequent handling operations. Inspection equipment may vary, depending on the needs. Frequently, custom and semi-custom inspection lines are used. The inspection equipment often involves gauges for sensing the lead positions. Vision systems for detecting bent leads or checking for lead planarity are also used.

11.4.1.8 Burn-in and testing

Screening premature failure at high temperature and high electrical loading is involved in this step. It is basically a reliability test of the packaged device to identify defective devices that would fail in the short term, a type of failure known as infant mortality. The devices are mounted on special circuit boards that apply electrical loading. The boards are then plugged into large oven chambers, where they are maintained at 120 °C for 9–24 hours at a particular voltage loading. The temperature, time duration and voltage loading vary according to device type and manufacturer. Typical conditions call for temperatures of 125–150 °C for approximately 24–48 h at a voltage of 6.2–7 V, well above the standard 5 V operating bias of most ICs. Burn-in is usually conducted on some fraction of the population, although some companies burn-in 100% of the devices. The newer burn-in ovens are computer controlled with intrinsic power and logic. Certain voltage and voltage cycles can be programmed for different circuit packs. A medium sized assembly and packaging operation will need ten to twenty full size burn-in ovens, each measuring approximately 6 cubic meters, as burn-in times are much greater than post-cure. The expense of burn-in is considerable and statistical sampling is an option often used to reduce it.

Generally, 100%, of the molded products are tested for electrical performance. The human and capital resources required for this testing contribute a significant fraction of the cost of assembly and packaging. Computer controlled machines which evaluate all functional parameters of the device, according to prepared testing algorithms, are generally used. The devices are fed to these machines either in tubes, cassettes, or manually. The test boards, essentially printed wiring boards, adapt different devices to the same test equipment. They are custom made to the type of device, its function and its testing procedure, and have to be made specially for a particular device code. An excessive drop-out rate after testing could indicate a serious reliability problem.

11.4.2 Injection molding process

Injection molding was developed for thermoplastic materials. Plastic, in the form of small pellets, falls from a feed hopper into a screw which forces the material forward through several heating zones. As the plastic is transported, it is heated to a highly viscous liquid state, arriving at the front zone fully melted, de-aired, dried and compressed up to 20 000 lb/in^2. The flights of the screw vary in each zone. When enough melt has accumulated, the screw is forced forward hydraulically, as a piston, pushing the melt through a nozzle, and down runners, to a mold, much as in transfer molding. This mold is cooled, usually with chilled water, to freeze the plastic as quickly as possible. The removal of heat from the plastic is the time-limiting factor, which determines the overall cycle time, from seconds to minutes. The thermoplastic molding materials are not altered substantially by this process, so sprues, runners, and scrap parts can be reground and blended with virgin material for reuse [8].

Injection molded parts are of extremely low cost because of the automated nature of the process and the very low cycle time. However, the molds are extremely complex, must withstand very high pressures and are very costly. Production must be considerable to justify the mold cost and setup expense.

11.4.3 Reaction injection molding process

This is a type of injection molding used to produce thermoset shapes, usually of large sizes, with polyurethane and polyester resins. Reactive liquid components are prepared separately and then pumped into a mixing head, where they are thoroughly combined. The mixed liquid resins are then forced into a heated mold which may also contain fiber-mat reinforcement and which cures like a compression molded part. Advantages include low cost and large parts. Disadvantages include the limited choice of materials, the rather low-performance physical properties, and the difficulty of controlling the reaction process [8].

11.5 MANUFACTURING DEFECTS

Among the more common manufacturing defects in plastic packaging are package cracking, paddle shift, wire sweep, defect location bias away from the gate, passivation layer cracking, delamination of interfaces, metallization deformation, ball bond liftoff, shearing, and fracture.

11.5.1 Package cracking

When the stresses generated by the constrained shrinkage of the molding compound exceed its strength, package cracking results. The shrinkage forces depend on the package geometry and the coefficient of thermal expansion of the leadframe and die. Finite element analysis can, to some extent, provide a mapping of the stress levels throughout the package, but it is still difficult to predict the occurrence of cracking. The presence of stress concentration points within the package significantly increases local stress levels. A severe stress concentration can magnify essentially any stress to above the strength of the molding plastic [1].

11.5.2 Paddle shift

The symmetry of the package is compromised by the presence of paddle shift, causing the stress gradients to be compressed and increased. The chip is moved closer to the surface of the package, shortening the moisture diffusion length. The displacement of the paddle stretches the wire bonds and places greater stresses on the ball bonds, increasing the chances for ball bond fractures. Paddle shift is most often realized with the paddle edge away from the gate sustaining the greater displacement. A mapping of the defects would indicate a higher incidence of failure in the pin numbers away from the gate, when this type of paddle shift is affecting reliability [1].

11.5.3 Wire sweep

The wires often remain stressed after sustaining flow-induced stress during molding, placing significant stresses on both the ball bond on the chip and the wedge bond on the lead. The term "wire sweep" is usually used to denote visible wire deformation, typically a lateral movement in the direction of molding compound flow through the cavity. Under this lateral deformation, both types of bonds can develop kinks at the attachment point, as the wire is pulled off the axis of the bond. This type of deformation is most often noted on the sides of the chip facing the gate [1].

11.5.4 Defect location

Defect mapping indicates that failures are biased either near or away from the gate. Biases near the gate are often attributable to high molding compound velocity and jetting that occur before the flow field has had an opportunity to stabilize and spread

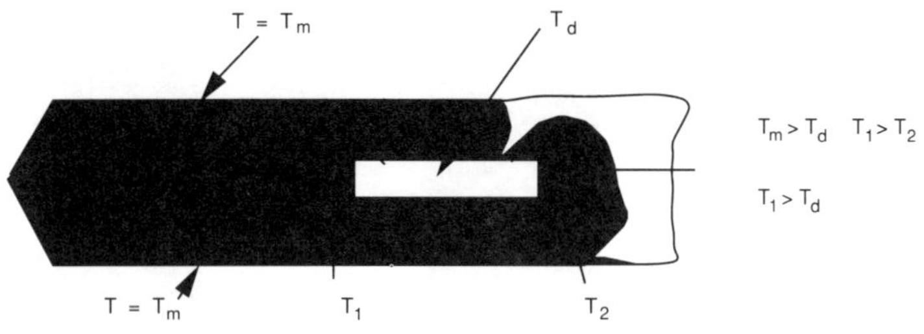

T_m : Mold temperature
T_d : Temperature of die and paddle assembly
T_1, T_2: Temperature of molding compound at points indicated

Figure 11.8 Flow effects that can promote a higher defect density away from gate of the cavity for the situation when the die and paddle assembly did not reside in the heated mold long enough to reach the mold temperature.

over the entire flow cross-section. A bias of high defect density away from the gate is more difficult to understand. One explanation may be that the chip and paddle are at a lower temperature than the sides of the cavity and the leads. The cavity walls are at the mold temperature and the leads are tightly clamped to the surfaces. The chip and paddle are not as well thermally interconnected since the pad assembly is supported by two to four thin cantilever beams, which span the entire distance to the center of the package. The pad assembly has a large thermal mass, compared to the thin leads, requiring a much longer time to heat. The consequence of this physical configuration is that the chip and paddle can be at a lower temperature than the cavity and leadframe. The molding compound can heat to near the mold temperature when it flows through the gate and along the leadframe, but on encountering a colder pad assembly, its temperature begins to drop and its viscosity increases. Therefore, the molding compound which flows up through the wires on the gate side of the paddle has a lower viscosity than the molding compound flowing through at the far side of the paddle, as illustrated in Figure 11.8 [1]. This results in greater flow-induced forces on the wires away from the gate.

11.5.5 Passivation layer cracking

Passivation layer cracking occurs when the stresses imposed by the shrinkage of the molding compound exceed the strength of the fragile passivating material [12]. Passivation layer cracking is often associated with ball-bond lift-off and ball-bond shearing because of the close proximity of these structures to the passivation layer damage on the edges of the chip.

11.5.6 Delamination of interfaces

Delamination at interfaces between the different materials within the package is a major cause of moisture ingress and subsequent premature package failure. The

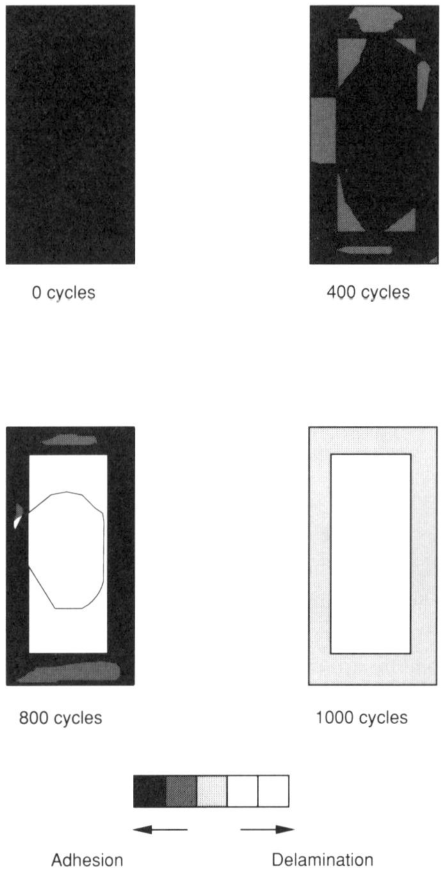

Figure 11.9 Mapping of the adhered and delaminated regions on the surface of a large chip with increasing number of temperature cycles.

principal areas for delamination are along the leads extending from the chip to the edge of the molded body, and along the die surface itself. Delamination is most likely to start at the corners of the chip, where stress concentrations exist. Present molding compounds can withstand hundreds of temperature cycles without showing delamination. Once delamination has begun, the line of stress concentration follows the adhesion edge on the chip surface, since this is the first constraint to molding compound movement encountered. The progression of delamination with temperature cycling is illustrated in Figure 11.9 [1].

11.5.7 Metallization deformation

The aluminum metallization on the device can be deformed by shrinkage forces imposed by the molding compound [13]. Damage often occurs at the edges of the chip, farthest from the zero tangential stress point in the center. The mechanism of deformation is the relative movement of the molding compound over the device

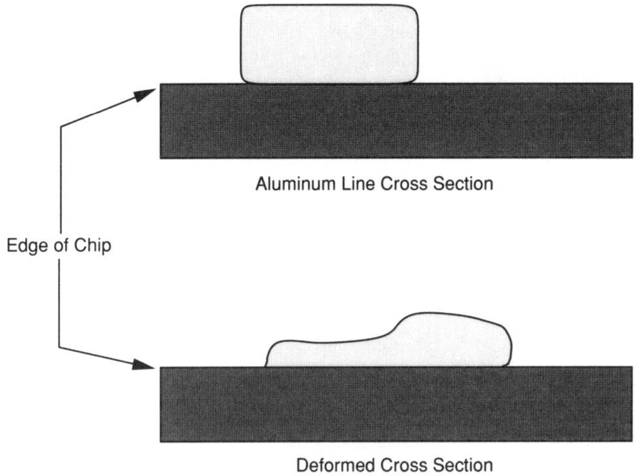

Figure 11.10 Aluminum metallization deformation that can be sustained during shrinkage of the molded plastic package.

surface, due to disparities in the coefficients of thermal expansion. The full extent of this deformation, in the absence of any constraints, can be surprisingly large. An estimate can be obtained by multiplying the difference in the coefficients of thermal expansion of silicon and molding compound by the temperature difference and one half of the length of the chip, as is shown in the equation below:

$$\begin{aligned}\Delta L &= (\alpha_p - \alpha_{si})(175^\circ - 25^\circ)\left(\frac{L}{2}\right) \\ &= (15 \times 10^{-6}/{}^\circ\mathrm{C})(150\,{}^\circ\mathrm{C})(3000\,\mu\mathrm{m}) = 6.8\ \mu\mathrm{m}\end{aligned}$$

In this relation α_p denotes the CTE of the molding compound, whereas α_{si} denotes the CTE of the silicon device, and L is the length of the chip.

Aluminum metallization deformation that can be sustained during shrinkage of the molded plastic package is illustrated in Figure 11.10.

11.5.8 Ball-bond liftoff, shearing, and fracture

Liftoff is the name given to the phenomenon caused by loss of contact between the ball bond and the bonding pad on the chip. Liftoff can be a result of either tensile or shearing forces induced by thermal stress. Open failures can also result from fracture of the ball bond or the attached wire. This can occur during molding as a manifestation of flow-induced stress, or as a failure mechanism after molding. Molded-in stresses, thermal induced stresses, and temperature cycling can fatigue the metallic bond and result in fracture failure. Figure 11.11 shows the types of ball-bond failures that can be caused by excessive flow-induced and thermomechanical stresses.

Ball bonds on large die plastic encapsulated ICs are subjected to high levels of shear stress, as the plastic packages are subjected to temperature extremes, due

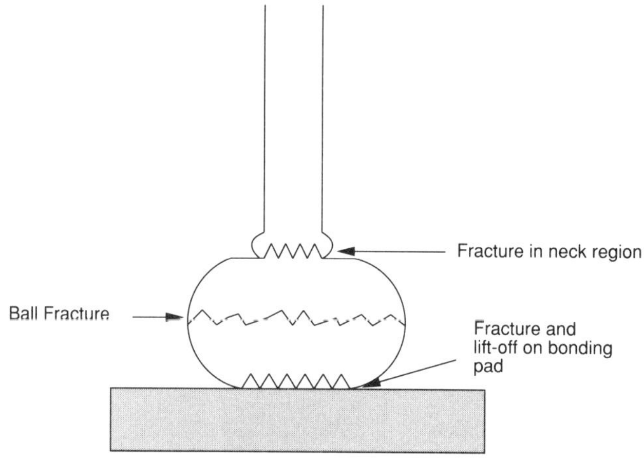

Figure 11.11 Ball bond failures caused by excessive flow and thermo-mechanical stress.

to the difference in the coefficients of thermal expansion. Surface mount plastic packages, such as PLCC and PQFP, are even more at risk, because of soldering temperatures during vapor phase reflow or infrared furnace board mount. As dies become larger, and with the more critical topography of the new generation of VLSI IC structures, these shear forces have made bond pad delamination and cratering a critical problem in industry [14].

Bond-pad cratering is the result of microcracks in the silicon and oxide layers under the bond pads, which may be caused by the bonder or bonder set-up problems, bond parameters, the characteristics of the Au wire and/or the material bonded, or a combination of these. The microcracks are the result of stresses introduced by the dynamic force of the gold ball at touchdown impact, the static force applied after touchdown, the ultrasonic energy, mechanical vibrations before or after bonding and the hardness of the gold ball. It has also been determined that doped oxide, MLO, polysilicon or other brittle layers under the bond pad, as well as bond-pad composition and metal thickness, make different devices more or less susceptible to cratering [15]. Residual stresses, occurring while cooling from cure temperatures, can, if sufficiently large, shear wire bonds from die pads, sever the wire at the neck region of the bond and induce cratering. Shear induced cratering typically results in damage to or separation of some portion of the silicon, glass, or other layers beneath the die bonding pads. In severe cases, the removal of the ball leaves a large pit on the surface of the chip. Complete separation of the ball from the substrate is rare. Typically cracks propagate through the oxide passivation layers and into the silicon itself, producing no visible damage, but degrading device performance [16].

Compliant coatings in the form of a gel or thin film are used on the die surfaces for the purpose of reducing stresses, providing moisture resistance, or protecting against alpha particle induced soft errors. The primary purpose of such coatings depends on the type of device being encapsulated. It is known that, while silicon gel coatings help reduce die surface stresses, they transfer greater stress to the gold wire, causing failures which depend on the thickness and extent of gel coat [14].

This is the reason why silicon gel coated devices in plastic packages may show continuity failures when subjected to extended temperature cycling. In most cases, a weak or broken bond wire neck is seen on decapping the package.

Bond-pad cratering, lifted metal and lifted balls are recognized as a serious quality and reliability problem in the industry, resulting in assembly yields loss, as well as potential reliability problems. It has been found that the silicon dioxide structure beneath the bond pads is greater for short duration dynamic forces than for static forces. The application of much higher bond forces during ball touchdown, combined with a reduction of static mash force, was found to reduce the damage caused to the sub-surface structure and improve reliability on ICs known to be sensitive to cratering problems [15]. Cratering effects can also be minimized with a thicker bond-pad metallization. A larger weld area (high ultrasonic energy and stage temperatures) and flatter bonds (high force) will produce bonds that are more resistant to any form of shear-induced cratering on SiO_2. Weak bonds are to be avoided since they can crater at much lower shear values. The combination of lower power, short dwell, hot stage temperature and high force will be required to produce reasonably strong bonds and reduce damage to the underlying dielectric layers [16].

High moisture-aging of devices with Au/Al bonds molded in epoxy, can lead to ballbond degradation in much the same way as has been observed previously for high temperature-dry aging. High moisture degradation appears to occur much faster at lower temperatu dry degradation. Antimony trioxide, Sb_2O_3, seems to be the overriding factor in most molding compounds in causing high moisture degradation [16]. Elimination of Sb_2O_3, and the use of ion scavenging additives with non-DGETA types of flame retardants, could markedly improve the moisture performance of epoxy molding compounds [16].

11.6 PACKAGE MOUNTING

11.6.1 Through-hole mounting

In a through-hole mounted board, all components and subassemblies must be leaded. The leads must be configured or mechanically formed to be inserted into the mounting holes of the board. For plastic PGAs or DIPs, the leads are already in the correct configuration for insertion. These components normally have their leads formed or bent into right angles for insertion by pickand-place equipment. Chip carriers and hybrids normally require preforming of the leads before they are put into the magazines that act as feeders to the pick-and-place equipment. After the components have been inserted, the leads coming out of the bottom of the board must be trimmed. Before the components are placed on the board, the board is plated with lead-tin. Once all the components are inserted, the board is sent through a wave solder. After soldering, the boards are cleaned and inspected prior to electrical testing. In through hole mounting, the temperatures encountered by the package are not as high as in surface mount components. Hence, the failure of devices due to thermal stresses is less in through-hole mounting compared to surface mounting.

11.6.2 Surface mount technology

The increased complexity of electronic systems in automotive and telecommunication applications necessitates the use of surface mount technology. Electronic systems are characterized by very high reliability requirements and, at the same time, must be cheap in high volume production. Hence, it is necessary to study the implications of surface mount technology on the overall reliabilityof the system. Surface mount technology imposes severe strains in plastic leaded chip carriers. Susceptibility to package cracking was observed in plastic packages after vapor phase or infrared solder reflow processes. The loss of integrity is a potential concern for all surface mount components. Surface mounting packages require different reliability requirements compared to through-hole mounting processes [17].

11.6.2.1 Soldering

The packages to be surface mounted are placed on the board and soldered. Brief summaries of the attachment options for surface mount components are presented below:

- *IR Reflow* This consists of prolonged exposure of the circuit board, with components, to high temperatures induced by absorption of radiation. It is a high-productivity method, since all components on the board are attached simultaneously. The peak temperature varies with thermal mass distribution.
- *Vapor phase reflow* Exposure of the circuit board with components to a high temperature solvent vapor, usually a chlorofluorocarbon (CFC). This is also a high productivity process since all components on board are attached simultaneously.
- *Solder dipping* The entire circuit board and components are dipped in molten solder. This gives high productivity, since all components on the board are attached simultaneously.
- *Laser heating* A laser beam, typically a YAG laser, is used to heat only the lead attach area. Low productivity since it is conducted one lead at a time.
- *Thermode pulse heating* A heated metal block formed to the lead attachment pattern contacts the leads of an individual package. A current pulse may also be used to rapidly reflow the solder. Productivity is low because it is a one component at a time method.

Several of these methods expose the entire molded body to the solder temperature for periods of 30 s to 5 min. The wave solder technique used with through-hole packages exposes only the leads of the package to the solder temperature. The introduction of surface mounting was the first time that the entire molded body was elevated to such temperatures. Yet, there have been no significant changes in the molding compounds to allow them to withstand this exceptional thermal loading. Solder dipping imparts the greatest stress on the molding, because it provides for the most rapid heating via a liquid immersion process. The other global heating techniques of IR reflow and vapor phase soldering also impart significant thermo-mechanical stresses. The last techniques, laser heating and thermode heating, impart the least thermal-induced stress, since they are equivalent to the localized heating experienced in the wave solder attachment of through-hole packages. Unfortunately, these localized heating techniques have far lower production rates than dipping, IR, or vapor phase reflow, so that they are not commercially important.

The mechanism of package cracking on a reflow solder attachment is related to the moisture absorbed within the molded body, both dispersed throughout the volume and adsorbed at the interfaces. This water is absorbed soon after manufacturing, if the packages are not kept in a humidity controlled environment. The molding compound is permeable to moisture even in the absence of cracks or delaminated surfaces. On heating to the reflow temperature above 200 °C, the plastic package softens rapidly, since it is at least 30–40 °C above its glass transaction temperature. The absorbed moisture, some of which may be in the form of clustered water in the pores of the molded body or at the interfaces, vaporizes and expands. A particular problem has been the delamination of the molding compound from the bottom of the die support paddle. This expansion severely distorts the package outline, popping it out and inviting the common label of "popcorn problem". Cracking can occur in the expanded state with the tearing of the molded body at the lower corners of the pad support, or it can crack on cooling and shrinking back to its normal dimensions. The lower pad surface is completely delaminated in most cases, effectively concentrating the thermal shrinkage forces on the paddle edges.

The continuing drive towards miniaturization means that more interconnections will have to be made in less space. Through-hole packages are restricted to 0.100 inch lead spacings because this is the practical minimum for drilling holes in a circuit board. Surface mount packages are not restricted to the through-hole spacing; they can accommodate much finer lead pitch and a higher interconnection density. In addition, components can be placed on both sides of the board, again increasing interconnection density, reducing system size, and lowering costs [1]. As plastic SMDs are fully compatible with cheap PCB material, their use appears to be inevitable. During the last ten years, performances of plastic packages have drastically improved. Looking at the end of the 1980s, LSI failure rates in plastic packages ranged from 1 to 10 FIT ($T_{\text{amb}} = 55\,°\text{C}$, CL = 60%, $E_a = 1$ eV). Such performance is closer to that achieved by ceramic packages. It is estimated that by the end of 1995, 50% of all the semiconductors will be in SMD packages and by the end of this century 70% of all the semiconductors will be in SMD packages. This conversion is accelerated by the fact that the manufacturing cost of plastic SMDs is becoming close to the manufacturing cost of plastic through-hole packages [18]. Figure 11.12 shows the transition in packaging technologies and the change in circuit board size and cost, corresponding to each packaging method [19].

"J" leaded PLCC packages have been registered in JEDEC and accepted as standard surface mount packages throughout the electronic industries. In the board mounting process, the entire semiconductor device is exposed to temperatures of 210 to 260 °C in surface mounting, while in PDIP mounting, only the leads of the device are exposed to the molten solder temperature (260 °C). This induces more stresses in surface mounted packages and causes cracks (Figure 11.13), which lead to environmental contamination [17]. The reduction cost of the surface mount technology at the printed circuit board is about 10% with plastic encapsulated

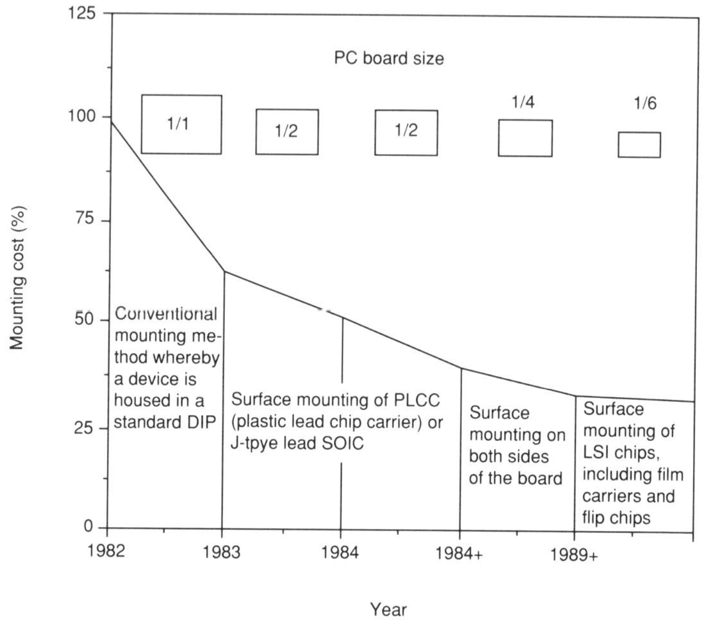

Figure 11.12 Reduction in PC board size.

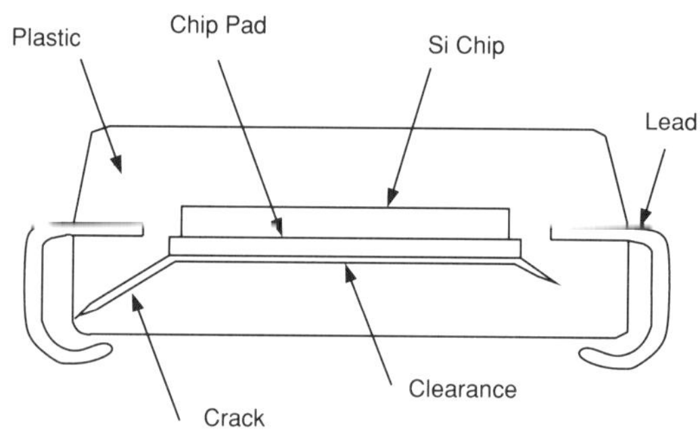

Figure 11.13 Cross-sectional view of cracked package.

integrated circuits instead of ceramic. Surface mount technology has begun to be applied to automotive electronic units recently and its use is expected to expand in the future for the purpose of obtaining smaller, lighter and more highly functional units [19].

Package cracks resulting from soldering stress can be divided into three types depending on the origin and directions of spread of the cracks [20]. The three types of package cracking defined by Omi *et al.* [20] depend on the location of origin and direction of spread. In Type I, the crack originates at one edge of the die pad and spreads to the bottom side of the package. In Type II, the crack originates on one edge of the die pad and spreads to the top surface of the package. In Type III, the crack spreads from an edge of the chip to the top surface of the package. Which type is predominant depends on the following factors:

- *Package structure* Chip size, relative resin thickness on the chip and under the die pad, leadframe material, and elasticity of the die bonding materials.
- *Soldering conditions* Soldering temperature.
- *Moisture absorption* Moisture absorption ratio.

Omi *et al.* [20] have also suggested three techniques for reducing cracks in the packages during surface mounting.

- *Molding compound with improved resistance to soldering heat*: A mold compound with drastically improved adhesion strength, mechanical strength, and moisture absorption ratio needs to be applied.
- *Polyimide coating on the back side of the die pad*: A polyimide film with excellent adhesion to both the die pad and the molding compound needs to be formed on the boundary between the two, in order to suppress possible mold compound exfoliation from the die pad, and to reduce the stress from moisture vaporization under solder heat.
- *Improved leadframe design*: The central portions of the longer sides of the die pad, where the maximum stress is known to be concentrated, needs to be scraped inward from the chip outline.

11.6.2.2 Methods of reducing package failures during surface mounting

One simple solution is to store the molded devices under a desiccant until the time they are attached to the circuit board. Although this is feasible, it is generally undesirable, because it imposes significant restrictions on component handling in the factory, and reduces efficiency.

Mold compound improvement. Low stress mold compounds have been proven to have less cracking than standard mold compounds. Parts which fail with old mold materials do not fail when using low stress materials [17].

Decreasing stress initiation points. Cracks in plastic packages are usually initiated at the leadframe burrs. Burrs which are created in the leadframe stamping process can be stress concentration points. Cracks propagate from the burr through the thinnest plastic region. By reversing the stamping direction the burrs could be formed on the top of the leadframe. Therefore, reversing the stamping direction could reduce the amount of cracking.

Improved die bond process. Package cracks sometimes extend from die bond voids at die edges. Therefore, a void-free die bonding process is crucial to preventing failures.

Reduce moisture content in the package. Insuring that the parts are mounted dry can be done by backing the parts immediately before the board mounting process, or by shipping dried parts in a sealed moisture barrier bag with desiccant.

Increase the glass transition temperature. One approach is to significantly increase the glass transition temperature of the molding compound until it nears the solder reflow temperature. Epoxies are capable of reaching this level of glass transition temperature. The problem is in maintaining the low stress properties that are important in surface mount components. The stress level goes up as the glass transition temperature increases, as shown in the following equation,

$$\sigma = (\alpha_{pg} - \alpha_i)E_{pg}(T_g - T_i)$$

where, σ = Thermal stress due to change in temperature,
α_{pg} = coefficient of thermal expansion of plastic at glass transition temperature,
α_i = coefficient of thermal expansion of silicon/leadframe,
T_g = glass transition temperature of plastic,
T_1 = initial temperature.

Reduce the glass transition temperature. One approach would be to keep the glass transition temperature below the solder reflow temperature to minimize stress. At the same time, one should increase the modulus of the molding compound in the rubbery region, so that the material does not sustain enough vapor-induced deformation to damage the package. Increasing the strength above the glass transition temperature would prevent tearing, as well.

11.7 NON-DESTRUCTIVE FAILURE ANALYSIS OF PLASTIC PACKAGES

11.7.1 Scanning acoustic microscope

The C-mode scanning acoustic microscopy (C-SAM) is a reflection acoustic technique with precision scanning for microscopic inspection, sophisticated signal analysis and display, and a broad-band acoustic transducer with a small numerical aperture for sub-surface imaging. The block diagram of C-SAM is given in Figure 11.14. The C-SAM is a pulse echo type acoustical microscope that generates an acoustical image of a package's internal layers by measuring the amplitude and phase of the reflected signal. The frequency ranges that can be used with

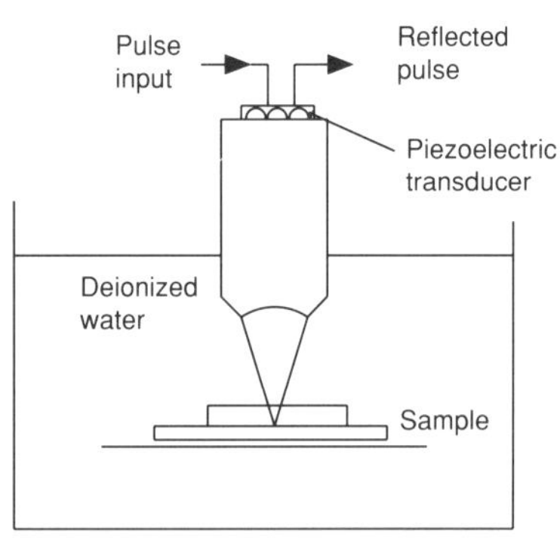

Figure 11.14 Block diagram of C-mode scanning acoustic microscope.

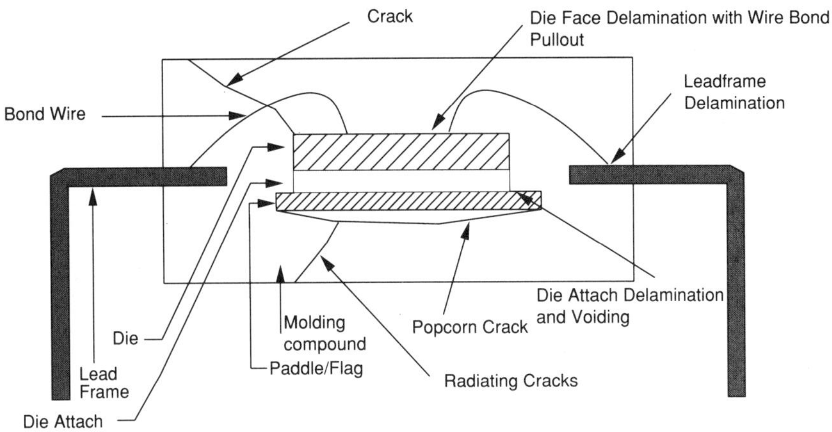

Figure 11.15 Typical plastic package defects that could be analyzed with scanning acoustic microscope.

the system are from 10 to 100 MHz. An internal layer of the package can be focused on by gating techniques, in which the detector is "on" only when the return signal for the desired layer reaches the detector. This excludes all other signals from the image. C-SAM can be used to monitor the integrity of the mold compound — substrate interface, mold compound — die interface, die-die paddle interface and any internal cracking of the die (Figure 11.15). A cross section image can also be obtained to locate the depth of the defects and also to measure the length of cracks in the package. Cracks, voids, or other interfacial delaminations not detected through other non-destructive techniques, such as X-rays, can easily be identified. The resolution of the image in plastic materials with a 100 MHz transducer is about 20 microns in the xy plane and 22.5 microns along the z-axis.

11.7.2 Scanning laser acoustic microscope

The scanning laser acoustic microscope (SLAM), on the other hand, is a transmission mode system that operates in the 10 to 500 MHz frequency range. SLAM produces images through the entire sample thickness in real time (Figure 11.16). In SLAM, a collimated plane wave of continuous-wave (CW) ultrasound at frequencies up to several hundred megahertz is produced by a piezoelectric transducer located beneath the sample. Because this ultrasound cannot travel through air (making it an excellent tool for crack, void, and disbond detection), a fluid couplant is used to bring the ultrasound to the sample. A laser beam serves as an ultrasound detector, by sensing the infinitesimal displacements created by ultrasound at the surface of the part. By rapid sweeping of the laser beam, the SLAM images are displayed on a high resolution video monitor. This instrument is widely used for non-destructive evaluation of surface mount components. Typical applications include the determination of molding adhesion in plastic IC packages, die attach integrity, and the analysis of TAB interconnection bonding.

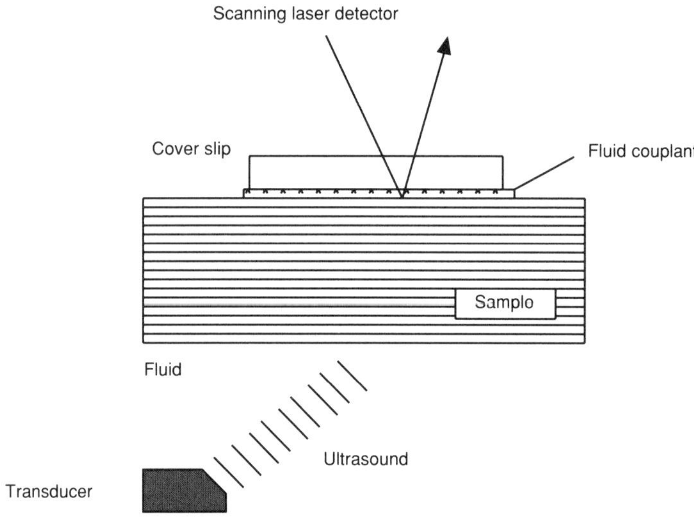

Figure 11.16 Block diagram of scanning laser acoustic microscope.

REFERENCES

[1] Manzione, L.T., *Plastic packaging of microelectronic devices*, Van Nostrand Reinhold, New York (1990).

[2] Edwards, D.R., Heinen, K.G., Groothuis, S.K., and Martinez, J.E., "Shear stress evaluation of plastic packages," *IEEE Transactions on Components, Hybrids and Manufacturing Technology*, CHMT-12(4), 618, IEEE Doc # 0148- 6411/87/1200-0618 (1987).

[3] Pittis, G., "High lead count/high reliability plastic packaging: issues and approaches", *Workshop on Moisture Measurement*, (1987).

[4] Liang, L., "High reliability molded plastic packaging through a systematic approach", *Workshop on Moisture Measurement*, (1987).

[5] Tummala, R., Rymaszewski, E.J., *et al.*, *Microelectronics Packaging Handbook*, Van Nostrand Reinhold, New York (1989).

[6] Lin, A.W. and Wong, C.P., "Encapsulation for non-hermetic mutchip packaging", *Electronic Components and Conference, Georgia*, May 11-16, 1991, 820-826.

[7] Cook, J.P. and Servais, G.E., "Corrosion failures in semiconductor devices and electronic systems, *Proceedings of the 2nd Automotive Corrosion Prevention Conference, Dearborn, Michigan*, 1983, pp. 187-197.

[8] Harper, C.A., *Electronic Packaging and Interconnection Handbook*, McGraw-Hill (1991).

[9] Yoshioka, O., Okabe, N., Yamagishi, R., Nagayama, S., and Murakami, G., "Improvement of moisture resistance in plastic encapsulants MOS-IC by surface finishing copper leadframe", *Electronic Components Conference, Texas*, May 22-24, 1989, pp. 464-471.

[10] *Electronic Materials Handbook, Vol. 1 — Packaging*, ASM International (1989), pp. 494-503

[11] Teat, W.H. and Bergstrom, J.S., "Severe environment packaging for integrated sensors, *SAE 870294*.

[12] Okikawa, S., Sakimoto, M., Tanaka, M., Sato, T., Toya, T. and Hara, Y., "Stress analysis of passivation film crack for plastic molded LSI caused by thermal stress", *Proceedings International Symposium on Test and Failure Analysis*, 275-280 (1983).

[13] Clatterbaugh, G.V., and Charles, H.K. Jr., "The effect of high temperature intermetallic growth on ball shear induced cratering", *Electronic Components Conference, Texas*, May 22-24, 1989, 428-437.

[14] Chidambaram, N.V., "A numerical and experimental study of temperature cycle wire bond failure", *Electronic Components and Conference, Georgia*, May 11-16, 1991, 877-882.

[15] McKenna, R.G. and Mahle, R.L., "High impact bonding to improve reliability of VLSI die in plastic packages", *Electronic Components Conference, Texas*, May 22-24, 1989, 424-427.

[16] Gallo, A.A., "Effect of mold compound components on moisture-induced degradation of gold-aluminum bonds in epoxy encapsulated devices", *International Reliability Physics, Louisiana*, March 27-29, 1990, 244-251.

[17] Cergel, L., "Considerations about the use of surface mount technology in automotive electronic applications", *Seventh International Conference on Automotive Electronics, London*, 1989, 51-60.

[18] Brizoux, M., Lerose, S., Chadebec, G., Potage, J., and Miet, J.M., "Plastic IC's for military equipment cost reduction challenge & feasibility demonstration", *1990 IEEE 40th Electronic Components and Technology Conference*, **2**, 918.

[19] Miura, N., "Automotive electronics — a japanese perspective", *Eighth International Conference on Automotive Electronics, IEEE*, 1991, 1-6.

[20] Omi, S., Fujita, K., Tsuda, T., and Maeda, M., "Causes of cracks in SMD and typespecific remedies", *Electronic Components and Conference, Georgia*, May 11-16, 1991, 771-776.

12

MANUFACTURING OF MICRO-ELECTROMECHANICAL SYSTEMS AND COMPUTER AIDED CONTROL OF MICROMACHINING

12.1 INTRODUCTION TO MICROELECTROMECHANICAL SYSTEMS

The recent evolution in microelectronics of combining electrical and mechanical functions has brought about an exciting new field — microelectromechanical systems (MEMS) [1, 2]. Miniature structures developed by new fabrication techniques on semiconductor wafers make possible new devices that have the potential to revolutionize instrumentation and control systems. At the University of Wisconsin (Madison), an air driven microelectromechanical generator has been developed by Henry Guckel. Using parts that are a fraction of the thickness of human hair, a generator was designed giving an output of five volts and weighing under five grams. In separate developments, a team at the University of Michigan (Ann Arbor), have built atomic-force microscopes that enable insights into surface science and produce miniature probes for use in advanced prostheses. At the Berkeley Sensor & Actuator Center at the University of California, researchers have built a microgripper capable of handling micron-sized structures. The heart of the gripper is a novel microstructure that consists of interdigitated fingers, or cantilevers, which are activated electrostatically to move the gripper arms. Such a device has potential applications in biomedicine and micro-telerobotics. These are just a few of the recent advances in the emerging field of MEMS, in which electronic and mechanical components are combined for applications as sensors, valves, actuators, and related systems.

Despite the vast variety of MEMS devices being proposed and fabricated, most of the devices fall into the categories of sensors and actuators, the most important constituents of instrumentation and control systems. The research on MEMS sensors has dated from the late 1960s, and extensive efforts have been made for the fabrication of MEMS sensors to become a mature technology application, as accelerometers which are used in automobile air-bag systems, temperature and pressure flowmeters, and neural microprobes for biomedical study. The promise of the MEMS sensors is that batch fabricated silicon wafers are very low in cost and very sophisticated in feature size, incorporating the use of on-chip circuitry. The micromachined sensors can be produced today with high yield and merged with integrated electronics both in monolithic chips and in hybrid multichip assemblies. For some types of sensors, accuracy can be as high as 16 bits and VLSI interface circuits are defined to allow features such as self-testing and digital compensation. On the other hand, consistent progress has also been made in the area of actuators. Microgrippers, piezoelectric micromotors, micropumps, magnetic microactuators, etc. have been successfully fabricated in laboratories. Though some problems remain to be solved in the area of microactuators all of the essential elements of simple MEMS are in place and complete microelectromechanical systems have been proposed with applications ranging from microrobots for security and medical appli cations to sophisticated positioning systems for assembly tasks at the submillimeter level.

The fabrication techniques used in MEMS are similar to those used by the electronics industry. The microminiatures are made using three micro-fabrication processes — surface micromachining, bulk micromachining, and LIGA processing. These processes employ methods such as photolithography, material deposition, chemical etching, electroplating, and X-ray radiation to shape the mechanical and electronic structures. Although micromachining has been used for over 20 years as a processing technique in the fabrication of sensors, only recently has it been extended to the commercial production of mechanical features from silicon. In the past five years there has been a heightened interest in the use of micromachining technology. Improved etching technologies, deep UV lithography, X-ray lithography for the LIGA process, and ion projection lithography make it possible to attain accuracy for features with high aspect ratios.

While progress has been made in MEMS there are still many technical challenges which must be solved. In the area of microactuators better driver mechanisms are required. Nonplanar technologies, with assembly techniques at the submillimeter and micron levels are also required. Workstation based design systems along with a database of material, structural, and performance information must be developed. These problems are being addressed and the progress achieved will be discussed in this chapter.

In the sections that follow, applications of MEMS are presented (section 12.2), and several MEMS technologies are discussed (section 12.3). The architecture and current status of CAD tools for MEMS processing simulation and design is also included.

12.2 MEMS DEVICES

The MEMS devices developed to date are sensors and actuators. The principles of these MEMS devices are presented including a description of the structures for sensors and actuators.

12.1.1 Sensors

12.1.1.1 Pressure sensors

The important category of MEMS devices is that of the silicon pressure sensor which has been identified to have applications in many areas including transportation, health care, and industrial process control. Substantial progress has been made on silicon-based pressure sensors in recent years as a result of advances in etching technologies to form thin silicon membranes. Most of the micromachined silicon pressure sensors contain silicon membranes. Depending on the pressure sensing mechanism they can be classified into two types of devices — piezoresistive sensors and capacitive coupled sensors.

Figure 12.1(a) shows the schematic of a piezoresistive sensor which contains a full bridge of diffused resistors to measure stress at four points on the diaphragm and to convert it to an electrical output signal. For a diffused resistor subjected to parallel and perpendicular stress components, $\sigma_{\parallel}$ and $\sigma_{\perp}$, the resistance change is [3]:

$$\frac{\Delta R}{R} = \pi_{\parallel}\sigma_{\parallel} + \pi_{\perp}\sigma_{\perp}$$

where $\pi_{\parallel}$ and $\pi_{\perp}$ are the piezoresistive coefficients parallel and perpendicular to the resistor alignment, depending on the crystal orientation. Thus by arranging the

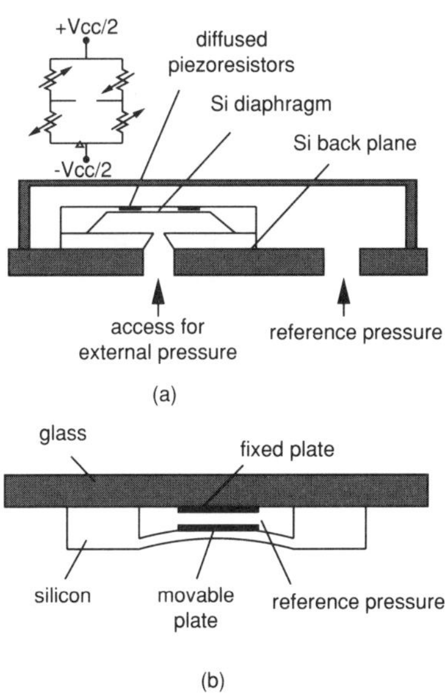

Figure 12.1 Schematics of (a) a piezoresistive and (b) a capacitive pressure sensor.

alignment of the diffused resistors in the bridge, two of them increase in value and the other two decrease in value upon application of the stress. A voltage signal is generated due to the resistance variation and the voltage signal is calibrated to obtain the pressure applied. In the capacitive type sensor shown in Figure 12.1(b), the diaphragm is used as a movable plate of a parallel plate capacitor. The pressure is determined from variation of the capacitance of the parallel capacitor.

By varying the diameter and thickness of the silicon diaphragms of 0–200 MPa have been fabricated. The bridge voltages are usually 5 to 10 volts and the sensitivity of silicon pressure sensors vary from 10 mV/kPa for low pressure to 0.001 mV/kPa for high pressure devices. Temperature compensation circuits are also added to the pressure sensor which can result in less than 1% sensitivity for sensors working in the range between 0 °C and 50 °C.

12.1.1.2 Accelerometers

Several silicon accelerometers based on the diaphragm or cantilever structure from micromachining have been fabricated [4]. One of the most interesting structures is a cantilever piezoresistive accelerometer which is illustrated schematically in Figure 12.2. The center layer of the glass–silicon–glass sandwich structure, which consists of a mass block and a cantilever, is the heart of the device. Fabrication of the accelerometer is a batch process utilizing standard IC photolithographic and diffusion techniques in addition to the special techniques required to shape the silicon and glass. The silicon element and the top and bottom glass covers are fabricated separately in wafer form and then bonded together. The micromachining technology will be discussed in more detail in section 12.3, while the details of the fabrication technology has been reviewed in Chapters 3–10.

The cantilever has a p-diffused path which forms a resistor of about 100 Ω/cm^2. Under acceleration, the mass block exerts stresses on the cantilever. The stresses will induce a change in the resistance of the diffused resistor. The magnitude of the change depends on the crystallographic orientations of the silicon and stress. The variation in the resistance is then detected by an on-wafer Wheatstone bridge circuit. By using of silicon micromachining technology the entire accelerometer can be made with weight under 0.02 gram and as small as $2 \times 3 \times 0.6$ mm. The sensitivity of such accelerometers can be as low as 0.01 g with accuracy under 1 percent over a range of 100 g, where g is the acceleration of gravity. The accelerometer also has good frequency response with a band-width of 100 MHz. The performance

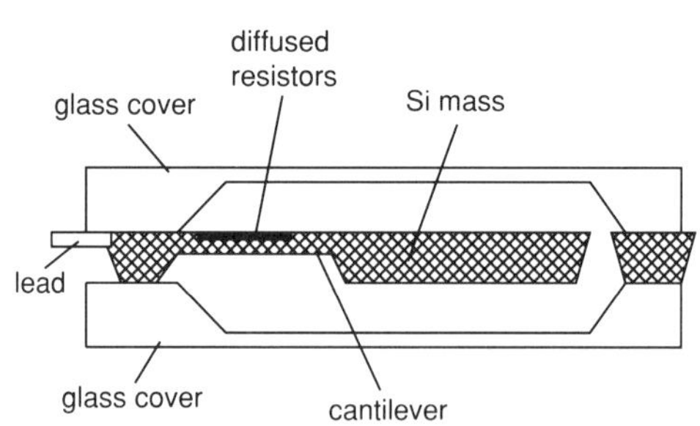

Figure 12.2 Structure of a cantilever piezoresistive accelerometer.

characteristics of such accelerometers can be varied over a wide range to meet the needs of different applications. One of the most interesting applications, for example, is in the automobile air-bag systems, which at the present time represents the largest application product.

12.1.1.3 Shear stress sensor

A second example of a MEMS sensor is a floating-element liquid shear stress sensor which has been recently developed at MIT by the group of Martin A. Schmidt [5], and has many potential applications such as flowmeters. The floating-element sensor ($120\ \mu m \times 140\ \mu m \times 5\ \mu m$) has been designed for high shear stress (1–100 kPa) and high pressure environments (up to 6600 psi) and utilizes a piezoresistive transduction scheme. Figure 12.3 shows the schematic diagram of the sensor, which consists of a plate ($120\ \mu m \times 140\ \mu m$) and four tethers ($30\ \mu m \times 10\ \mu m$). The tethers function both as mechanical supports for the plates and as resistors in the transduction scheme. The plate and the tethers are constructed from a 5 micron thick lightly doped silicon layer and are suspended 1.4 μm above another surface. A flow over the floating-element structure and parallel to the length of the tethers generates a shear stress on top of the suspended plate. Assuming the plate moves as a rigid body, the shear stress forces the suspended plate in the direction of flow. Two of the tethers experience compressive stresses and the other two tensile stresses. These stresses generate axial strain fields throughout the tether structure, which introduce a change of resistance of the tether due to the piezoresistive properties of single crystalline silicon. The change in resistance is then converted to a change of a voltage signal change by applying bias to the structure, which reflects the shear stress.

12.1.1.4 Biosensors and neural probes

Micromachined structures are emerging as useful instruments in medicine, where precision sensing is critical. An electromechanical sensor array that is small enough to fit inside a blood vessel has been recently developed. The device is inserted into an artery within a catheter that has an inside diameter of 650 microns. It measures the levels of oxygen, carbon dioxide, and PH in the blood. The firm also developed

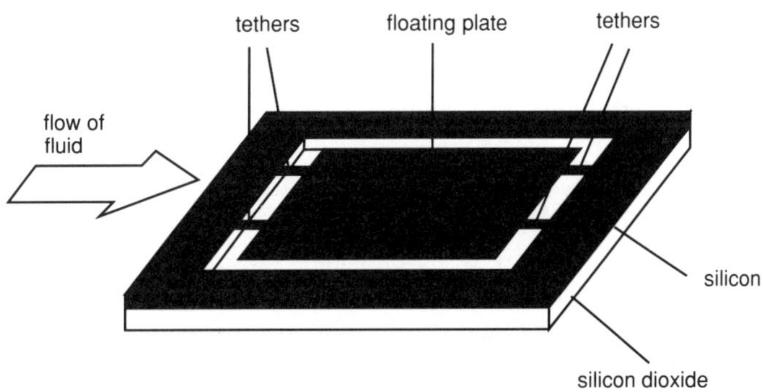

Figure 12.3 Schematic of a shear stress sensor.

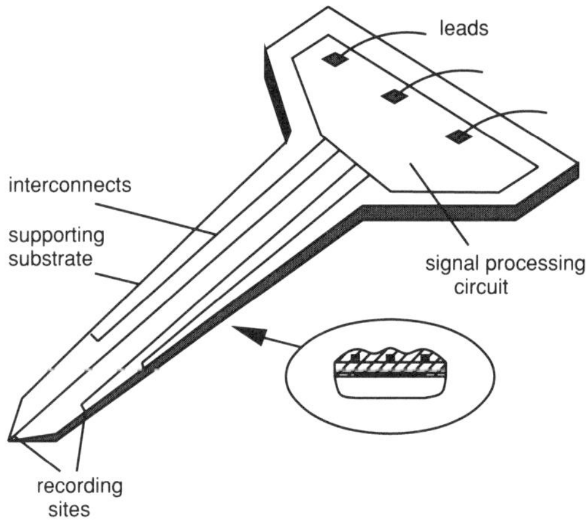

Figure 12.4 Schematic of a neural probe.

a sensor for monitoring the carbon dioxide and oxygen that a patient inhales during surgery.

Even more significant to the medical profession may be the development of micromachined neural probes at the University of Michigan. In the past, sensors have been used to record impulses from one site in the brain at a time. The new neural probe developed is able to record single brain cell activity from 30 sites simultaneously. The probe data opens the possibility for examining structures in the brain at the circuit level, which would help scientists learn how to treat neural disorders. Such a probe will also help to develop a neural-electronic interface which will control auditory, visual, and neuromuscular prostheses. The schematic of the neural probe is illustrated in Figure 12.4.

12.2.2 Microactuators

12.2.2.1 Microgripper

Recent progress has been made in the MEMS area in the construction of micro-grippers, which are capable of handling micron-sized objects and have applications in biomedicals as well as in micro-telerobotics. A typical microgripper consists of a fixed closure driver and two movable jaws. The jaws are closed by an electrostatic force from an electrostatic voltage applied across them and the closure driver.

Figure 12.5 shows the schematic of a microgripper developed at the University of California [6], by Richard S. Muller. The microgripper consists of a silicon die (7 mm $\times$ 5 mm), a 1.5 mm long support cantilever, made from boron-doped silicon substrate material, and a 400 micron long polysilicon overhanging gripper extending from the end of the support cantilever. The support cantilever, heavily doped with boron, is defined geometrically by etch-stopping in ethylenediamine and pyrocatechol (EDP). It is about 12 microns thick, 1500 microns long, and tapered from 400 microns width at the base to 100 microns width at the end. This support

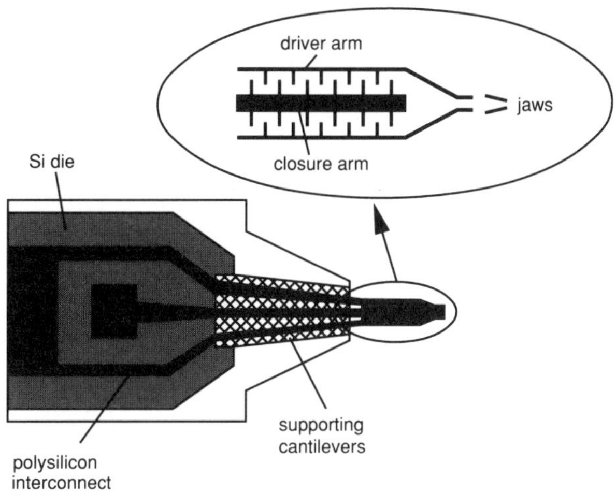

Figure 12.5 A silicon-fabricated microgripper.

cantilever accurately locates the overhanging polysilicon microgripper and provides a thin extender to the unit.

The silicon die is formed by anisotropic etching in EDP and is 5 mm × 7 mm in area and 0.5 mm thick. The silicon die is snapped free along a backside V-groove with a portable vacuum pen. The microgripper and its foundation die is mounted on a positioner and electrically connected using two large contacts that are provided for wire bonding. The microgripper is reported to exert 40 nN of force on the specimen with an applied voltage of 40 V.

12.2.2.2 Piezoelectric micromotor

Another application of MEMS in microactuators is micromotors for microrobots. Interest in microrobots has been driven by recent success in developing intelligence architectures for mobile robots, which can be compiled efficiently into parallel networks on silicon. Various types of micromotors have been fabricated. Figure 12.6 shows the top view and cross-section of an electrostatic stepping motor. A silicon center rotor is free to move around a center pole that is bonded to the glass substrate. A stopper is located between the glass substrate and the silicon center pole hub to prevent the rotor from separating from the center pole. During operation the oppositely placed stators are sequentially stepped with applied voltages, an electrostatic force is exerted on the rotor, which tends to align the teeth of the rotor with the activated stator, and cause the rotor to rotate around the center pole.

In addition to electrostatic motors, micro-magnetostatic motors have also been developed. In general, electrostatic motors are preferred over magnetostatic motors because electrostatic forces scale favorably as dimensions shrink. Also, because dielectric materials are more easily patterned and processed than magnetic materials, especially in the realm of silicon devices. The three-dimensional windings required in magnetostatic motors would be very hard to fabricate in silicon. However, the small gap sizes that allow electrostatic motors to withstand increased electric fields before breakdowns are easily fabricated using photolithographic techniques.

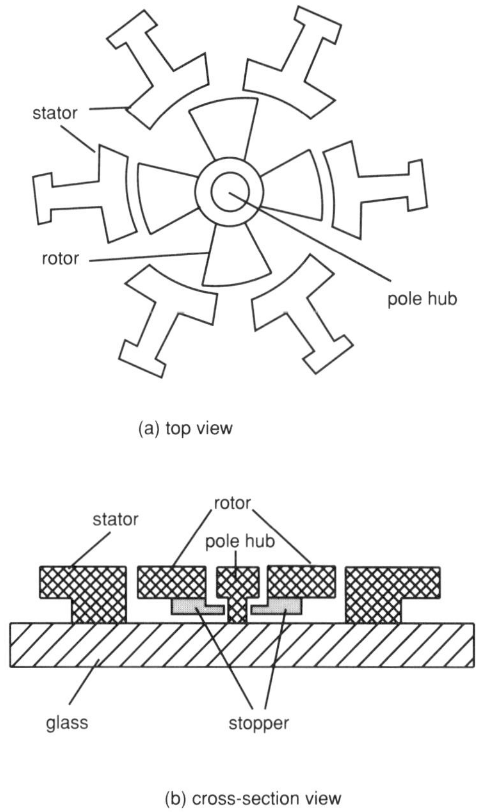

Figure 12.6 Top view (a) and cross-section view (b) of an electrostatic stepping motor.

Although electrostatic micromotors have been successful, there are also limitations in fabrication aspect ratio constraints and torque-to-speed characteristics.

12.3 MEMS TECHNOLOGY

12.3.1 Bulk micromachining

Early silicon products were fabricated using bulk micromachining technology with isotropic wet chemical etchants such as hydrofluoric, nitric, and acetic acid mixture. The isotropic etchants have no preferential etch rate to crystal orientation and attack different planes at the same rates. Isotropic etching has problems with etch control, selectability, and precision. Most of the chemicals used today are wet etching anisotropic etchants which selectively etch the $< 100 >$ and $< 110 >$ directions of silicon. Etching in a $< 111 >$ direction is much slower than the $< 100 >$ and $< 110 >$ directions (typically 50 times slower). The use of anisotropic etchants results in vertical sidewalls for the $< 110 >$ silicon substrate and at an angle of 54.7° for the $< 100 >$ orientation. Lateral geometries and etch profile can be precisely controlled through anisotropic etching.

In a typical bulk micromachining process, photolithography is used to transfer a two-dimensional pattern from an etch mask to an underlying silicon wafer. Ultraviolet light waves irradiate a mask, which admits a selected pattern of light. The waves pass through the mask, altering the molecular makeup of the photoresist on the surface of the silicon wafer. A chemical etchant then selectively attacks the wafer.

In this manner, a predetermined wafer shape is micromachined. The etching process can also be stopped in midstream to control the shape of the microstructure, using thermal diffusion or ion implantation of boron, an etchant resist material into the wafer. In fabricating a typical pressure sensor, for example, etch-stops are used to prevent the chemicals from eating right through the wafer. When the etching process is stopped, a thin layer of silicon is formed — usually between 1 to 20 microns, which functions as the diaphragm of a sensor. Figure 12.7 illustrates a bulk micromachining process.

Dry etching technology has also been used as an alternative to the conventional wet-etching technique. According to the etching mechanisms dry etching can be classified into ion-etching (i.e., sputter etching and ion beam milling) and reactive etching (i.e., plasma etching and reactive ion etching). The ion beam milling technique [7] uses the kinetic energy of etchant ions beams. Etchant ions are generated from an ion source and accelerated by an electric field. The ion particles (Ar, CF_4) of the etchant hit the wafer with high energy. The wafer is precoated with ion beam resist and pre-patterned. The portion of the wafer under the ion beam without resist coating is etched away. What makes ion beam milling attractive is that it is completely anisotropic and provides high accuracy fine machining. Moreover, ion beam milling provides a greater flexibility in the control of milling conditions. By selection of the ion source (chemically active or inactive) and control of the ion beam density, one can get various geometries of interest. In the reactive etching technologies, a radio-frequency or electroncyclotron resonance generated plasma produces neutral atoms, molecules, and radicals that react with the surface of the part to be machined and produce volatile compounds. Common gases for silicon etching including freon (CF_4), carbon tetrachloride (CCl_4), and Cl_2. In the etching process fluorine or chlorine produced in the plasmas react with silicon to form SiF_4 and $SiCl_4$.

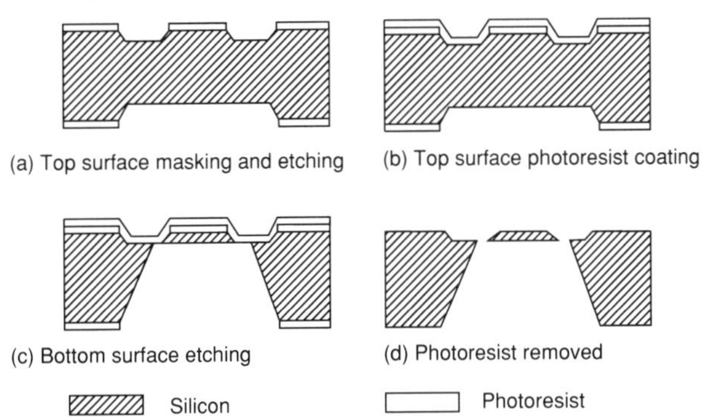

Figure 12.7 Typical bulk micromachining process.

12.3.2 Surface micromachining

In a typical surface micromachining process, a silicon wafer is first oxidized in a low pressure chemical vapor deposition (LPCVD) system, which grows silicon dioxide on the surface. Chemicals are then used to etch away parts of the silicon dioxide and to define a required shape. The structure is put through a second LPCVD process, in which a silicon layer grows on top of the silicon dioxide and adheres to the wafer in the areas where silicon dioxide has been etched away. In subsequent etching steps, the shape of the layer is defined, and the film of silicon dioxide is dissolved. Figure 12.8 illustrates a typical surface micromachining process.

In the above technique, the end component is a suspended mass, or a miniature bridge, a geometric feature common to many surface micromachined components. The silicon dioxide is known as the sacrificial layer for the sake of fabricating a suspended structure. In addition, chemical vapor deposition, sputtering and direct evaporation are used to deposit thin films on a chip. Polysilicon, silicon nitride, and tungsten, among other materials, are used for the structures of surface micromachined parts while phosphorus-doped silicon functions as the sacrificial layer. There has been strong interest in polysilicon micromachined parts recently, however some problems still exist. Polysilicon structures of thickness larger than several microns are difficult to fabricate due to the larger intrinsic stress which results from the thicker polysilicon layers.

The advantage of surface micromachining is that the elements are fabricated in preassembled form. This avoids assembling individual mechanical parts by hand. The bulk micromachining technique, in contrast, cannot provide such integrity of micromachined parts. However, the surface micromachining process tends to introduce intrinsic stress as the structure layer is grown on the sacrificial layer. Such stress causes structure warp and limits the thickness of structure to be fabricated with in several microns. Recently an alternative process has been developed for

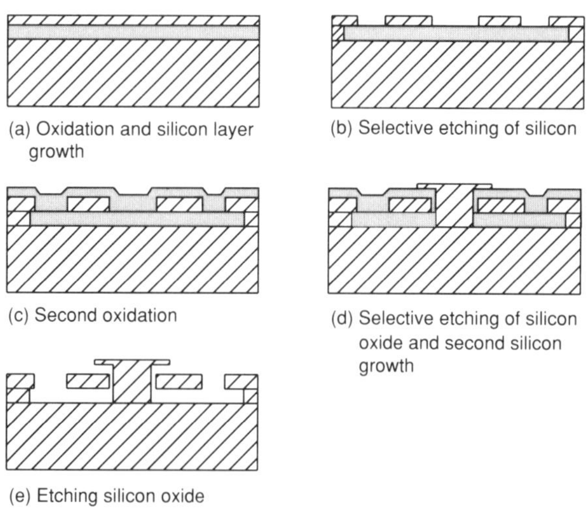

(a) Oxidation and silicon layer growth

(b) Selective etching of silicon

(c) Second oxidation

(d) Selective etching of silicon oxide and second silicon growth

(e) Etching silicon oxide

Figure 12.8 Surface micromachining process.

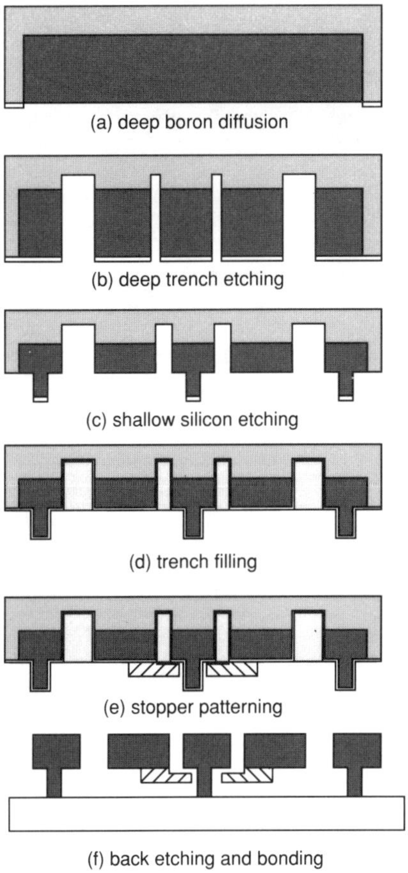

Figure 12.9 Overall process of a micromotor.

single crystal silicon micromotors [8] based on the combination of both bulk micromachining and surface micromachining techniques. The new approach utilizes deep etch stop (boron) diffusion and deep trench etching to define the vertical structure of the parts and employs the surface micromachining technique (i.e., sacrificial layer and epilayer growth) to form the joint parts. The new process provides up to 20 microns mechanical structure with higher production yield. The overall process of the micromotor is illustrated in Figure 12.9.

12.3.3 LIGA process

To date, most of the micromachined devices are made using either surface or bulk micromachining. A micromachining technology called the LIGA (an acronym for the German term for lithography, electroforming, and plastic modeling) process is less compatible with integrated circuit manufacturing, but is showing great promise for the development of microstructures. The technique provides well defined, thick microstructures, that have extremely flat and parallel surfaces. These characteristics

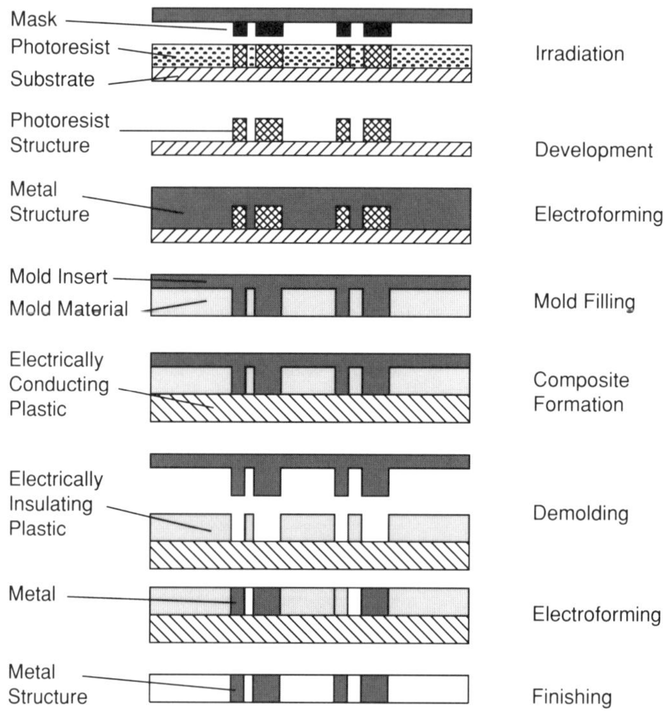

Figure 12.10 Typical LIGA process.

are particularly useful for fabricating motors, gear trains, and generators having spinning parallel parts that come in contact.

In a typical LIGA process as shown in Figure 12.10 [9], polyamide is first deposited on a silicon wafer as a sacrificial layer. The shape of the layer is then defined by passing ultraviolet light waves through a mask. Next the substrate is covered with a plating base made of a thin film of titanium under a layer of nickel. A photoresist layer over the plating base is deposited, this usually consists of polymethyl methacrylate, a photopolymer similar to Plexiglas.

A mask is placed over the photoresist material and X-rays are radiated through it. The X-rays are powerful enough to penetrate a photopolymer as thick as 300 microns. The X-ray mask is a membrane, typically of silicon nitride, which is only 1 to 1.5 microns thick. It is mostly transparent to allow the X-rays through, but gold is deposited in defined regions of the mask to absorb certain X-rays.

X-ray radiation changes the polymer's molecular weight in selected areas, and chemical etchants dissolve away the regions of material that have been exposed. The resulting geometry is a mold into which nickel is electroplated. In subsequent steps, chemicals are used to remove the mold, plating base, and sacrificial layer. Use of a sacrificial layer allows us to fabricate structures that are semisuspended or free of the substrate. These components are assembled into parts that are attached to the substrate to make interacting microstructures. In a motor, for example, the shaft and the rotor are fully attached parts, while the spinning rotor is machined free of the substrate and mounted to the shaft.

12.4 COMPUTER AIDED DESIGN OF MICROMACHINING

12.4.1 CAD tools

As the field of MEMs develops there is an increasing need for CAD/CAE to assist device and system designers in a variety of tasks, for example, mask design, process planning, design verification, device simulation, and optimization. A better understanding of the fabrication process is required to ensure reproducibility for commercial products. A fundamental aspect of such a requirement is to construct a three-dimensional solid model from a description of the mask set and process sequence used in the fabrication of a microelectromechanical device. Equally important to the designers, the material properties for each component of the device to be fabricated must be determined after each step of processing since these material properties may be process dependent. The designer thus can construct the solid model of the final product and perform mechanical and material simulation to ensure that it meets the requirements. Several such CAD systems have been presented or are under development, such as IBM's OYSTER [10], the crystal etching program of Sequin [11], MIT's MEMSCAD [12], and Michigan's CAEMEMS [13].

The OYSTER developed at IBM is able to construct a three-dimensional polyhedral-based solid model mask set and primitive process description. In another approach, the CAEMEMS (acronym for Computer Aided Engineering Workbench for Micro-Electro-Mechanical Systems), developed at the University of Michigan, provides a database of material properties and process model parameters, a process modeler that takes process information as input and produces a solid model of the device to be fabricated, a solid modeler that allows for visualization and design verification, and a device modeler that performs finite-element simulations of components and systems. The block diagram of the CAEMES system is shown in Figure 12.11. The MEMS database will incorporate existing knowledge of the isotropic and anisotropic etch rates of crystalline and polycrystalline silicon, silicon dioxide, silicon nitride, quartz, and other materials such as metallic thin films. This database contains etch-rate information as a function of material, doping, orientation, etchant, and temperature. The database can be updated as new information becomes available. The MEMS CAD takes mask information and process steps as input. By using appropriate diffusion models and etching simulators, an accurate three dimensional representation of the microelectromechanical device, including its detailed composition, is generated as output. The MEMS CAD will have the capability of capturing the three-dimensional geometries of devices in the form of finite-element meshes. The user then can apply appropriate boundary conditions and forcing functions to the finite-element meshes to perform finite-element simulation, including displacement, stress, and thermal analysis, both static and dynamic.

12.4.2 CAD architecture

To better understand the CAD programs discussed earlier, we will discuss the architecture of the MEMS CAD being developed at MIT in this section. The CAD

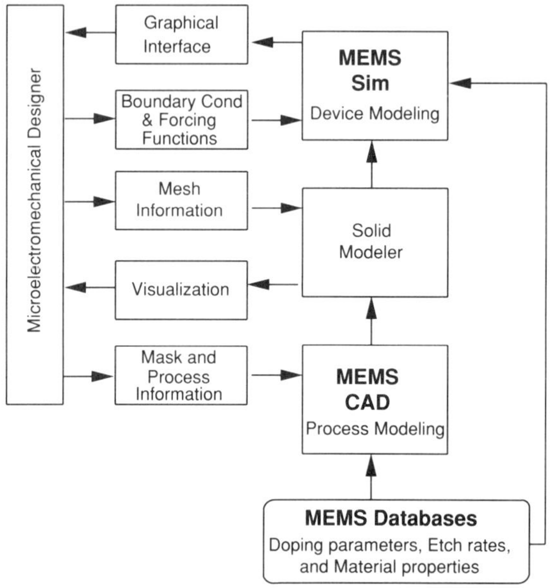

Figure 12.11 The block diagram of the CADMEMS system.

system consists of three sections, the microelectronic CAD section, the mechanical CAD section, and the material property simulator, as outlined by dashed blocks in Figure 12.12. The interactions among sections and their various constituents are shown by arrows and the flow of information is denoted by the direction of the arrows. The primary interface for mechanical modeling is through PATRAN, a mechanical CAD package which provides interactive construction of three-dimensional solid models, graphical display, and interfacing with FEM packages.

12.4.2.1 Microelectronic CAD

In the microelectronic CAD section, the mask is created in CIF [14] using KIC [15] and the process sequence is created in the MIT Process Flow Representation (PFR) using a standard editor. SUPREM-3 [16] and SAMPLE [17, 18] are installed to provide depth and cross-sectional modeling capabilities. The structure simulator is the critical part in this section, mask layout and process information are combined in the structure simulator to construct a three-dimensional solid model. Information about the geometry of the structure, the material type and associated process conditions must be tracked. Physically, a process step causes a change in the wafer. The structure simulator performs its functions in a process-step by process-step bias, making appropriate modifications to the solid model for each process step (Figure 12.13). The geometry generated is directly passed to the mechanical type and process conditions for each component are passed to the Material Property Simulator through the Process History File.

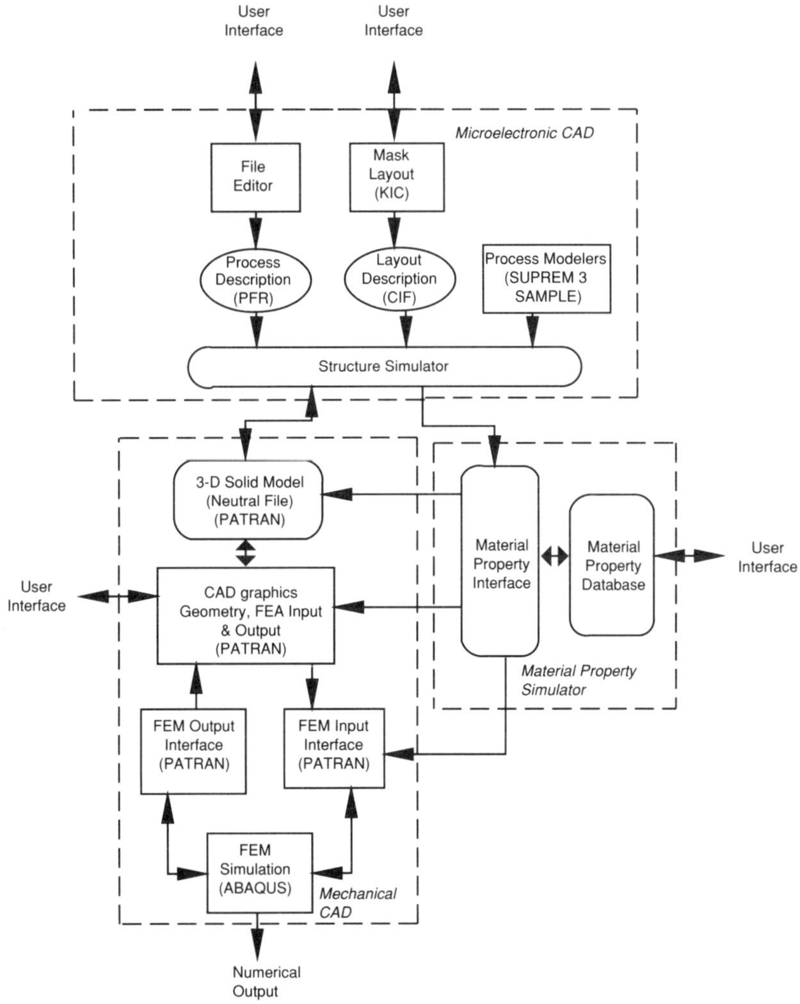

Figure 12.12 Outline of the MIT MEMSCAD.

12.4.2.2 Material property simulator

The Material Property Simulator reads the sequence for each component of the solid model from the Process History File and generates a set of material property data. The material properties are passed to the mechanical CAD section, either into the PATRAN Neutral File, the PATRAN interactive section, or the FEM input file itself. The purpose of the FEM input file path is to allow for the use of the FEM simulator independent of the PATRAN graphics software. The material property simulator generates both intrinsic stress and thermal stress data for the solid model, which are passed to the mechanical CAD section.

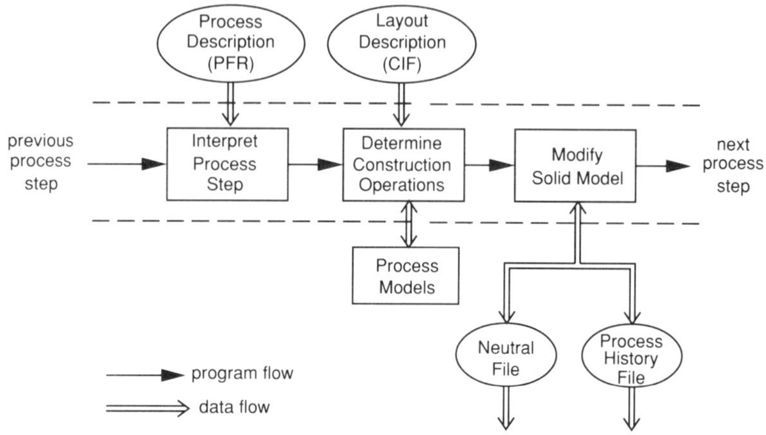

Figure 12.13 Process step simulation.

12.4.2.3 Mechanical CAD

In the mechanical CAD section, the geometric information from the microelectronic CAD section is discretized into finite elements. Material properties, geometries, loads, and boundary conditions are specified. The solid model geometry and material property data are read and manipulated in PATRAN graphs. A complete input model for finite element analysis is then generated and optimized interactively. The FEM model is translated into an ABAQUS [19] input file for FEM analysis. The results of FEM analysis are then passed to PATRAN to examine the mechanical behavior of the device.

REFERENCES

[1] O'Connor, L., "MEMS: microelectromechanical systems", *Mechanical Engineering*, February 1992, 40–47.

[2] Wise, K.D., "Integrated microelectromechanical systems: a perspective on MEMS in the 90s", *IEEE MEMS Workshop*, 1991, 33–38.

[3] Middelhoek, A.S. and Audet, S.A., *Silicon Sensors*, Academic Press, 1989, pp. 118–122.

[4] Roylance, L.M. and Augell, J.B., "A batch fabricated silicon accelerometer", *IEEE Trans. Electron Dev.*, **ED-26**(12), December 1979, 1911–1916.

[5] Shajii, J., Ng, K.Y. and Schmidt, M.A., "A microfabricated floating-element shear stress sensor using wafer-bonding technology", *J. Microelectromechanical System*, **1**(2), June 1992, 89–93.

[6] Kim, C.J., Pisano, A.P. and Muller, R.S., "Silicon-processed overhanging microgripper", *J. Microelectromechanical System*, **1**(1), March 1992, 31–35.

[7] Arimatsu, K., Hashimoto, I., OOishi, S., Tanaka, S., Sato, T. and Gejyo, T., "Development of large scale ion beam milling machines", *Nuclear Instruments and Methods in Phys. Research*, B37/38, June 1989, 833–837.

[8] Suzuki, K. and Tanigawa, H., "Single crystal silicon rotational micromotors", *IEEE MEMS Workshop*, January 1991, 15–20.

[9] Harmening, M. *et al.*, "Modeling of three dimensional microstructures by the LIGA Process", *IEEE MEMS Workshop*, 1992, 202–207.

[10] Koppelman, G.M., "OYSTER, A 3D structural simulator for micro electromechanical design", *Proc. IEEE MEMS Workshop*, 1989, 88–93.

[11] Sequin, C.H., "Computer simulation of anisotropic crystal etching", *Digest of 1991 Int. Conf. Solid-State Sensors and Actuators*, June 1991, 801–806.

[12] Maseeh, F., "A CAD architecture for microelectromechanical systems", *IEEE MEMS Workshop*, 1990, 44–49.

[13] Crary, S. and Zhang, Y., "CAEMEMS: an integrated computer aided engineering workbench for micro-electro-mechanical systems", *IEEE MEMS Workshop*, 1990, 113–114.

[14] Conway, L. *et al.*, *Introduction to VLSI Systems*, Sec. 4.5, Addison-Wesley, 1980.

[15] Billingsley, G.C., "Program reference for KIC", Memo No. UCB/ERL M83/62, U.C. Berkeley, October 1983.

[16] Ho, C.P. *et al.*, "VLSI process modeling SUPREM 3", *IEEE Trans. Electron. Dev.*, **ED-30**, November 1983, 1438–1452.

[17] Oldham, W.G. *et al.*, "A general simulator for VLSI lithography and etching processes: Part 1 — Application to projection lithography", *IEEE Trans. Electron. Dev.*, **ED-26**(4), April 1979, 717–722.

[18] Oldham, W.G. *et al.*, "A general simulator for VLSI lithography and etching processes: Part 2 — Application to deposition and etching", *IEEE Trans. Electron. Dev.*, **Ed-27**(8), August 1980, 1455–1559.

[19] Hibbitt, Karlsson, and Sorensen, Inc., Providence, RI.

13

MANUFACTURING OF ELECTRONIC PACKAGES BY INDUSTRIAL ROBOTS

13.1 INTRODUCTION

Electronic package assemblies consist of many electronic and non-electronic components which must be assembled in a precise fashion within a single facility. Each electronic package type follows its own process flow. For example, an insertion-mounted printed wiring board (PWB) can be built in the same facility as a surface-mounted PWB. Although different equipment and processes are used, both boards may be assembled by the same type of robots [1].

For mass produced PWB, purpose-built insertion robots are used for the package assembly. With insertion rates of up to 3000 per hour, these robots are designed to accommodate large volumes of small but uniformly shaped components. The robots are flexible and have the ability to carry out simultaneous insertions. Robots are generally seen to be applicable in the 1800 components per hour region [2]. They can therefore be faster than manual assembly methods, but can also accommodate a number of shapes and sizes, especially when fitted with multiple grippers or through interchangeable tooling. One of the factors that directly affects insertion rate is the type of component being handled. This will also affect the case with which the robot will be able to grasp and manipulate the component, and hence will determine the critical forces.

In this chapter, we analyze the assembly of electronic packages with industrial robots. Surface mounted devices related to robots are presented in Section 13.2. General assembly processes for electronic packages are described in Section 13.3. Section 13.4 covers the assembly system configurations. Section 13.5 deals with the industrial robot configurations. End effectors for the robots are discussed in Section 13.6. Finally, Section 13.7 analyzes the problems of assembly with micro-robots. The assembly processes must be designed for high throughput assemblies and hence for high volume products.

13.2 SURFACE-MOUNT TECHNOLOGY BY ROBOTS

Surface mounted devices are particularly relevant for assembly by robots. They were first used on commercial products in the late 1970s, and are increasingly used for PWB applications [2]. These devices differ from the conventional electronic components because they do not have the long wire leads which connect to the board of the PWB. The conventional component leads are inserted through drilled holes in the board and soldered to the copper. Surface mounted components, however, are soldered directly on to the board, thus removing the need of drilled holes. Consequently, a more densely populated and less expensive PWB can be obtained. In addition, such boards have higher density, greater reliability, better electrical performance, and lower costs than conventional ones.

The major significance of robots is in the capability for achieving automatic assembly. This is due to their external similarity in appearance, which enables one or only a few end effectors to handle the range of devices required to populate a PWB. The sensors and computerized counting mechanisms also allows the routing pattern to be recorded for future reference. This similarity also simplifies part presentation to the robot. Surface mounted devices are generally much smaller than their conventional counterparts. Therefore it is more difficult to handle them manually for high density assembly. Various automatic soldering techniques such as wave soldering can be used in conjunction with robot placement equipment to give fully automated board production. The micro-robotic operation is typically a pick and place operation.

Pick-and-place robots offer speed and accuracy for component placement. This is the advantage for applying robots to the packaging of surface-mount technology packages (SMT). These robots not only can perform specialized tasks efficiently, but also have additional assembly capability available from machine options that enhance SMT. Each available robot configuration has advantages and disadvantages so that, when purchase is contemplated, a careful review of the limitations is necessary. The robot selection process begins with the determination of the following requirements [3]:

- Ease of integration for device types and assembly techniques.
- Operational equipment, such as adhesive application, component electrical test verification, or alignment system.
- Operational complexity and software support.
- Quantity of device feeder mechanisms and styles available.
- Desired accuracy.
- Ease of technology upgrade.
- Safety of equipment.
- Cost.

The advantages of incorporating robotic techniques have been recognized for many years throughout industry. Some robotic surface-mount technology advantages are:

- High quality product which is also uniform in appearance.

- Reduced component stock areas resulting in higher component density.
- Enhanced process controls programmed into the insertion process.
- Reduced cost of assembly per board.

Having introduced the concept of robotics in packaging, the processes related to assembly will be introduced.

13.3 GENERAL ASSEMBLY PROCESSES

There are many levels to the assembly of electronic packages for a given technology. The final system can contain many subassemblies. Each of these subassemblies can also contain its own subassemblies, and so on. Most subassemblies are manufactured as deliverable, self-contained units and can be built at various facilities. The assembly system obtains all of these subassemblies and integrates them into the final system. The parts used within a system are commonly listed according to the subassemblies. This list starts with the final system and the major subassemblies are listed as branches connected to the final system. These subassemblies then branch out further with their own subassemblies and components.

Figure 13.1 shows the electronic package manufacturing process from the raw silicon wafer through the mask and wafer processing steps, which is sometimes referred to as front-end processing [4]. The back-end processing includes:

- Wafer dicing.
- Die attachment or die bonding.
- Wire or tape bonding.
- Package placement and sealing.

Figure 13.1 therefore shows packaging steps as a continuation of wafer processing.

The simplest subassemblies are the die structures which have been discussed extensively in the previous chapters. The die is at the lowest point of the package assembly hierarchy but does contain the most sensitive elements which must be protected during the assembly operation. The die-substrate assembly may be either hermetically sealed or protected by a polymer coating prior to placement on the printed circuit board [1].

There are three primary ways to connect the die to the package: alloy or eutectic bonding, solder attachment, and adhesive bonding. Dice that have been separated from the wafer may be in waffle pack form for manual bonding to a package or hybrid circuit. Dice may also be mounted on a releasable plastic substrate which keeps them in an uniform position after the dicing operations for attachment with pick-and-place robots. A heated vacuum tip picks up the die and moves it to the package, which is on a heated platen [4].

Assembly processes for hybrid circuits are generally applied to both thin-film and thickfilm interconnect substrates. Differences lie primarily in the choice of attachment materials, interconnect techniques, and sealing methods. The assembly sequence selected depends on whether the substrate is attached to the inside of the package with alloy (solder) or with epoxy. If alloy attached, the substrate should be connected to the inside of the package first, since this is the highest-temperature operation [3].

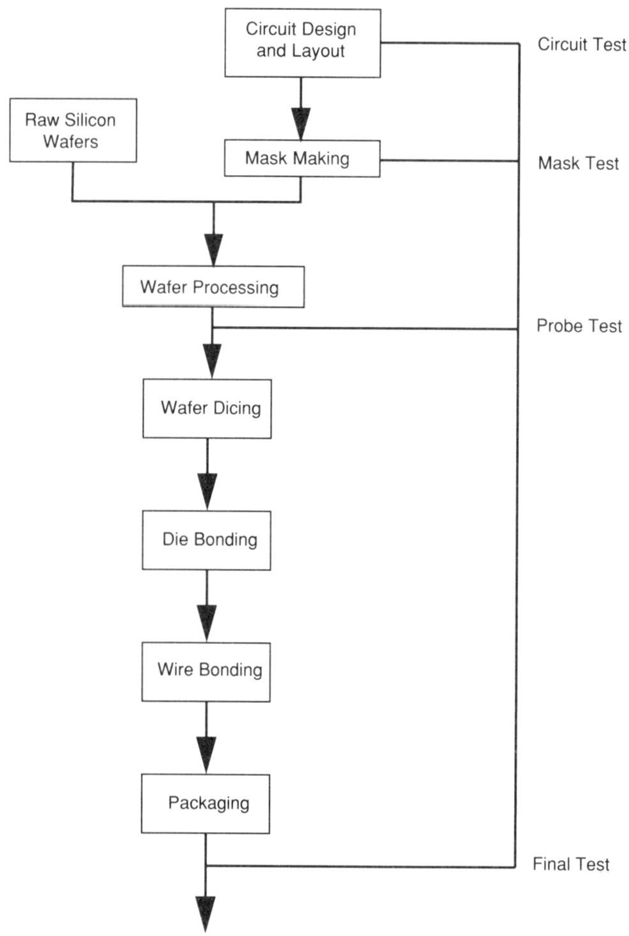

Figure 13.1 Electronic package manufacturing process steps [4].

Subsequent die and component attachments should be performed in order of decreasing processing temperatures. Thus, if some components are to be attached with soft solder and others with epoxy, the higher temperature solder attachments should be completed first. It is most desirable and less costly to use a single attachment process, preferably epoxy adhesive bonding which is the least expensive and easiest for rework.

If epoxy is used as the attachment material, two sequences may be followed:

- The substrate may be attached to the inside of the electronic package first, then the die attached to the substrate.
- The die may be attached to the substrate first, then the substrate to the electronic package.

After attachment, the dice are interconnected, generally by bonding gold or aluminum wire, to the device pads and to the metal bonding sites on the substrate.

The last step involves connecting the outer bonding pads of the substrate to the in-board pads or pins of the package. Throughout the assembly sequence, numerous cleaning steps are required. Cleaning should be thorough to remove all contaminants which may cause degradation of die, wire bonds, or substrate metallization. These processes are accomplished by the automatic transfer of die cassettes from process centers to process centers.

13.4 ASSEMBLY SYSTEM CONFIGURATIONS

There are three basic configurations of assembly systems for electronic packages: a single workstation assembly system, series assembly systems, and parallel assembly systems [5]. Combinations of these three basic types are also possible. For example, it is sometimes beneficial to design a series configuration with certain stations in parallel.

13.4.1 Single workstation assembly systems

In single workstation assembly systems, all essential parts for completing the desired assembly of electronic packages are presented to the robot at a single workstation. All of the joining tasks for the assembly are completed at the single robotic workstation. A single robotic workstation assembly system is typically used for small and medium quantities of electronic packages, in which there is a limited number of assembly tasks and parts to be handled. This means that the electronic package is of low to medium complexity.

The single workstation assembly system is not very fast, even for assemblies of relatively low complexity. The production rate of the cell is inversely proportional to the number of parts in the assembly. The more parts that must be assembled by the system, the less reliable it will be. A large number of distinct parts will present more difficult problems to the designer of the robot gripper.

In the case of the gripper, either the design must be more sophisticated to handle the variety of different electronic devices; the gripper must be instrumented with sensors to accommodate the electronic device differences; or a tool-changing mechanism must be devised for exchanging the various specialized grippers and other end effector tooling. The single workstation assembly system has one attractive merit: it requires the least cost for small quantities of electronic packages. In addition, the CAD/CAM software may be used as an application tool in such a system.

13.4.2 Series assembly systems

This configuration is used in assembly situations with medium and large quantities of electronic packages. The assembly line consists of a series of robotic work-stations, at which only a few operations are performed on the electronic package. Each robotic station works on a different package, the packages gradually building up as they move down the line.

Typical hardware used for a series assembly system includes intermittent transfer systems and non-synchronous transfer systems [5]. In both cases, the work flows in a straight line. Another possible configuration for a series robotic assembly system is the dial indexing machine. This consists of a round table with a number of workstations evenly spaced around the periphery. At the end of each work cycle, the table indexes the work to the next station. The phases of the design and manufacturing processes serve to define the entire configuration of the series assembly system.

13.4.3 Parallel assembly systems

The parallel robotic assembly system for electronic packages is shown in Figure 13.2. In essence, the assembly work can take either one of the two or more routes to have the same operations performed. The parallel robotic workstations would normally be considered under two conditions. First is the situation where electronic package cycle times, at a particular workstation or group of workstations, are too long to keep up with the other sections of the line. The other robotic stations are forced to wait for the slow workstations. In this case, the use of two parallel stations effectively reduces the cycle time, and permits the workload to be more evenly distributed among the workstations.

The second reason for considering a parallel configuration is when reliability of a certain station or group of stations is a concern. Production requirements may be such that a shutdown of the line cannot be tolerated. For example, it is likely that certain robotic stations will have greater chance of breakdown, and those would be provided with a duplicate station in parallel. This would improve the proportion of uptime of the line. However, the duplication of equipment adds cost to the system.

The workstation concept to electronic assembly seems to form the basis of the next generation of robotic systems. The style of distributed commands and computing offered by the control assembly workstations is preferred over that of control computing offered by a mainframe. The argument for such a preference is typically environments provided by the latter systems, the networking potential of workstations, and low entry costs.

13.4.4 Other assembly system configurations

In addition to the three traditional assembly configurations presented above, it is possible to develop other systems that take advantage of the programmability and

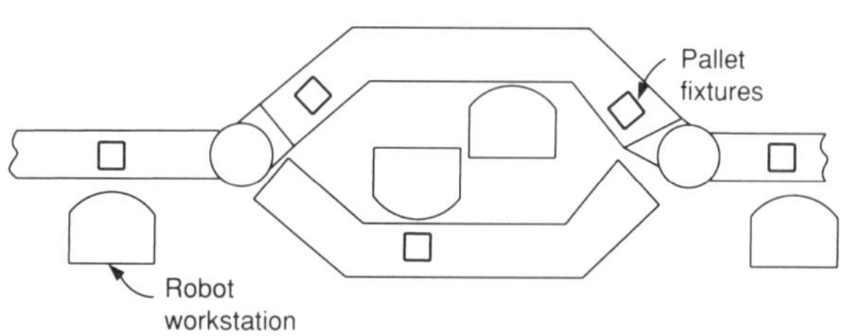

Figure 13.2 Parallel workstations on an assembly system [5].

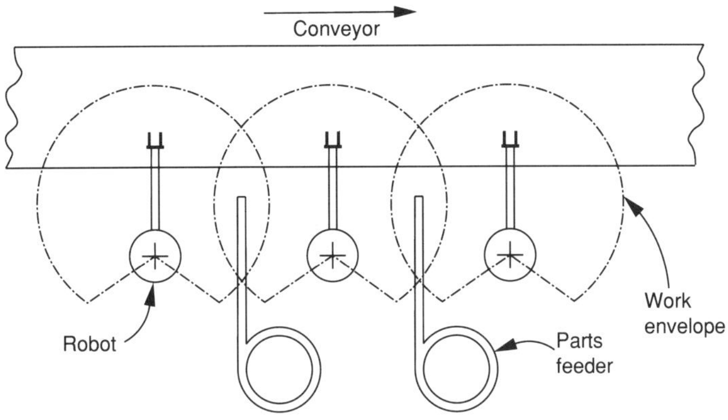

Figure 13.3 Series layout with overlapping working envelopes [5].

adaptability of robots. Based on the previous configurations, increasing the number of robots in the system increases throughput but decreases reliability. A desirable objective in the design of any robotic production system is to provide the greatest throughput with the highest reliability at the lowest possible cost.

An advantage that robotic systems have over most other forms of automation is their capability to adapt to changing conditions in the workplace which may be introduced by the variability of the die size. If required, the robot can alter its programmed cycle in response to the changes. One cell concept which utilizes the adaptability of the robot is the series layout with overlapping working envelopes, shown in Figure 13.3 [5]. This is basically a series configuration, in which the robot working envelopes overlap each other significantly. The tooling of each robot is configured so that they can perform the tasks of the robots in adjacent stations. The reason for this redundancy is to allow the system to tolerate failures of some of the workstations and to continue to operate, although at a reduced production rate. When a station failure occurs, the workcell controller commands the robots on either side of the failed station to share its work cycle, such as the pick and place of passive components in addition to active transistors.

Each robot on the line would be programmed to perform its primary tasks. In addition, when a failure occurs at an adjacent station, a secondary program would be downloaded which instructs it to perform a portion of the work of the broken robot next to it. Although the throughout of the system would be decreased, the line would remain operational in this mode long enough to repair the failure. The advantage of this concept is that it provides the redundancy of the parallel system without the additional expense.

13.5 INDUSTRIAL ROBOT CONFIGURATIONS

Although industrial robots take many different physical forms, there is no absolute best configuration for the assembly of electronic packages. The best configuration will be determined by the electronic packages concerned. Different configurations have different advantages which make them more suitable for certain assembly

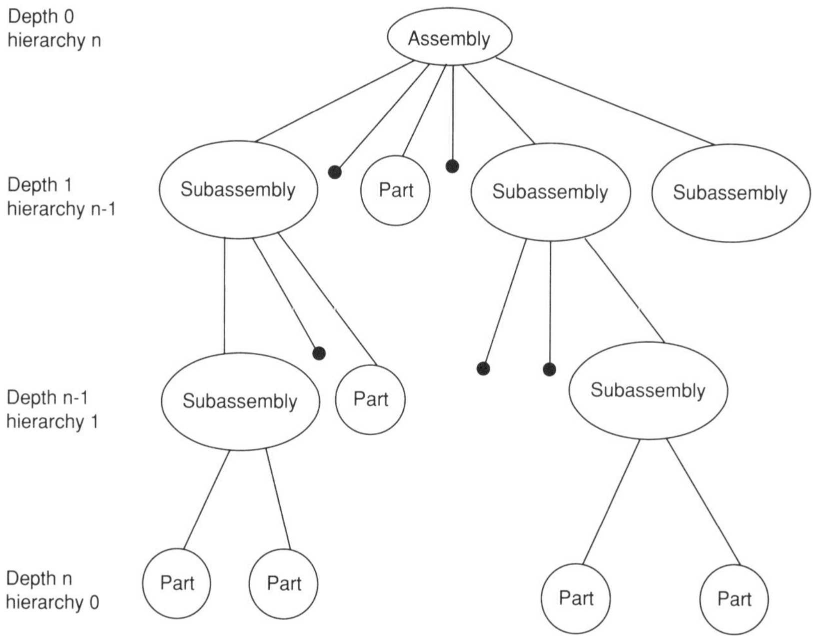

Figure 13.4 Generalized assembly tree.

tasks. A detailed study should always be conducted, since the result could indicate which robot is inappropriate for a particular application. Such result is very valid and helpful as it may prevent large cost on an ineffective resource item.

Industrial robots vary greatly in the number and types of movements they possess, their geometrical arrangement, and their physical make-up. These factors together determine the final design of the robot. The configuration of a robot determines to a large extent those tasks to which it is best suited. A robot configuration typically comprises a number of rigid, jointed members that together form the manipulator or arm. These members are made from a material strong enough to withstand a variety of applied forces and loads. There are two basic industrial robot configurations that may be used for the assembly of electronic packages, described as follows: articulated and SCARA configuration.

The two robot configurations are part of an overall assembly tree for the electronic package. The assembly tree shown in Figure 13.4 shows that the assembly procedure is divided into several subassemblies and each subassembly at depth $(n - 1)$ is composed of various parts. The robot configurations are designed to assemble parts at depth $(n - 1)$. The levels of the tree represent individual parts, its nodes represent subassemblies and its root represents the assembly itself. Figure 13.5 shows a printed wire board assembly, consisting of a microprocessor, a coil and a DC power supply forming a microcomputer system board. The tree shown in Figure 13.6 represents an assembly sequence by which the microcomputer assembly can be produced. The assembly tree is not unique as it is possible to generate other valid assembly sequences.

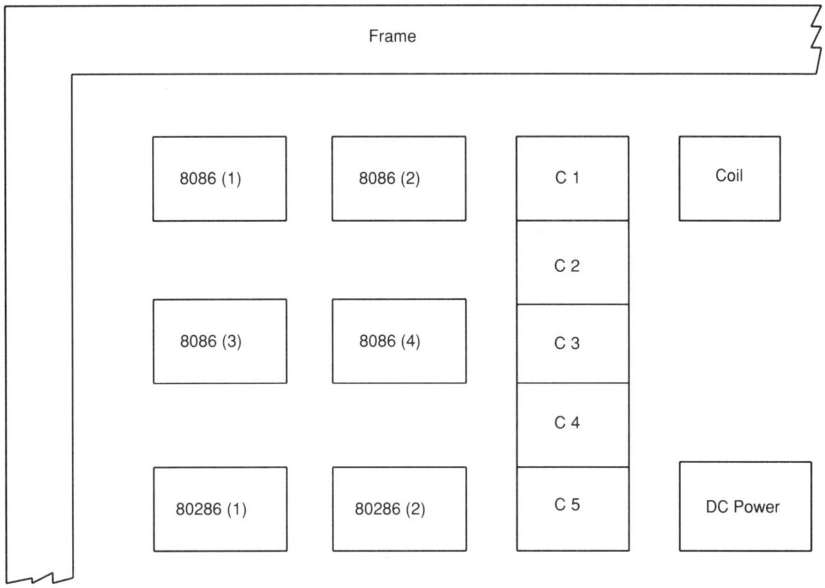

Figure 13.5 Microcomputer system board assembled by micro-robotic operations.

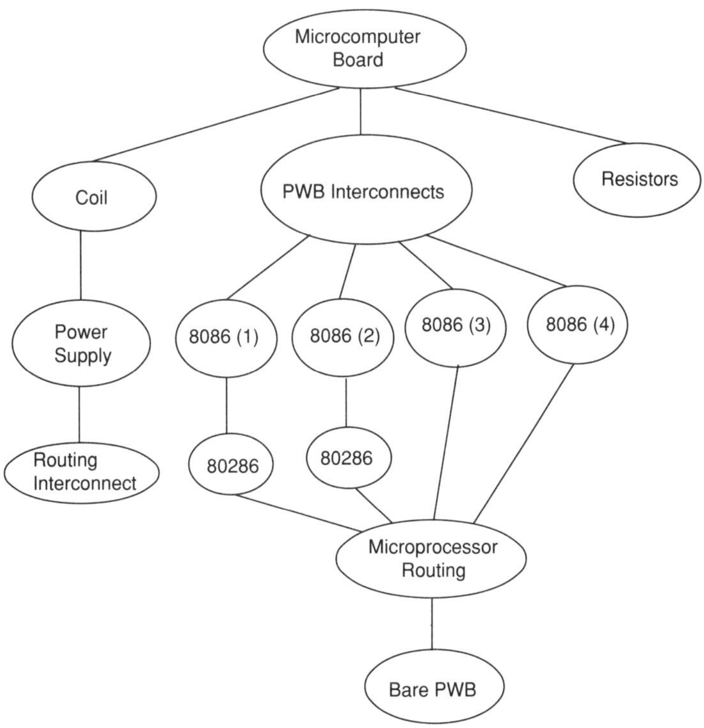

Figure 13.6 Assembly tree of the microcomputer board.

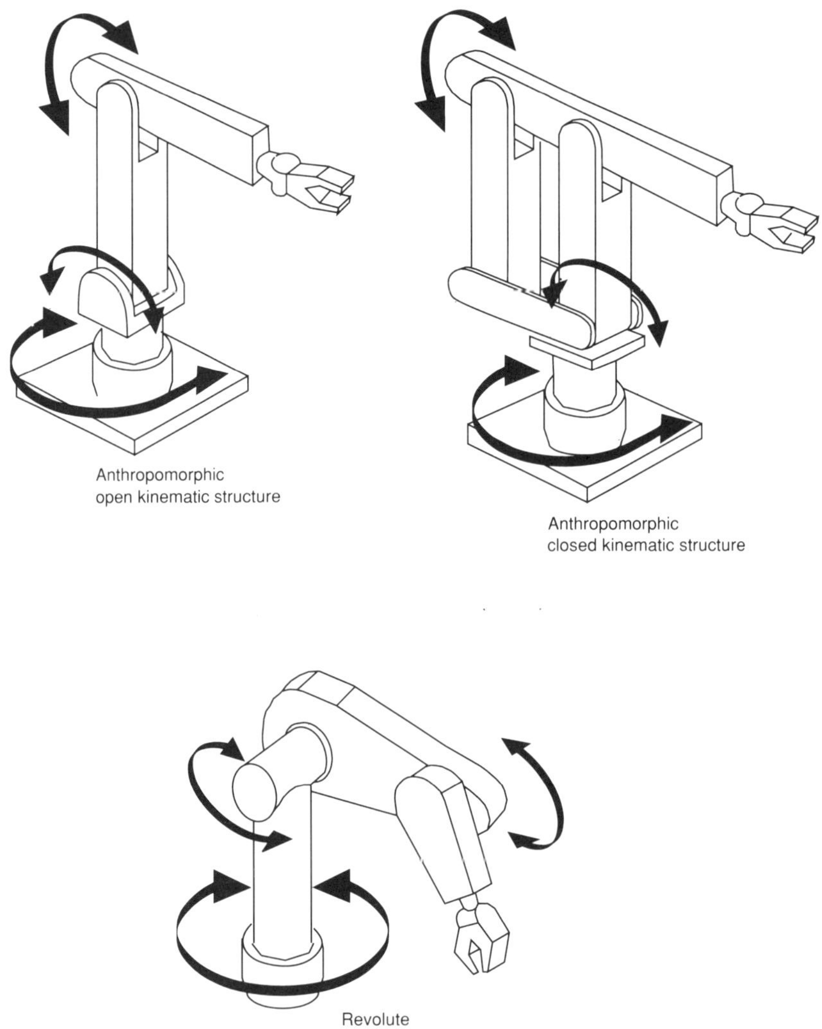

Figure 13.7 Three implementations of the articulated robot configuration [6].

13.5.1 Articulated or revolute configuration

The articulated configuration consists of a number of rigid arms connected by rotary joints. Rotary movement around the base is also provided. It is sometimes referred to as a jointed-arm configuration because it resembles the movements of the human body. The reach is determined by the length of each arm member in the configuration. Certain configurations may have either an open kinematic structure or a closed kinematic structure. These are shown in Figure 13.7. Since it is not easy to visualize three-dimensional free-form movement, programming the arm is often troublesome. Particular advantages of this configuration include [6]:

- Ability to reach over obstructions.
- Easy access to front, sides, rear, and overhead.
- Large reach for small floor area.
- Slim design allowing easy integration into restricted workplace layouts.
- Fast operation due to rotary joints.
- Ability to traverse complex continuous paths

The articulate configuration must be programmed so that it transforms the coordinates of the parts to be assembled from its local coordinate system to the global coordinate system of the PWB. This operation requires computer interfacing through a transformation matrix achieved by specifying the location of the local coordinate system of a part relative to the assembly global coordinate system and by forcing the orientation of the local system to coincide with the orientation of the proper WCS (working coordinate system).

13.5.2 SCARA configuration

The selective compliance assembly robot arm (SCARA) is a combination of the cylindrical configuration and the revolute configuration operating in the horizontal plane. A three-linked arm with two rotary joints provides movements in the horizontal plane. Vertical movement is provided at the end of the arm. A typical SCARA robot configuration is shown in Figure 13.8. Particular advantages of this configuration include [6]:

- Extremely good maneuverability and access within its programmable area.
- Fast operation.
- High accuracy.
- Relatively high payload capacity due to stiff structure in the vertical direction.

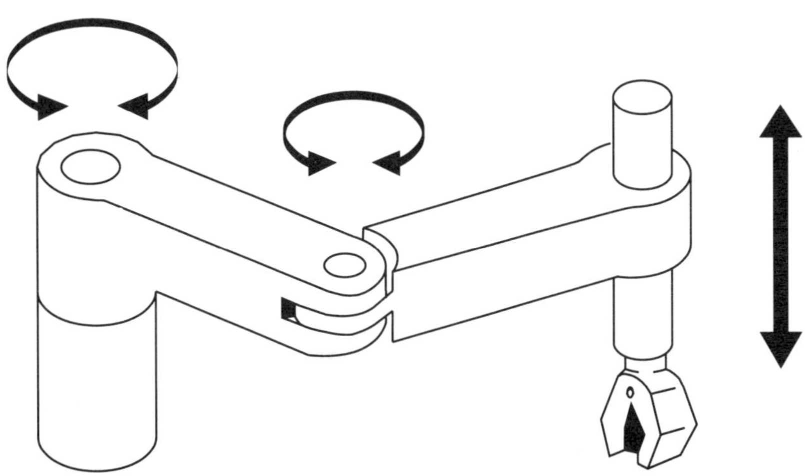

Figure 13.8 A typical SCARA robot configuration [6].

The SCARA configuration was developed primarily for assembly type operations. This configuration can be used as the pick-and-place robot for the assembly of electronic packages. Assembly tasks predominantly require movement in the horizontal plane coupled with simple vertical movement for picking, placing, and insertion operations.

13.5.3 Working coordinate system (WCS) for robotic assembling

The WCS is defined relative to the assembly global coordinate system and forms the basis for assembly motions in order to achieve robotic assembly. The proper WCS used to merge an electronic component into the board assembly can be defined such that its XY plane coincides with the XY plane of the part MCS (model coordinate system). In effect the transformation matrix relates the part MCS to the host MCS (assembly). This method of merging parts into their assemblies is commonly used in various existing CAD/CAM systems. The transformation equation can be written as

$$P = [T]P_W$$

where P is the position vector of a point relative to the MCS and P_W is the vector of a point relative to the active WCS. Each vector is given by

$$P = [xyz]^T$$

The matrix $[T]$ is the homogeneous transformation matrix and is given by

$$[T] = \begin{bmatrix} t_{11} & t_{12} & t_{13} & : & t_{14} \\ t_{21} & t_{22} & t_{23} & : & t_{24} \\ t_{31} & t_{32} & t_{33} & : & t_{34} \\ .. & .. & .. & + & .. \\ 0 & 0 & 0 & : & 1 \end{bmatrix} = \begin{bmatrix} & {}^M_W[R] & & : & {}^M P_{W,\mathrm{orig}} \\ .. & .. & .. & + & .. \\ 0 & 0 & 0 & : & 1 \end{bmatrix}$$

where ${}^M_W[R]$ is the rotation matrix that defines the orientation of the WCS relative to the MCS and ${}^M P_{W,\mathrm{orig}}$ is the position vector.

For example, if a microprocessor (B) is to be assembled on to a pre-routed PWB (A) then the basic assembly consists of two parts, A and B. The user first creates the data bases of A and B with the MCS of each part. We may take part A as the base part or the host and merge part B into it. To merge the case of B on top of A, the X, Y, Z, WCS is defined by the user, allowing for the transformation matrices to be derived. The other alternative is to represent an assembly based on specified spatial relationships between its individual parts as mating conditions (faces and centerlines).

13.6 END EFFECTORS

In order for the robot to perform useful work and to conform to the definition of an industrial robot, it must have some way of manipulating tools and materials. This is done by means of an appendage attached to the end of the arm, i.e. an end effector. The motion of the tool may be programmed so that the routing geometry may be simulated.

13.6.1 Description of end effectors

The end effector may adopt an infinite variety of configurations, but basically it can be classified into two district types: (a) process end effectors, and (b) grippers which can range in sophistication from simple two fingered pneumatic grippers to complex intelligent hands with their own sensory systems and microprocessors [7]. The distinction between the process end effector and the gripper may appear to blur in some instances, e.g. when interchangeable tooling is used with end of arm adaptors, or special purpose vacuum grippers are employed for workpiece manipulation. However, the method of differentiation here is that, if the end effector is a tool directly used to change the state of the workpiece, then it is a process type. If it is used to grasp and directly manipulate or transport a workpiece, material, or tool, then it will be referred to as a gripper [2].

As robot end effectors are considered, they are found to be usually much less general purpose and often custom designed. Process types will naturally be specific to the job at hand, and if it is of the gripper type it will often be specially designed for the component being manipulated. General purpose grippers are widely available but will usually be purchased with a range of movement and grasp to suit a specific job. Thus when a robot is purchased, the end effector will come under the classification of additional equipment and will not be included in the basic robot cost. The end effector costs are usually relatively low. It may cost up to 20% of the price of the robot if it requires specially design, special complexity, or is a particularly expensive process type.

The end effector will be bolted on to the robot wrist, which will be provided with a faceplate for this purpose. Because there is little standardization in this area, an adaptor plate to fit between the wrist faceplate and the end effector may have to be made. It is possible in some cases to have compatibility. There are, in fact, designs that enable the robot to change its own end effector using couplings which allow for pneumatic, hydraulic, and electrical connections between the wrist and effector. The end effector for handling components was originally in the form of a multiple rotary indexing gripper, the indexing being effected by an integral stepper motor. In this particular situation, the user might decide that it was more efficient to use the previously mentioned exchangeable tool system, which increased accuracy, but also increased cycle time.

End effectors may be broadly classified as grippers, tools, or sensors, depending on their design and intended purpose [6]. Grippers represent, by far, the majority of end effector types. They are characterized by being able to grasp, manipulate, transport, and position an object. Although they do not necessarily have to be mechanical in operation, many grippers do incorporate tactile sensor elements to feed back information regarding the operation or status of the gripper, or to discriminate between different components.

Mechanical grippers are the most common form of gripper devices for the assembly of electronic packages. The majority of designs are modelled on clamping mechanisms. Figure 13.9 shows two different configurations of parallel motion mechanical grippers. Figure 13.10 shows three different configurations of pivoted jaw mechanical grippers.

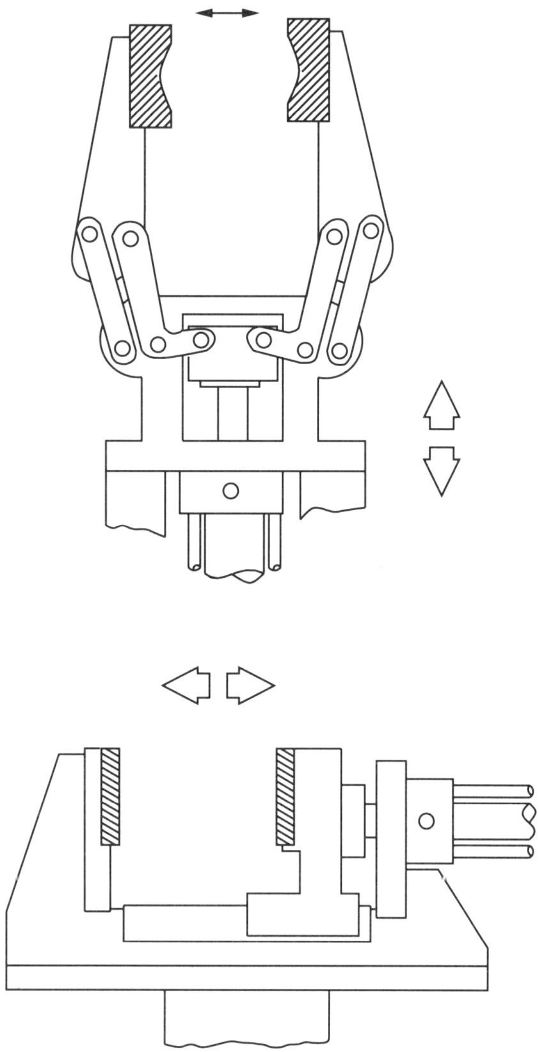

Figure 13.9 Two different configurations of parallel motion mechanical grippers [6].

13.6.2 End effector path generation

Critical to the successful assembly of electronic components on PWBs is the motion of the end effectors in a pick-and-place operation utilizing CAD routing databases. These are recognition of the PWB surfaces (feature recognition), total path generation, tool path verification, and height recognition or collision detection. Automatic feature recognition for electronic component manufacturing is a crucial step in integrating robotics with CAD/CAM. Collision detection is concerned with finding if the robotic arm may collide with other pre-assembled components. One method to solve the collision problem is to find the swept volume generated by the arm motion in space and intersect it with any of the models of the packaged components.

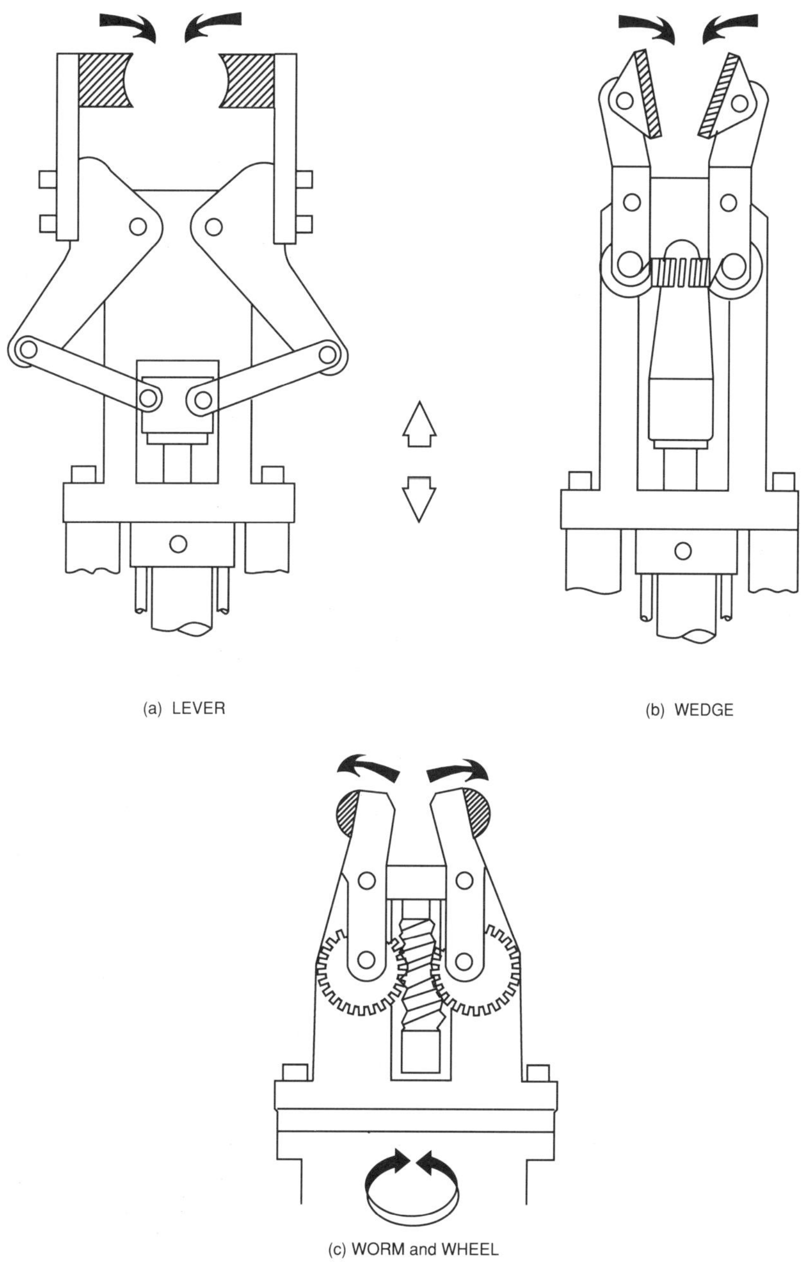

Figure 13.10 Three different configurations of pivoted jaw mechanical grippers [6].

If the intersection is not null, collision occurs and the components would have to be repositioned.

The motion of the robotic and effector is programmed using advanced NC programming languages such as APT which employ surface geometry techniques.

Robotic motion programming based on APT uses only unoriented, unbound single surfaces and as such is ideal for 3D module assembly. In order to place an electronic component on a surface parametrically, the arm location points are generated using the surface definition and the stated robot parameters. These points are stored within a memory file known as the CL-file. The robotic arm tip is directed to move in straight line segments to each point. For example, if we choose a step size of 0.1 in both the u and v directions of a parametric surface, and assuming both u and v have a range from 0 to 1, 11 points are created along each of the u and v directions. Thus 121 points are needed to define the surface, and these points include the location of bond regions for robotically placed electronic components.

The robotic arm path generation algorithm for the placement of electronic components on a board may be written as follows:

(1) Select the robotic motion direction of the PWB surface. Assuming the u direction is selected, set $v = 0$. The surface representation is assumed to include the tolerance for placement.

(2) Set the first placement point $u_C = 0$ and evaluate the position, unit tangent and unit normal vectors and the radius of curvature at $u_A = 0$ (the initial point).

(3) Evaluate an extra point D at a parameter value u_D.

(4) Select the variable w.

(5) Update the parameter value u.

(6) If $u = u_{\max}$, then let $u = u_{\max}$; otherwise go to step 7.

(7) Evaluate the position, unit tangent and unit normal vectors and radius of curvature at the new u value.

(8) Calculate the end effector path point at u.

(9) If $u < u_{\max}$, set $u_C = u_D$ and go to step 3; otherwise go to step 10.

(10) If $v < v_{\max}$, set $v = v + D_v$ and go to step 2; otherwise go to step 11.

(11) Stop.

The above algorithm is based on the fact that the surface patch to be bonded to occupies the range described by $0 \leq u \leq u_{\max}$ and $0 \leq v \leq v_{\max}$. Further enhancements and extensions to the above algorithm are possible such as the inclusion of collision detection mechanisms.

13.7 ANALYSIS OF THE ASSEMBLY PROCESSES

In order to establish transferable processes in the assembly of electronic packages, an inquiry has been carried out and selected assembly problems examined on the basis of workplace analyses shown in Table 13.1. The aim of the inquiry, conducted in an industrial area with emphasis on precision engineering, was to narrow down the product range with respect to quantity, variety of types and models, batch sizes, part characteristics, extent of assembly and obstructions to automation [8].

The most significant obstruction for the assembly, in addition to small batch sizes and a wide variety of types and models, is the minute size of parts with

Table 13.1 Method and aim of the analysis.

	Analysis		
	Part range	Assembly job	Tolerances
Objects	Insertion parts Basic parts	Assembly processes Assembly parameters	Parts Assembly system
Results	Volume classes Geometry classes Joint position Cross-section Materials	Joint directions Joint lengths Insertion plays Joint forces Position deviations	Insertion part tolerances Basic part tolerances Handling device Loading equipment
Aim	Determination of weaknesses of assembly		

fitting tolerances. To determine the requirements for tolerance compensation, a tolerance study needs to be conducted and the tolerance chain for an assembly system with industrial robots needs to be established. In the assembly of electronic packages, all tolerances which lead to position deviations must be considered. The part and assembly system tolerances result in a chain in which all tolerances and deviations are added. The tolerance chain relevant to the assembly of electronic packages is based on measured values, manufacturer specifications and experience.

A substantial part of the tolerance chain consists with the components, handling systems, and loading equipment, as well as insertion part tolerances. The position deviations caused by the assembly system depend strongly on the choice and type of equipment. The position deviations in the fitting cross-section plane are decisive for the success of the fitting procedure and the requirements of a tolerance compensation. External influences, such as temperature, vibration, and air currents, must also be taken into consideration in assembly tasks. In a typical ROM based CNC system, all of the tolerance information may be stored as a section of the part program.

The assembly part program is a combination of robotic tool codes, specific instructions and tolerances. Since part programmers usually prepare programs for NC robotic tools (numerically controlled), they must have knowledge of all tolerance requirements and manufacturing requirements.

13.7.1 Description of robotic tools for electronic assemblies

Figure 13.11 shows a schematic of an NC robotic assembly tool which contains the MCU (machine control unit) and the machine tool itself. The MCU is the brain of the effector, it reads the part program and controls the machine tool operations. After reading the part program, the MCU decodes it to provide commands and decoded instructions to the various control loops of the robot's axes of motion. The MCU contains a data processing unit (DPU) which reads and decodes the part program statements and provides data to the CLU (control loop unit). The CLU receives the data from the DPU and converts it to control signals. For instance in

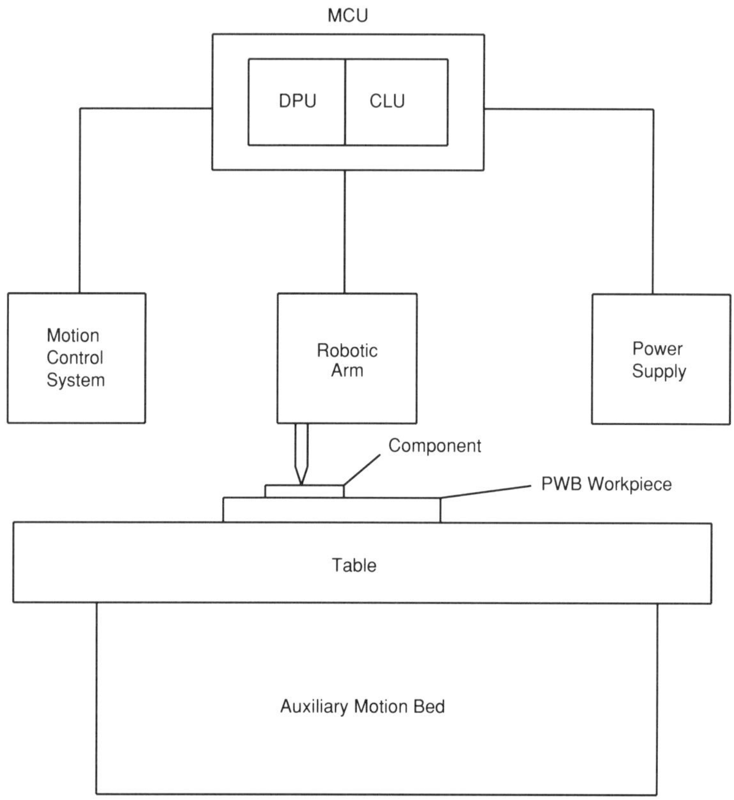

Figure 13.11 Schematic of a NC robotic tool for electronic package assembly.

a 2-axis machine motion, the Z axis control plane is parallel to the XY plane. In multiple axis tools, the tool position is simultaneously controlled in X, Y, and Z directions.

13.8 SUMMARY

Assembly tasks occupy 53% of the time and 22% of labor costs in manufacturing an electronic product [9]. Assembly requires skill in both selecting and orienting parts and in fitting joint components, sometimes using serpentine motions to align the parts. Joining parts into a composite involves welding, sealing, or fitting together.

Automated assembly involves the use of robots to select and assemble components, and must also be matched with a system of designing components in order to ensure that they can be assembled by robots. They offer flexibility in assembly by being able to transition from small batches to millions of parts. This flexibility makes them ideal for advanced manufacturing facilities utilizing just-in-time (JIT) concepts.

REFERENCES

[1] Pecht, M., *Handbook of Electronic Package Design*, Marcel Dekker, New York (1991).
[2] Mair, G.M., *Industrial Robotics*, Prentice-Hall, New Jersey (1988).
[3] Harper, C.A., *Electronic Packaging and Interconnection Handbook*, McGraw-Hill, New York (1991).
[4] Giacomo, J.D., *VLSI Handbook*, McGraw-Hill, New York (1989).
[5] Groover, M.P., Weiss, M., Nagel, R.N., and Odrey, N.G., *Industrial Robotics: Technology, Programming, and Applications*, McGraw-Hill, New York (1986).
[6] Jones, B.L., *Elements of Industrial Robotics*, Pitman, London (1987).
[7] Lammineur, P., and Cornillie, O., *Industrial Robots*, Pergamon Press, New York (1984).
[8] Schweigert, U., "Precision assembly with industrial robots," *Industrial Robot*, **19**(2), 1992, 28–34.
[9] Bedworth, D.D., Henderson, M.R., and Wolfe, P.M., *Computer-Integrated Design and Manufacturing*, McGraw-Hill, New York (1991).

14

MANUFACTURING OF MICROWAVE SUBSTRATES FOR PACKAGING OF MONOLITHIC MICROWAVE CIRCUITS

14.1 INTRODUCTION

With the advancement of the microwave microelectronics industry, the requirement for faster, denser, thinner, lighter, reliable and low-cost packages which can dissipate heat, is growing rapidly. To meet these growing demands, thick film hybrid ceramics are used, since they are compatible with multilayer structures. The number of dielectric layers in a multilayer substrate can vary between 5 and 40, depending on the application. Each layer is electrically connected to its adjacent layers through vias. Depending on the material constituents and firing temperature of the substrates, the multilayer ceramic substrates can be classified as hightemperature co-fired ceramic (HTCC) and low-temperature co-fired ceramic (LTCC) systems [1].

If the ratio of ceramic to glass is high (9/1), the green substrate layer (a substrate in the unfired state is called green tape) can only be sintered at firing temperatures of approximately 1500 °C. In such a materials system, the dielectric consists of glass fillers in a ceramic matrix and is referred to as a high-temperature co-fired ceramic. Consequently, thick-film pastes which are co-fired with the substrate layer also have to withstand high temperatures. Usually, tungsten and molybdenum-manganese are used as the metallization materials.

Material systems with a much lower fraction of ceramic (3/1) as the dielectric, and which can be sintered at much lower firing temperatures (850–900 °C), are generally referred to as low-temperature co-fired ceramics (LTCC), or multilayer glass-ceramics (MGC). With lower temperatures, low resistivity precious metal conductors, including gold and silver, can be used. These offer significant performance advantages, compared to the refractory metals in high-temperature co-fired systems [2]. Alumina has a high dielectric constant (ε = 9–10) which causes delay in the switching speed, since this varies proportionally. As the demand for a new packaging technology to meet high frequency and high interconnect density requirements increases, new glass-ceramic materials with low dielectric constant are being developed. Low resistivity conductors on glass-ceramic substrates may be obtained by controlling the silver and palladium particle shape. The multilayer glass-ceramic (MGC) substrates, which are sintered at about 900 °C in air, have four characteristics: (i) the new MGC substrate has a small propagation delay, due to its low dielectric constant material; (ii) low electrical resistivity and low cost is realized by using a silver-palladium conductor system; (iii) improved electrical design may be used to control characteristic impedance and crosstalk coupling noise; and (iv) a highly accurate fabrication process has been developed to take advantage of the new materials [3]. The ability to achieve very high circuit densities in a relatively small area is an appealing feature of low-temperature co-fired ceramics [4]. The typical constituents of high-and low-temperature co-fired ceramic systems are given in Table 14.1. Two basic processes related benefits of LTCC are that it can be fired in a modified thick film furnace, and gold plating is not required.

In LTCC, even though gold conductors may be used for interconnections, silver-palladium dominates since it provides low resistivity at a cost lower than gold. The dielectric materials for the MGC substrates are 55 wt % Al_2O_3–45 wt % lead borosilicate glass (G-C(I)), 45 wt % cordierite ($2MgO{\cdot}Al_2O_3{\cdot}5SiO_2$)–55 wt % borosilicate glass (G-C(II)), and 35 wt % SiO_2–65 wt % borosilicate glass (G-C(III)). These have controlled particle size, and dielectric constants of 7.8, 5.0 and 3.9, respectively. Low conductor resistivity achieved with Ag-Pd reduces the voltage drop along the line length. The line resistance increases with decreasing line width and thickness. The higher line resistances produce voltage drops and localized heating, which can affect device switching behavior. The MGC substrate is advantageous in achieving high wiring density, due to the dimensional control. It is therefore compatible with high speed signal propagation. A reduction in the processing temperature to around 900 °C facilitates the use of many low-resistivity metals such as Ag–Pd, Ag, Au, or Cu [5–8].

Table 14.1 Typical constituents of high- and low-temperature co-fired ceramic systems.

	High-temperature co-fired	Low-temperature co-fired
Dielectric		
Matrix	Ceramic (96 wt% alumina)	Glass/crystallizable glass
Filler	Glass (4 wt% glass)	30 to 50 wt% ceramic
Organic binder	Poly vinyl butyryl	Acrylates/methacrylates
Solvent	Toluene	Xylene
Plasticizer	Polyethylene glycol	Dicotyl phthalate
Metallization	Tungsten, polymanganese	Ag, Au, Cu, Ag-Pd

14.2 MANUFACTURING PROCESSES FOR LTCC

The various steps involved in the manufacture of low temperature co-fired ceramics are discussed below (Figure 14.1).

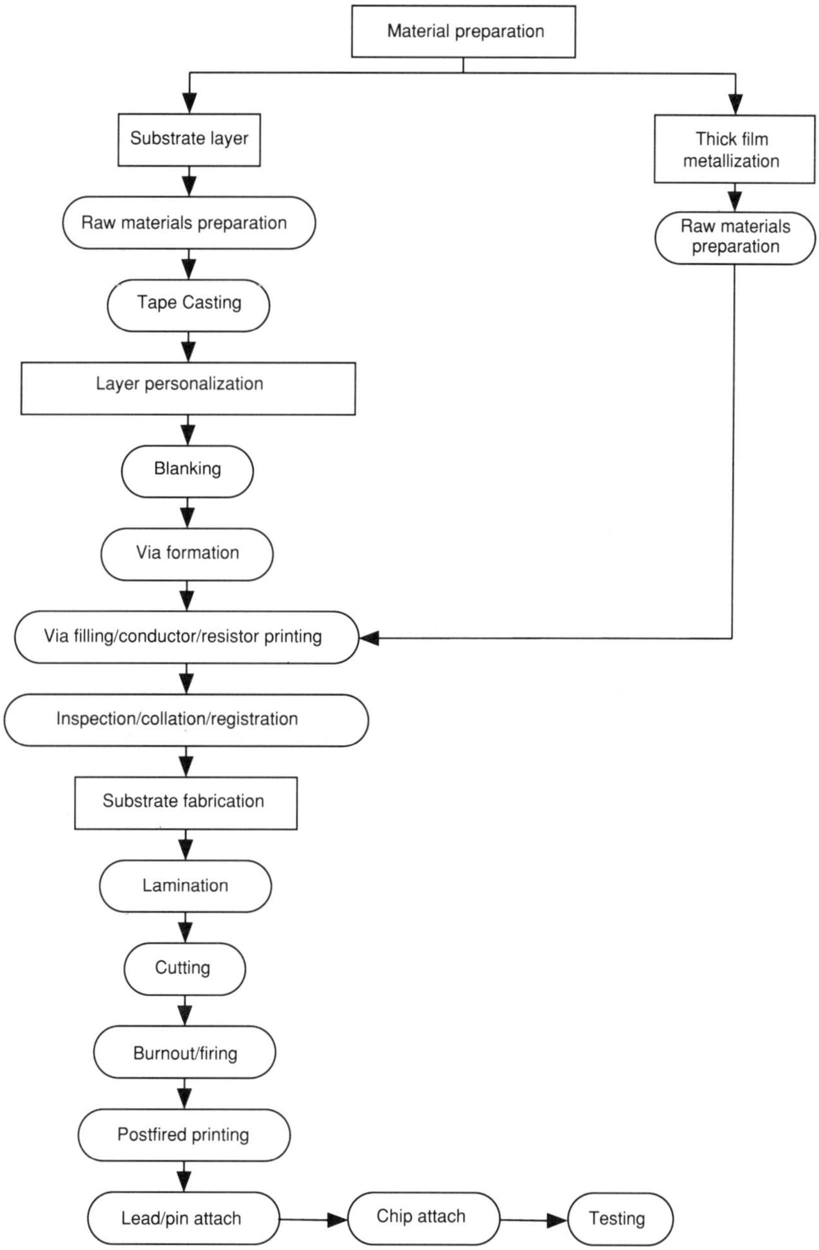

Figure 14.1 Process flow for low temperature co-fired ceramic.

14.2.1 Material preparation

The initial material system consists of a dielectric in the form of a green tape. Subsequent processing involves the preparation of the dielectric substrate layer and thick film paste to form the conductor traces, via fills, and passive components such as resistors and capacitors. The raw materials for the substrate layers consist of a mixture of organic and inorganic constituents. The inorganic components are various ratios of ceramic and glass powders [1]. For high temperature co-fired processes, the inorganic components consist of a blend of alpha-alumina and borosilicate glass (9/10), with controlled particle size fabricated by various types of jet milling operations. This ensures homogeneity and uniformity of the required particle size. Inorganic components for low temperature co-fired processes are similar, but have lower ratios of ceramic to glass (3/1). The choice of ceramic depends on the coefficient of thermal expansion (CTE) match desired [1]. For low CTE, alumina, quartz, calcium zirconate and forsterite are preferred, versus fused silica, mullite, cordierite, and zirconia for high CTEs. The amount of glass composition depends upon the dielectric constant required. The glass transition point must be high enough so that the organic burnout takes place prior to densification, but low enough to ensure sintering, in order to achieve a high density. Hence, the difference between the deformation temperature and the softening point of the glass is critical. A difference of 50 to 70 °C is ideal.

The organic components provide strength and flexibility to the green sheets during substrate personalization and fabrication. These consist of a polymeric binder and plasticizers dissolved in a solvent, which are added and blended in the dispersion operation, for both high and low temperature co-fired processes. The organic binders should have a low glass transition temperature, a high molecular weight for toughness and strength, good burnout characteristics, and solubility in volatile and preferably non-flammable solvents. The organic components consist of polymers, such as vinyl butyrals or alcohols, polylactones and copolymers of lower alkyl acrylates. Also, they consist of methacrylates that can be cleanly burned in air or in inert atmospheres between 300 and 400 °C. For optimum binding efficiency, it is preferable to use at least 5% binder by weight. Higher contents increase the amount of binder which must be removed during burnout, and therefore reduce the particle packaging. This results in an increase in shrinkage during firing. Plasticizers such as dioctyl phthalate and solvents such as acetone, xylene, methanol, and ethanol are often used. The solvent is chosen to obtain a complete solution of the binder and to have sufficient volatility to evaporate at low temperatures and atmospheric pressure. After all the constituents are blended, milled and homogenized, they acquire the form of a paint-like suspension. The dielectric, composed of glass and filler materials carefully combined in suitable proportions, can produce dense structures for hermetic applications, since a low glass content may lead to poor hermeticity. On the other hand, a very high glass content can lead to entrapment of organics and blistering during firing. The filler materials are selected in order to optimize the heat transfer coefficient and dielectric constant of the dielectric tape. The material system also contains conductor paste compositions for signal line and via fill printing. These are designed to shrink at the same rate as the dielectric tape, in order to prevent warpage. The materials used for inner layer and via filling conductors are Ag, Pd/Ag, and Au. The top layer consists of thick film conductors and resistors which are printed and fired after the co-firing operation. They provide excellent soldering or bonding, without the requirement for plating.

The dielectric green tape is first cast on a Mayer polyester film carrier coated to facilitate stripping. Table 14.2 lists the general properties of a typical dielectric tape whose material composition is glass and filler. The tape is cast by spreading the slip on to a moving belt, which is usually a polyester film carrier sheet, passed under doctor blades, and moved into a drying zone, where the solvents are driven off (Figure 14.2). The casting machine consists of a smooth aluminum plate on which the polyester film carrier moves. The film is supplied from a spool and enters the chamber through a slot at its end. Electrostatic brushes are mounted to remove dust particles from the polyester film carrier, before the film carrier enters the chamber. The doctor blade at the exit side of the chamber is adjusted by means of micrometer screws. The layer of slip passes through a carefully adjusted gap between the blade and film carrier. A thickness monitor, mounted beyond the doctor blade, measures the thickness of the freshly cast layer of slip. Filtered air is blown in the opposite direction, and passes through the entire length of the casting machine, as it is saturated with solvent. The concentration gradient above the tape is low, thereby allowing the solvent to redistribute itself, preventing curling of the slip as it dries.

The thickness of the cast tape depends on the doctor blade gap, belt speed, drying temperature, depth of slip behind the reservoir, slip viscosity, and tape shrinkage. Greater precision casting can be obtained by a two-bladed doctor blade (Figure 14.3). This consists of a curved bed of tempered glass to ensure good contact between the carrier film and the bed, and a two-bladed doctor blade. Shrinkage occurs in the dielectric tape during firing, with the removal of porosity.

Table 14.2 Typical green tape properties.

Thickness	0.114 mm
Width	254 mm
Carrier tape thickness	27 mm
Color	white
Density	2.08 g/cm^3
Tensile strength	0.13 kg/mm^2
Young's modulus	19.87 kg/mm^2

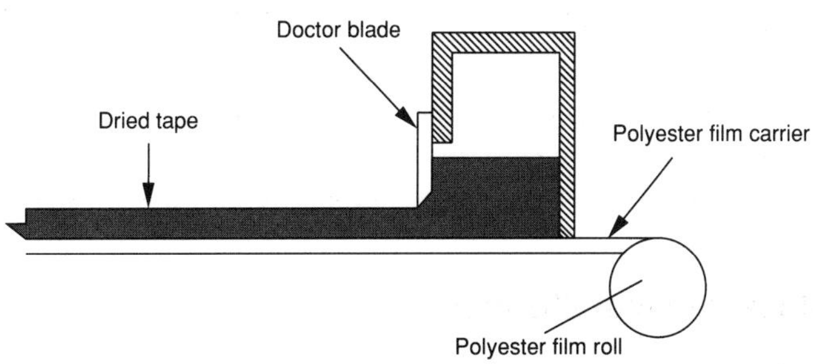

Figure 14.2 Doctor blade casing machine.

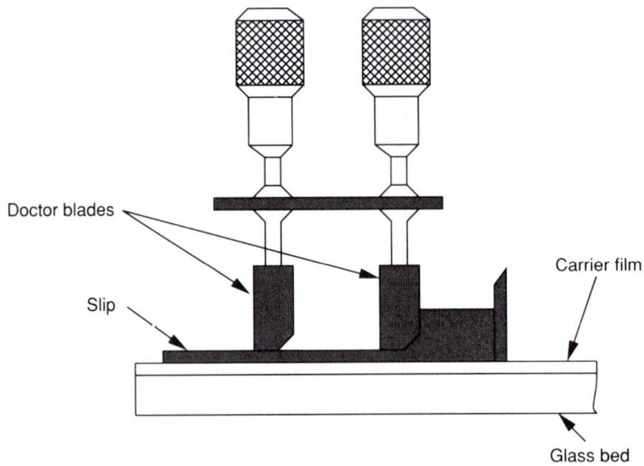

Figure 14.3 Two blade doctor blade for precision tape.

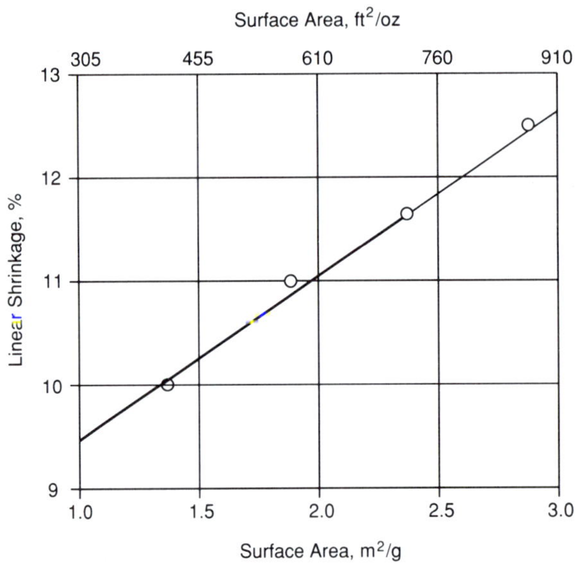

Figure 14.4 Linear shrinkage versus glass surface area.

This process depends on the pore distribution before firing, and the particle size distributions of the glass and ceramic fillers. The shrinkage variation due to changes in the surface area of the glass is shown in Figure 14.4 (graph).

14.2.2 Thick-film metallization

Thick-film metallization processing on the substrate layer is accomplished by screen printing thick film pastes, in order to form conductor traces and fill vias. A screen

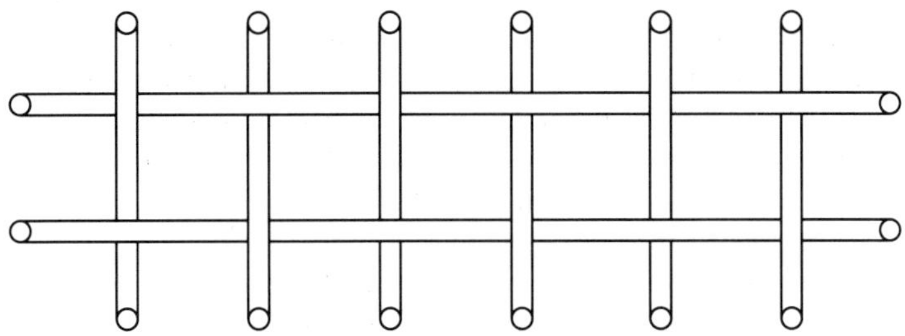

Figure 14.5 Mesh opening in a screen.

consists of a wire mesh stretched over a metal frame. The mesh comes in various diameters and mesh counts [9], i.e., number of wires contained in a linear inch of mesh. The finer the wire diameter and the higher the mesh count, the finer the obtainable print resolution. The wire used for fabricating thick-film screens is stainless steel (Figure 14.5). Thick film pastes for conductor traces and via fills consist of a conductive phase, a binder, a vehicle, and additives. The conductive phase of the paste is usually metallic, commonly used materials being Ag, Au, Cu, Ag–Pd, or Pt–Au. These constitute 50–70 wt% of the paste. The particle size distribution and particle morphology influence the final electrical and physical properties of the paste.

The binder consists of low-melting glasses, which facilitate densification and adhesion to the substrate layer. They constitute 10–20 wt% of the paste. There are primarily three methods used to bond a thick film to a substrate layer: frit bonding, flux bonding and reactive bonding. Frit bonding is the most common of these and involves the addition of 2 to 10% glass powder (frit) relative to metal. During firing, the glass wets the substrate and penetrates the metal, providing mechanical, as well as chemical bonding between the phases. For optimum bonding, the glass should have the proper surface tension and viscosity during firing. In addition, the sintering rate of the metal powders must be controlled, so that the glass can flow and bond to the metal and substrate before the metal powder sinters. Flux bonding involves the addition of 1 to 5% of an oxide, which forms a liquid phase with the substrate layer at the firing temperature. Bismuth oxide is generally used, since it forms low-melting phases which wet the metal particles and the substrate, in order to produce effective bonding. Reactive bonding utilizes 0.1 to 1% of an oxide such as CuO or CdO, in order to react with the substrate layer, and to form a Cu or Cd spinel, which promotes adhesion. The advantage of this method is the small amount of oxide required for bonding, which keeps electrical resistivity as low as possible. This is especially critical for top metallization, where a "pure" metal surface can enhance wire bonding.

The vehicle consists of 12–25 wt% of conductor paste and influences the substrate printing and burnout characteristics. As the paste is screen printed, it must exhibit a proper degree of pseudoplastic and thixotropic behavior. It should have a low viscosity at the high shear rate produced when the squeegee traverses the screen, so the paste can be squeezed through the screen mesh on to the substrate layer. This viscosity should increase rapidly to its low shear value so that the paste does not spread on the substrate layer once the screen is removed. The rheological properties of the paste are generally controlled by the proper selection of the organics. Usually a high molecular weight polymer (ethyl cellulose or an acrylic) is dissolved in a low vapor pressure solvent, such as terpineol or butyl carbitol. Vehicles also control burnout characteristics. They must burn out completely (decompose at elevated temperatures), and should not affect the inorganic constituents before their removal. Via filled pastes are similar to conductor pastes, but have a higher viscosity. Thick-film pastes are also used to make passive components, such as resistors and capacitors.

14.2.3 Layer personalization

Layer personalization includes blanking, via formation, and via filling. Blanking consists of removing the tape from the polyester film carrier and then cutting it into blanks by using a precision blanking die [10]. The die simultaneously cuts the tape and punches holes. As there is a slight shrinkage differential between the casting direction and the transverse direction of the tape, each sequential layer of tape is rotated by 90° in order to achieve a uniform shrinkage.

Via holes are used to provide a conducting path between the substrate layers. They are formed by creating small holes through each layer and filling them with a conducting paste. The via pattern must be generated with $x-y$ coordinates referenced to a common origin and must be compensated for green tape fired shrinkage. Via diameters usually range from 0.005 to 0.008 inches. Vias can be formed by a number of methods, such as die set rigid punch arrays, numerically controlled punches, drills, or lasers. A rigid punch array (Figure 14.6) is suitable for high volume production, in which all vias can be formed in one operation. Numerically controlled equipment is satisfactory for design and prototyping. The repetition rate of the equipment and the movement of the table holding the substrate are typical parameters which must be controlled during this operation. Typical hit rates are of the order of 2 to 3 per second, although hit rates of 50 to 100 vias per second are possible with laser drilling. The CO_2 laser, which has good coupling characteristics with ceramic-filled glasses, is extensively used. The laser drilling method can achieve vias faster, with greater precision and lower damage, but is limited by high equipment costs.

Vias are filled by a standard screen printer with a modified printing stage, using thick pastes. Complete filling of the vias in a single printing pass is accomplished by using contact printing through a stencil made of brass foil on a polyester film carrier. Via fill compositions are designed to match the shrinkage rate of the dielectric. Overfilled and underfilled vias can result in shorts or opens, respectively. A special printing stage is required for printing vias or printing conductors. A porous stone plate through which a vacuum is drawn is used to hold down the tape and to provide a flat, distortion-free substrate for printing (Figure 14.7). The machine setup for conductor printing is similar to that for via filling, but the conventional off-contact printing mode is used. The substrate layer with the conductor print pattern is then dried. Thick film resistors can also be printed on the substrate layer, either in the top layer or buried between the layers. For obtaining precise values of resistance, the printed resistors are laser trimmed after metallization.

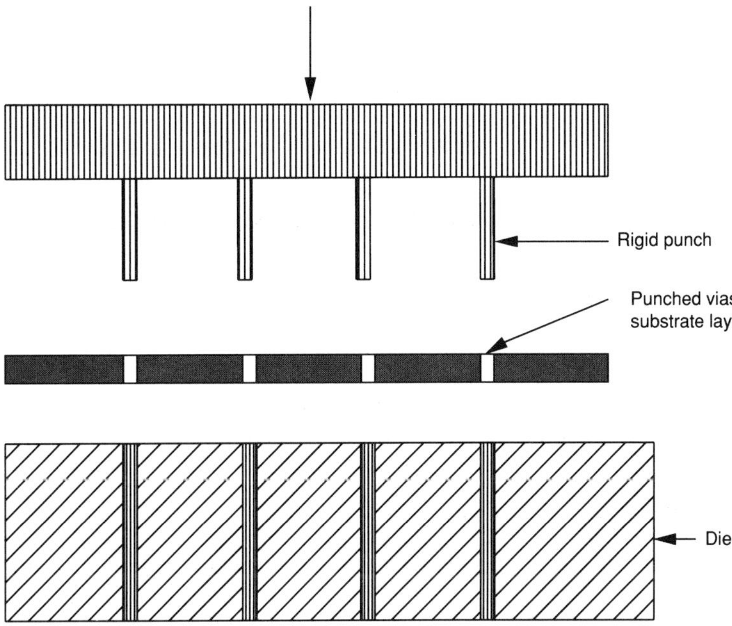

Figure 14.6 Rigid array punch for generating vias in tape.

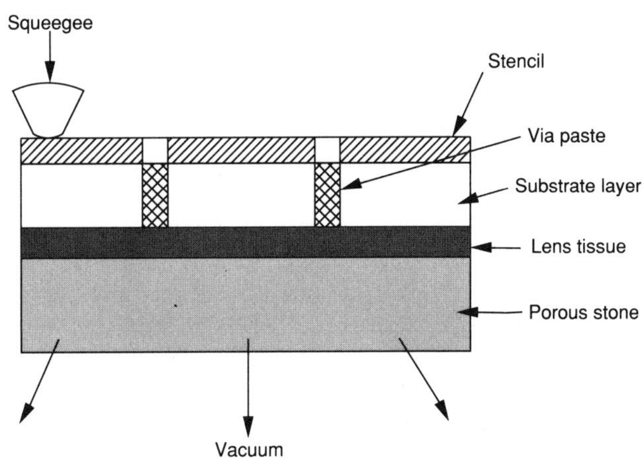

Figure 14.7 Filling vias in prepunched tape.

After drying, the substrates are inspected for line continuity and via registration. Registration of the substrate layers must be held to less than ± 0.1 mm tolerance through the via forming, conductor printing, and lamination steps. After inspection, the tape layers are collated (stacked) and aligned, so that the corner registration holes are concentric, for placement of the registration pins in the lamination die (Figure 14.8).

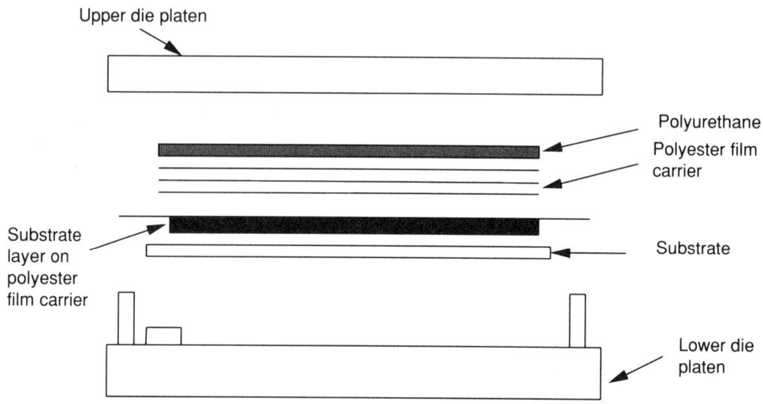

Figure 14.8 Lamination stack for tape laminated on a substrate.

14.2.4 Substrate fabrication

Substrate fabrication includes lamination, cutting, firing, post-fired printing, and lead or pin attachment. The stacked and registered layers are placed in a die having cavity dimensions identical to those of the blank sheet. The die walls prevent lateral distortion of the sheets during lamination, and the registration pins ensure mechanical registration and circuitry alignment. Shrinkage can be controlled by local laminating conditions. The distribution of porosity can be changed by controlling the lamination pressure. Once the vias have been formed, the substrate layer is placed on a premetallized substrate which has been set in the lamination die. Compressible materials, such as polyurethane, are placed over the part for an even distribution of pressure. Parts are cut to their final size in either the fired or unfired condition. A large number of tools, such as a hot razor or steel rule die, a dicing saw, and laser scribing are available in order to complete the process.

The objectives of burnout/firing are to remove organics from the substrate layers and then to sinter the green ceramic to a dense, rigid structure. The burnout and firing operations are done separately, since they use conventional thick-film belt furnaces. With proper furnace design, both processes may be combined. In the burnout, the laminated green substrates are heated in box furnaces to 350 °C for air-fired metallizations, such as Ag or Au and to 550 °C for Cu metallizations. The substrates are heated for one hour in the case of Ag and Au, and for five to six hours in the case of Cu, to completely remove the organic binders and plasticizers. After burnout, the parts are too fragile to be handled individually, but can be transported to the belt furnaces on thin alumina setters. For Ag or Au metallization, the burnout parts are sintered in a belt furnace at a temperature of about 850 °C for 30 minutes. The complete process for burnout and firing takes about two or three hours. Additional firing steps are required for all metallizations, except for copper. One common approach is to use copper oxide as the starting powder in the conductor paste. This eliminates cracking of the dielectric because of volume expansion in the oxidation of pure Cu during burnout. In this case, burnout is done in air in order to remove the organics completely. Reduction is subsequently

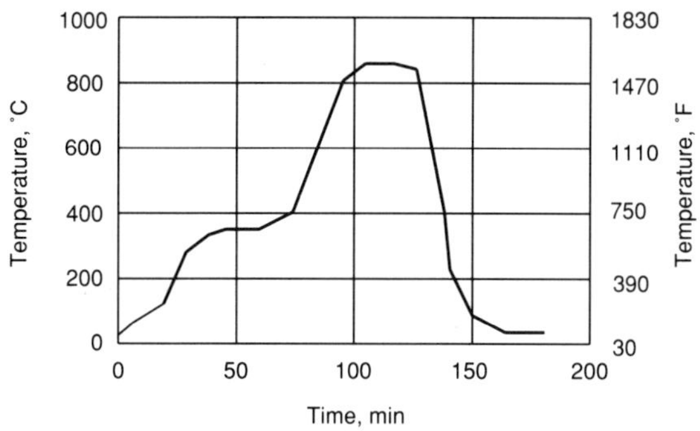

Figure 14.9 Burnout and firing profile for low temperature co-fired ceramic.

conducted in an atmosphere of nitrogen/hydrogen or carbon monoxide/carbon dioxide mixtures at 300 to 400 °C, to convert the oxide back to metallic Cu before firing. The final sintering is done in a conventional nitrogen belt furnace, at temperatures between 900 to 1050 °C for 20 to 30 minutes. The entire burnout and firing for Cu metallized parts takes about 12 to 14 hours (Figure 14.9).

After firing, the top layers are ready for personalization. A post-fired printing process is conducted, which involves the printing of pads, conductor traces, or precision resistors. In the case of low temperature co-fired ceramics, it is possible to have high-conductivity innerlayer metallization, such as Ag, and then post-fire a solderable Cu conductor on the top layer. Here, the post-fire has to be done in a nitrogen belt furnace.

The leads or pins are attached to the substrates to allow for connection to the next level of packaging. Low temperature co-fired ceramics use soldering as the preferred lead or pin attach method. Low temperature brazing and thermocompression may also be used.

Electrical testing is carried out after personalization. When defective wiring is found, rewiring is performed by adding lines on the top surface, wire bonding to join one pad to another, or separating pads from vias by cutting conductor lines with a CO_2 laser. Chip replacement is accomplished by mechanical removal, thermal rework of the substrate pads, and replacement with a new chip. The selected removal and reworking techniques depends on the chip attachment method used.

14.3 ELECTRICAL PROPERTIES

The various electrical properties of LTCC substrates are discussed and tabulated in Table 14.4. The electrical properties of LTCC substrates are also compared to those of HTCC substrates.

14.3.1 Dielectric constant

The dielectric constant, which is an important property for substrate materials, controls the signal transmission speed. The propagation delay time (T_{pd}) is formulated as follows:

$$T_{pd} = \frac{\sqrt{\varepsilon} l}{C}$$

where c = speed of light
ε = dielectric constant
l = interconnect length

The propagation delay for many megabits of information, which must travel through large computer and supercomputer systems, cannot be disregarded. Therefore, it is desirable to reduce the dielectric constant as much as possible. Any technique

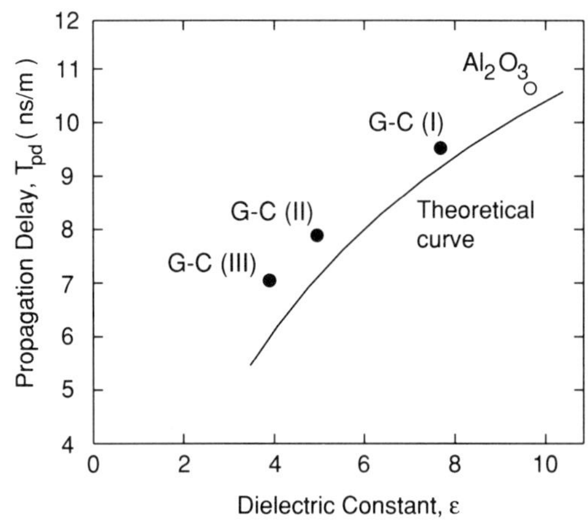

Figure 14.10 Relation between propagation delay and dielectric constant. (G-C(I): Al_2O_3-borosilicate Glass; G-C(II): cordierite-borosilicate glass; G-C(III): SiO_2-borosilicate glass.)

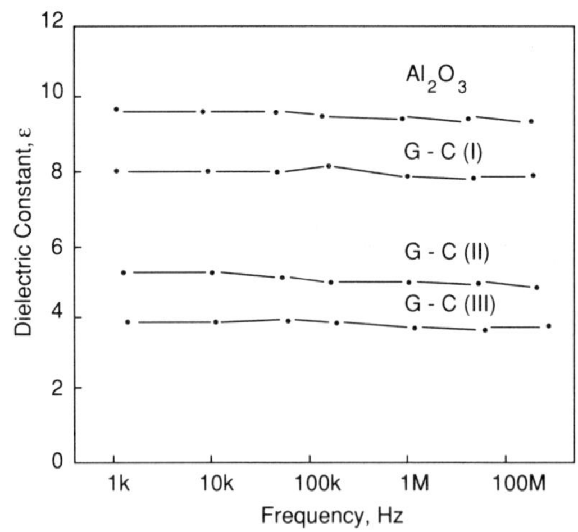

Figure 14.11 Dielectric constant frequency dependence.

which increases the specific volume of a material, and any lower dielectric constant material would be beneficial (graph). A graph for the propagation delay for various compositions of LTCC substrates is given in Figure 14.10. The frequency dependence of the dielectric constant for LTCC substrates and for 96% alumina is given in Figure 14.11.

14.3.2 Characteristic impedance

Characteristic impedance depends on the dielectric constant, the geometry of the signal line, and the distance between the signal line and the ground plane (Figure 14.12). The characteristic impedance (Z_0) is formulated according to the following relationship:

$$Z_0 = \sqrt{\frac{L}{C}}$$

where L = inductance of the line,
C = capacitance of the line.

14.3.3 Crosstalk coupling noise

Crosstalk coupling noise is an important factor, because the packaging density is restricted by the limited space and length of parallel lines. While near-end crosstalk is quite significant in a homogeneous medium, far-end crosstalk is zero. However, in the multilayer ceramic substrate, the presence of conductors in adjacent channels, orthogonal lines, and via holes have little effect on the mutual inductance. Therefore, the inductive coupling is larger than the capacitive coupling, contributing to the far-end crosstalk coupling noise.

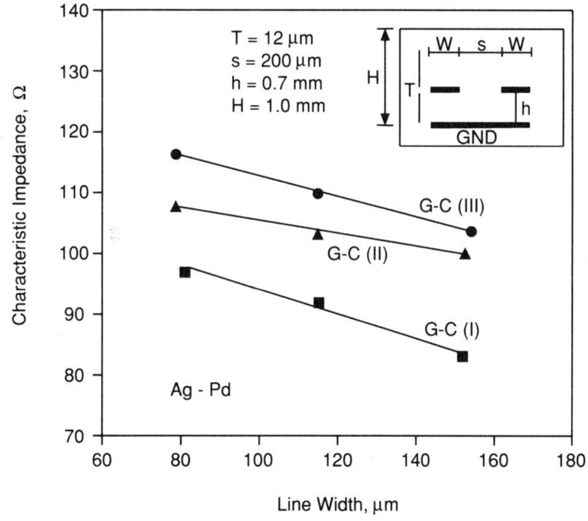

Figure 14.12 Characteristic impedances for LTCC materials.

14.3.4 Loss tangent

The loss tangent is also referred to as the dissipation factor, and is defined as the ratio of the loss index, k'' to the relative permittivity, k' [11].

$$\tan\delta = \frac{k''}{k'}$$

Loss index, $k''(\varepsilon_r'')$ is the magnitude of the imaginary part of the relative complex permittivity and is given as [12],

$$k'' = \frac{p}{E^2 f V \times \text{constant}}$$

where p = power loss,
E = electric field intensity,
f = frequency,
V = volume.

The value of the constant is 5.556×10^{-13} when the power loss is expressed in watts, electric field intensity in volts per centimeter, frequency in hertz and volume in cubic centimeters.

Relative permittivity k' is the real part of the relative complex permittivity. It is also the ratio of the equivalent parallel capacitance C_p of a given configuration of electrodes with a dielectric, to the capacitance C_v of the same configuration of electrodes with vacuum (or air for most practical purposes):

$$k' = \frac{C_p}{C_v}$$

The loss index should be small in order to reduce the heating of the material and to minimize its effect on the rest of the network. In high frequency applications, a low value of loss index is particularly desirable, since for a given value of loss index, the dielectric loss increases directly with frequency. The loss tangent may be used to indicate the characteristics of a material in other respects, such as dielectric breakdown, moisture content, degree of cure, and deterioration from any cause. However, deterioration due to thermal aging may not affect the loss tangent unless the material is subsequently exposed to moisture. While the initial value of the loss tangent is important, the change in loss tangent with age may be much more significant. The dissipation factor for a typical LTCC substrate should be approximately 0.002.

The reciprocal of the loss tangent is called the quality or storage factor, Q. The quality factor of a materials system is inversely related to the transmission line power loss of a microwave signal in the medium. The conductor loss depends on frequency. The Q-factor can also be defined as the ratio of the ability of a materials system to store energy, to the signal loss in the medium. When two materials have the same dielectric constants, their dielectric losses per unit volume, at a given frequency and applied voltage gradient, are directly reflected by their respective loss tangents. Since the dielectric loss in an insulating material results in the generation of heat, it is desirable in most cases that these losses be kept as low as possible. This is important not only from the standpoint of the overall

Table 14.3 Loss tangent for LTCC substrates.

LTCC substrates	Loss tangent (1 MHz)
55 wt % Al_2O_3–45 wt % lead borosilicate	0.003
45 wt % cordierite–55 wt % borosilicate glass	0.005
35 wt% SiO_2–65 wt % borosilicate glass	0.003

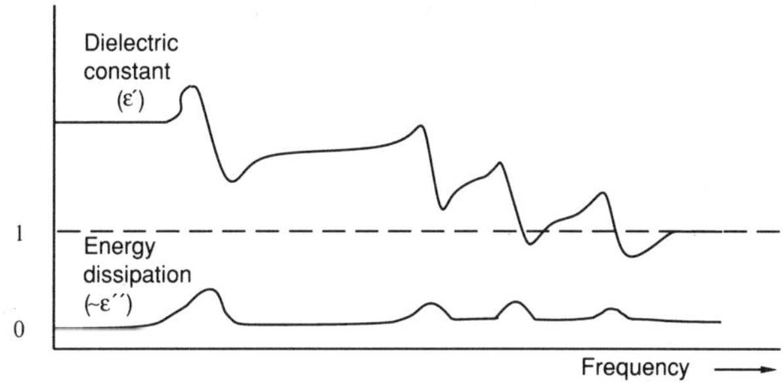

Figure 14.13 Variation of dielectric constant and dissipation factor with frequency.

efficiency of an electrical system, but also since the increased temperature generally results in significant changes in both the relative permittivity and loss, and therefore contributes to circuit mismatch which may occur at high frequencies. The dielectric loss, as measured by the loss tangent and loss index, may serve as a quality control criterion and as a means of determining batch-to-batch uniformity of a product. It is also an excellent means of measuring the effects of weathering, aging, and absorption of moisture by the substrate, since these result in substantial increases in the dielectric loss index. The dissipation factor, in combination with the dielectric constant, is useful in estimating the contribution of the dielectric loss to the total attenuation characteristics of microwave systems. The relative permittivity, in combination with the loss tangent, determines such transmission characteristics as velocity of propagation, attenuation distance, dB loss per meter, phase factor, complex index of refraction, index of absorption, and dielectric conductivity. The dielectric constant and dissipation factor are sensitive to changes in chemical composition, impurities, and homogeneity. Hence, measurement of these properties is useful for quality control and for determining the effect of environments with moisture, heat, or radiation on substrates. In microwave circuitry, the loss tangent (along with copper losses) influences attenuation and power loss [13]. The dissipation factors for various compositions of LTCC are given in Table 14.3. The general variation of dissipation factor and dielectric constant with frequency is given in Figure 14.13 [14].

14.3.5 Measurement of dielectric constant and dissipation factor

The dielectric constant of a material is defined as the ratio of the capacitance between two conductors when embedded in the material, to the capacitance between the same configuration of conductors in a vacuum (or air). The dissipation factor is

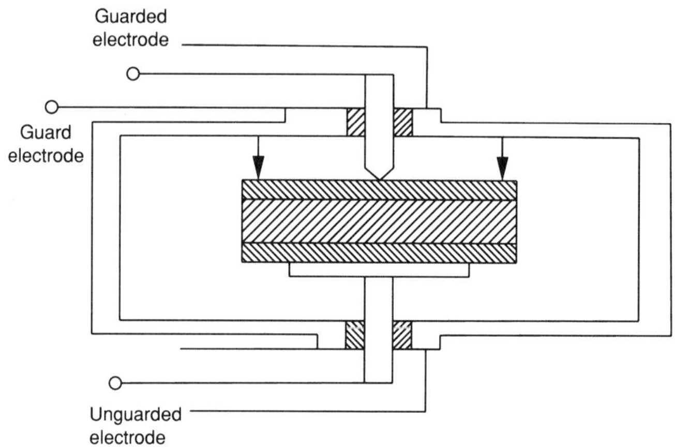

Figure 14.14 Three-terminal cell for solids.

defined as the ratio of resistive to capacitive currents in the dielectric. The dielectric constant and dissipation factor for devices operating at room temperature can be determined by the standard ASTM test method D-2149-79 [15]. In this method, the dimensions of the test specimen play a very important role. The geometry of the specimen is either that of a sheet or a cylinder, but usually a sheet form is preferred since dimensions can more accurately be defined for a sheet than for a cylinder. The test method consists of placing the specimen in an electrode system, with a vacuum capacitance that can be either calculated accurately or determined by calibration in the absence of the solid material. The material whose dielectric constant has to be determined is placed between two parallel plate electrodes to measure its capacitance. In addition, there are fringing and stray capacitances, which can be eliminated to a large extent by the addition of a guard electrode (Figure 14.14). If the test specimen and guard electrode extend beyond the guarded electrode by at least twice the thickness of the specimen, and if the guard gap is very small, the field distribution is identical with that existing when vacuum replaces the dielectric. The ratio of these two capacitances is the permittivity. The common test frequencies are 60 Hz, 1000 Hz, and 1 MHz.

14.3.5.1 Measuring circuits

For measurements at 1 MHz and above, a resonant-circuit susceptance variation method is recommended (Figure 14.15) [11]. The Q of the circuit should be at least 200 except for very low loss materials, for which a Q of 500 or higher is desirable. For measurement from 50 Hz to 100 kHz, a substitution method using a low-voltage capacitance bridge is recommended.

14.3.5.2 Test enclosure

The test enclosure is shown in Figure 14.16. The material used in the specimen holder should have good thermal and electrical conductivity, yet be oxidation resistant and have sufficient strength to maintain its mechanical dimensions after

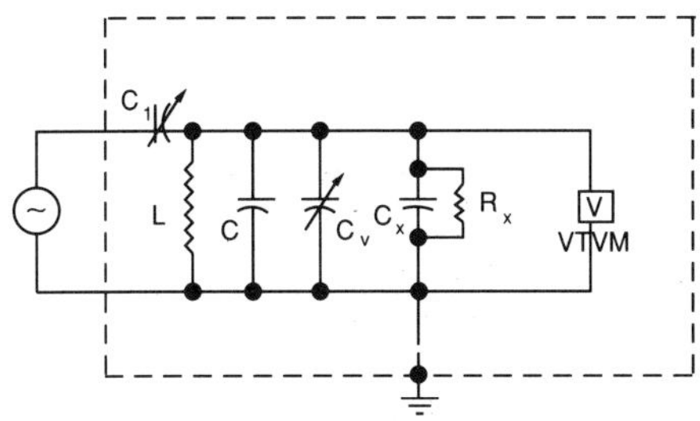

Equations

$$C_x = \Delta C_v$$
$$\Delta C_v = C'_v - C_v$$
$$(C_{v1} - C_{v2})$$
$$D_x = [(C'_{v1} - C'_{v2})/2C_x] \cdot (V' - V)/V$$
or
$$D_x = [(C_{v1} - C_{v2}) - (C'_{v1} - C'_{v2})]/2C_x$$

Figure 14.15 Susceptance-variation method.

repeated heating. AISI stainless No. 316 is the most commonly used material. It satisfies all the requirements except that of thermal conductivity. The insulators may be aluminum oxide, beryllium oxide, or polytetrafluoroethylene. The electrodes should be 50 mm in diameter and at least 5 mm thick, with sharp corners. They should be parallel to within 0.01 mm. The length and cross-section of the tubes should be such that the temperature of the insulators does not exceed 100 °C when the oven is at 500 °C. A dial gauge is used to determine the electrode separation, and to monitor the specimen expansion. This should have a precision of 0.005 mm.

14.3.5.3 Electrode systems

There are various systems of contact and non-contact electrodes. The choice of the electrode system depends on the material, application, and convenience. The micrometer non-contact electrode (Figure 14.17) is generally used, as it eliminates errors caused by the series inductance and resistance of the connecting leads [16].

14.3.5.4 Test specimen

The test specimen is a disk 40.00 ± 0.01 mm in diameter and 2 or 3 mm in thickness, with surface finished to 32 μ in. or better and parallel within 0.01 mm, free of bubbles and other defects.

14.3.5.5 Temperature control

Temperature measurements are taken at frequent intervals not exceeding 20 °C. The temperature is reduced to the lowest required test temperature and allowed to reach equilibrium. Then it is increased at the rate of 2 ± 0.5 °C/min, until the required temperature range is transversed. Prior to the test, the electrodes are conditioned at room temperature (23 ± 1 °C) and 50 ± 2% relative humidity for about 40 hours.

14.3.5.6 Test procedure

The specimen to be tested is clamped between the electrodes as shown in Figure 14.16, and the drive nut is rotated until a sudden decrease in friction is observed. A micrometer reading is taken, and compared against the setting at

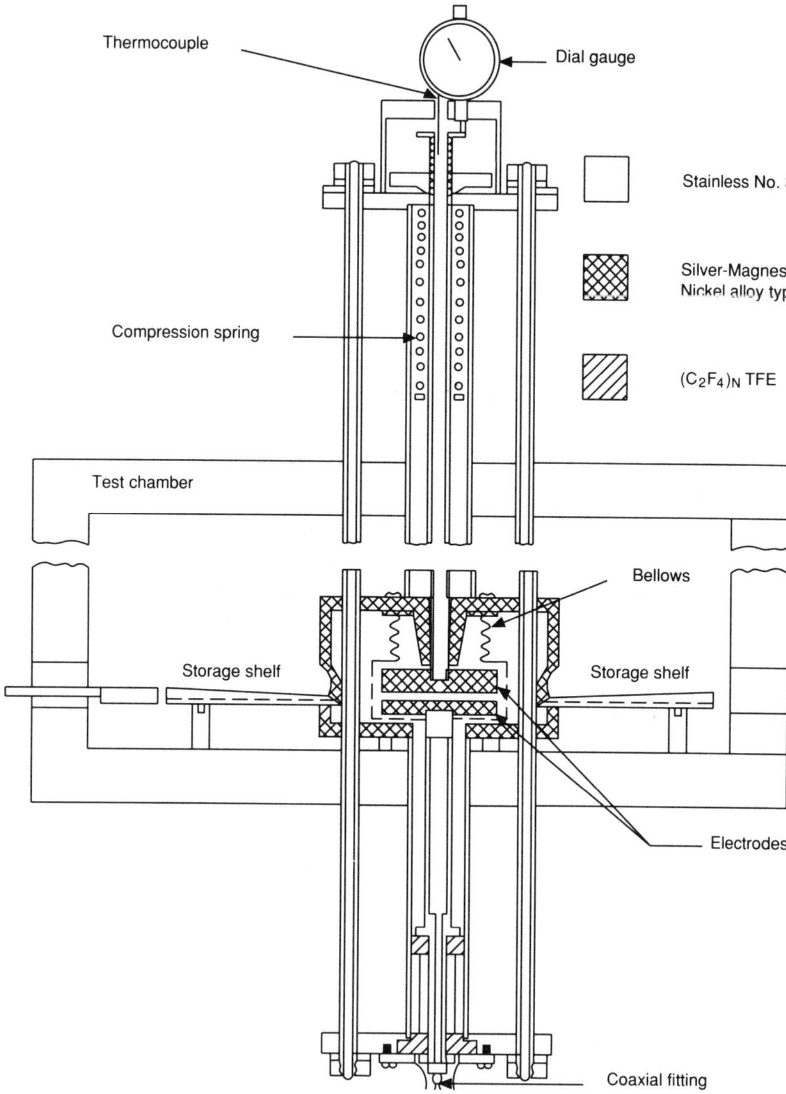

Figure 14.16 Specimen holder.

which friction first increases with increasing electrode spacing. The Q of the measuring circuit is peaked. Next the electrodes are opened and the specimen is taken out. The balance of the measuring apparatus is restored without changing its capacitance setting, by reducing the spacing between the electrodes and adjusting the measuring circuit to balance the loss component. The dissipation factor and micrometer settings are noted. This is repeated at each test temperature and each required frequency. Using the micrometer method, the capacitance of the specimen can be calculated as follows:

$$C_s = C_0 - C_t + C_v$$

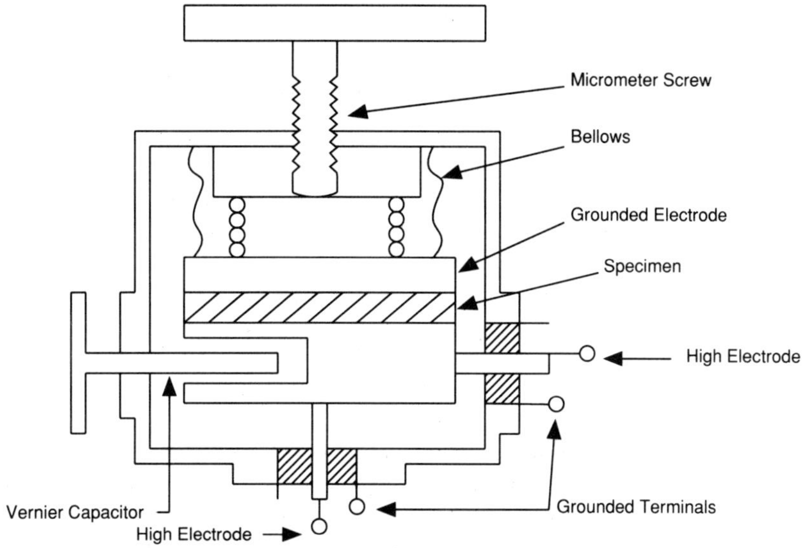

Figure 14.17 Micrometer-electrode system.

$$C_s = \Delta C + C_v$$

where C_0 = capacitance of the specimen holder with the specimen out,
C_v = equivalent geometric vacuum capacitance of the specimen,
C_t = total capacitance at the unknown terminals of the measuring circuit.

The dielectric constant, k'_x and dissipation factor, D_x of the material are determined using the following relations,

$$k'_x = \frac{1}{1 - \dfrac{\Delta C}{C_1}\dfrac{t_0}{t}}$$

$$D_x = D_c + Mk'_x \Delta D$$

where ΔC = capacitance change when specimen is inserted,
C = capacitance with specimen in place,
D = dissipation factor change when specimen is inserted,
ΔD_c = dissipation factor with specimen in place,
t_0 = parallel plate spacing,
t = average thickness of specimen,
$M = t_0/t - 1$.

14.3.5.7 Lead length correction

The length of the leads should be kept to a minimum between the measuring instrument and the specimen holder. The correction in ΔC for a dissipation factor, when its value is less than 0.10, is given by

$$D_s = \left(\frac{C_t}{C_s}\right)(D_t - D_0)\left(\frac{C_H}{C_m}\right)^2$$

where C_H = capacitance of the specimen holder,
C_m = capacitance indicated by the measuring circuit,
D_t = dissipation factor of the measuring circuit when the specimen is between the electrodes,
D_0 = dissipation factor of the measuring circuit when the circuit is rebalanced with the specimen removed from the electrodes

If C_H/C_m is not less than 0.99, no correction for lead length is required.

The dielectric constant and the dissipation factor are sensitive to changes in chemical composition, impurities, and homogeneity. Hence measurement of these properties is useful for quality control and for determining the effect of environmental factors such as moisture, heat, or radiation on substrates.

Dielectric properties may vary over the 2–18 GHz range and be significantly different from those measured at the traditional 1 MHz [17]. The dielectric constant and loss tangent of the substrate are critical to microwave circuit design, where controlled impedance is necessary. At microwave frequencies, the LTCC substrate was found to have a lower dielectric constant than that measured at 1 MHz. There are various dielectric measurement techniques at frequencies above 1 GHz. The microstrip resonator (ring resonator) technique is the most widely used. Other measurement techniques are the waveguide resonator, end-coupled resonator, microstrip lines, and multilayer filters. The ring resonators consist of a microstrip ring and two microstrip lines (Figure 14.18) [17, 18]. Ring resonators have been made fundamentally resonant at about 1 GHz, but can be used satisfactorily over a

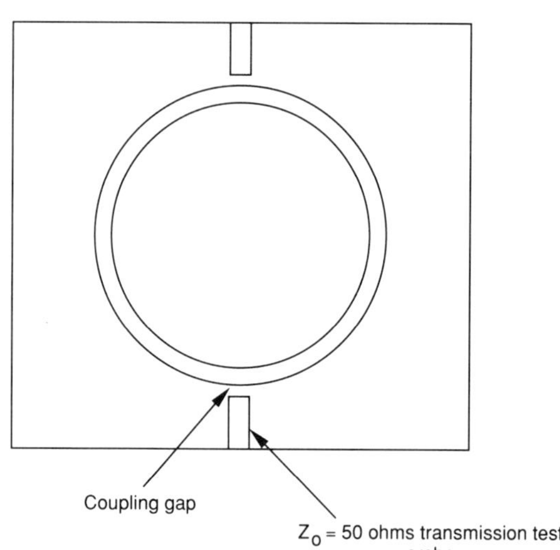

Figure 14.18 Plan view of ring resonator and transmission-test probes.

wide range of frequencies. However, ring resonators can result in inaccuracies due to ring curvature and uncertainty in the exact electrical length of the ring. The end-coupled resonator does not suffer from the curvature effects of the ring resonator. Curvature effects may cause inaccurate estimates of the electrical length of the ring from the mean physical circumference. But the end-coupled resonator may suffer from radiation effects at the open ends. This radiation may be eliminated or minimized if the structure is enclosed in a cavity whose dimensions are such that it is below cutoff to any waveguide or cavity type modes. The microstrip samples are the metallized substrates with thin-film gold on both sides, processed and diced. A multilayer elliptic function filter was designed to operate with a center frequency of 8 GHz [19]. The process flow chart for the elliptical filter is given in Figure 14.19.

The shorted transmission and resonant cavity perturbation methods can also be used to measure the dielectric constant and dissipation factor, and are discussed in ASTM standard test method D 2520-86 [15, 20].

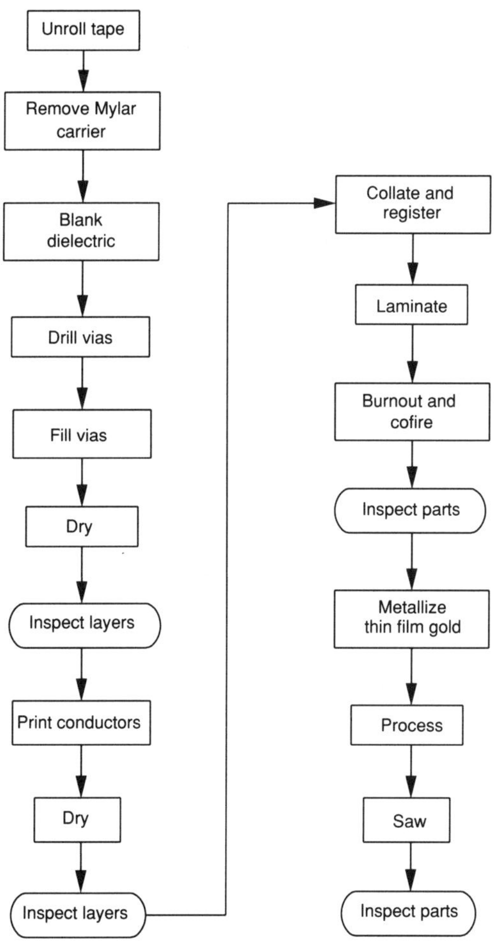

Figure 14.19 Process flow for elliptical filter.

14.3.5.8 Shorted transmission method

The shorted transmission line method is used for specimens precisely formed to the inside dimensions of the waveguide. This method can be used at any frequency where suitable transmission lines and measuring equipment are available. Transmission lines capable of withstanding temperatures of up to 1650 °C in an oxidizing atmosphere should be used to hold the specimen.

Figure 14.20 gives a block diagram of equipment components. The equipment consists of a generator, square-wave modulator, frequency meter, isolator, slotted section, probe, VSWR meter, cooling sink, waveguide specimen holder, and tube furnace.

Generator. A generator with stable power and frequency, and with low harmonic output is required.

Square-wave modulator. In a square-wave modulator, a 1.0 kHz output frequency is needed for a voltage standing wave ratio (VSWR) meter.

Frequency meter. The frequency meter can either be heterodyne or cavity absorption, with an uncertainty not to exceed 1 part in 10^4.

Isolator. A 30 dB isolation, and having an output VSWR of less than 1.15 dB isolation is required.

Slotted section. The apparatus also consists of a slotted waveguide section and carriage capable of measuring gross distances to 0.025 mm and small distances to 0.0025 mm. A micrometer head is required and it should move parallel to the axis of the line.

Probe. The probe must be adjustable for depth. The detector must be operated in the square law [21] region, if the voltage-decibel scale of the standing wave ratio (SWR) meter is used.

VSWR meter. The VSWR meter should be readable in decibels.

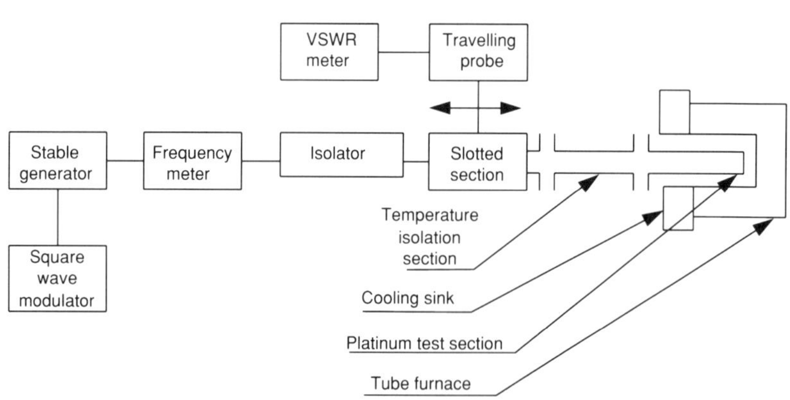

Figure 14.20 Block diagram of apparatus used to perform the measurement of dielectric properties by the shorted transmission line method.

Cooling sink. A cooling sink is used to provide sufficient conduction to a water or air stream, so as to maintain suitable temperature and waveguide dimensions.

Waveguide specimen holder. The waveguide specimen holder is platinum-20% rhodium for a temperature of 1650 °C, platinum for 1300 °C, copper or silver for lower temperatures, to withstand thermal damage and corrosion. The length should be sufficient for a main temperature transition region of the order of the guide wavelength in the line and to still keep the temperature uniform to 5 °C.

Tube furnace. The tube furnace is platinum wound to accept the test section, and to maintain a 50 mm length at a constant temperature of ±5 °C, up to 1650 °C. The slope of the attenuation characteristic of the slotted line should be normal, that is, the VSWR should change by the expected amount in going from one node to another, while looking into a shorted termination. The dielectric holder should slope downward at 45 ° to 90 °, to maintain the specimen against termination. Termination should be flat to 0.010 mm and perpendicular to the axis of the waveguide within ± 0.05 °. The cross-section of the holder should meet the Electronic Industries Association (EIA) specifications. At 9 GHz this requires a ± 0.075 mm tolerance on transverse dimensions.

Test specimen. The transverse dimensions of the specimen should be 0.05 ± 0.025 mm, less than those of the transmission line. The front and back faces are parallel within 0.01 mm and perpendicular to the axis of the transmission line within ±0.05 °.

Procedure. The procedure consists of measurements of impedances with and without the specimen, by keeping the frequency of the source and temperature distribution of the line the same for both observations.

14.3.5.9 Resonant cavity perturbation method

The resonant cavity perturbation method covers the measurement of the microwave complex permittivity of dielectric specimens in the form of rods, bars, sheets, and spheres. The measurement frequency depends on the available resonant modes of the cavity, which limits the freedom of selection to a few values. Resonant cavities exhibit very high Q values (2000 to 5000 or more) and are therefore inherently sensitive for low loss measurements. Although dimensions are important and must be measured accurately, the specimen does not need to have a close tolerance fit within the specimen holder. A test cavity, which is a short section of a rectangular waveguide, is shown in Figure 14.21. The block diagram of a typical microwave system for measurement of permittivity by the perturbation method is given in Figure 14.22. The main apparatus includes a microwave frequency meter (6 digit precision), VSWR meter, or equivalent power indicators (0.1 dB or less sensitivity), crystal diodes and holders (or bolometers), directional couplers (or a waveguide to coax adopters), microwave signal generator with adjustable frequency, square wave modulator, variable attenuator (precisioncalibrated, 0.1 dB or less uncertainty, range 10 dB or more), variable attenuator, isolators, waveguide hardware, coaxial cables, and connectors. A 100 MHz digital frequency counter plus a transfer oscillator are often used by mixing a harmonic of the oscillator with

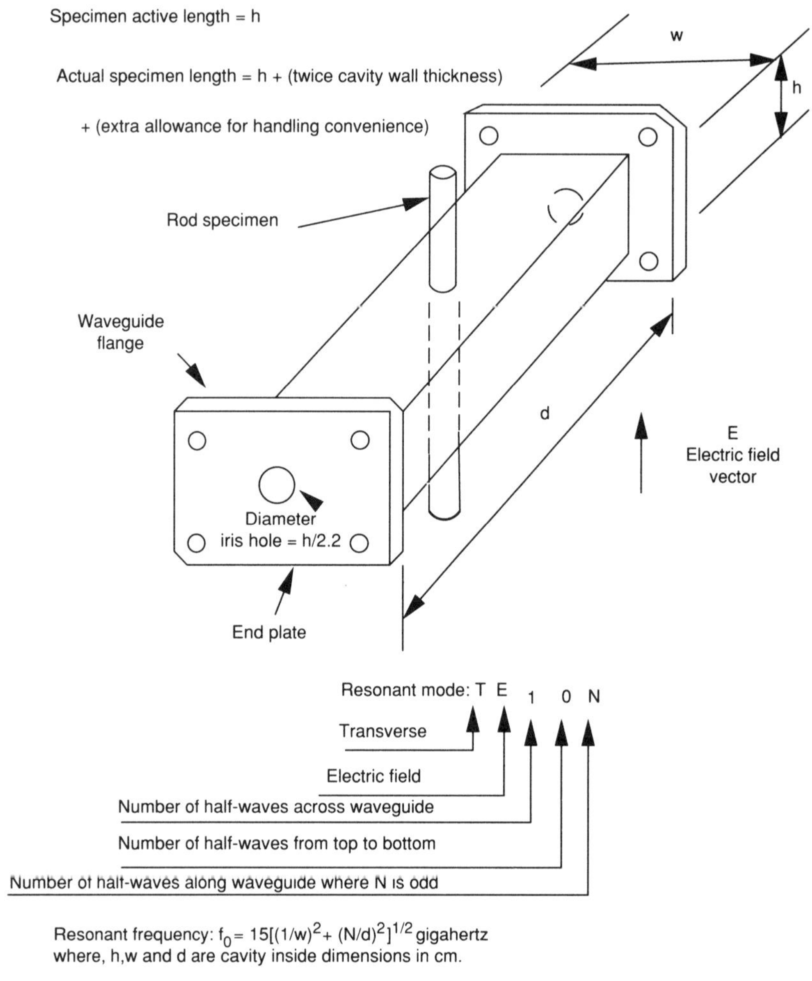

Figure 14.21 Rectangular microwave cavity for permittivity measurements by perturbation method.

the microwave signal to get a null beat frequency on an oscilloscope. The diameter of the rod specimens in X-band frequency (8 to 12 GHz) should be 1.04±0.00 mm. The active length is the inside length of the waveguide cavity, 10.16 mm. At S-band frequencies (2.6 to 3.9 GHz) the rod diameter should be 3.2 ± 0.00 mm. The diameter of spherical specimens or the thickness of strip and sheet specimens should be small (approximately 10% or less) compared to the wavelength. The rod specimen diameter should be measured by a micrometer to within ±0.0025 mm.

14.3.6 Skin effect

The skin effect phenomenon was first mathematically described by Maxwell in 1873, and followed up by Rayleigh in 1886. It is defined as an effect characteristic

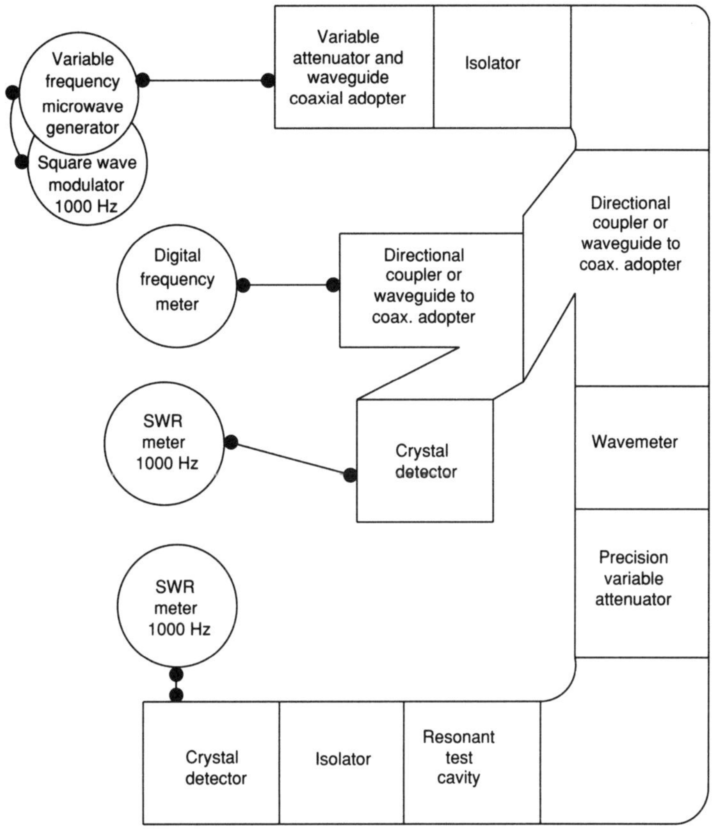

Figure 14.22 Block diagram of typical microwave system for measurement of permittivity by perturbation method.

of current distribution in a conductor at high frequencies, by virtue of which the current density is greater near the surface of the conductor than in its interior. The current concentration near the conductor surface increases with the frequency, the conductivity, and the permeability of the conductor. This results in a corresponding increase in resistance and a decrease in internal inductance at frequencies for which this effect is significant [22, 23]. The skin depth is the depth inside the conductor in which the current density is concentrated and is given by

$$\delta = \frac{c}{\sqrt{2\pi\mu\sigma\omega}}$$

where c = velocity of light,
μ = permeability of the medium,
σ = electrical conductivity,
ω = angular frequency.

As the frequency increases, the skin depth decreases. The skin depth for copper and silver is approximately 0.002 mm at 1 GHz and decreases by a factor of 10 at 100 GHz [22]. In the case of materials having a high dielectric constant, the characteristic decay distance is given by [24],

$$\Delta Z = \frac{c}{2\pi\sigma}\sqrt{\frac{\varepsilon}{\mu}}$$

where ΔZ = decay distance,
c = velocity of light,
ε = dielectric constant,
μ = permeability of the medium,
σ = electrical conductivity.

14.4 MECHANICAL PROPERTIES

The thermal conductivity of fired LTCC tape is about 15–20% that of HTCC. The thermal expansion coefficient of fired LTCC (7.9 ppm/°C) tape is more than that of

Table 14.4 Electrical and mechanical properties of HTCC and LTCC.

	HTCC	LTCC
Electrical characteristics		
Conductor resistance	Tungsten (15 mΩ/square)	gold (5 mΩ/square)
Thick film resistors	Not possible	100 Ω/sq and 10 kΩ/sq
Dissipation factor (1 MHz)	15×10^{-4} Ω-cm	16×10^{-4} Ω-cm
Dielectric strength	550 V/mil	808 V/mil
Volume resistivity at 20 °C	> 1014 ω-cm	> 1014 ω-cm
Narrow well-defined conductors	5 mil and less	5 mil and less
Buried layers as with PC boards	Yes	Yes
Vias interconnecting layers	4 mil diameter	4 mil diameter
Parallel plate capacitors	Yes	Yes
Inter-digital capacitors	Yes	Yes
Dielectric constant	9.7	5.8
Mechanical characteristics		
High power dissipation	7 W achieved to date	8 W achieved to date
Heat sink metal	Kovar	Tungsten-nickel
Heat sink thermal conductivity	0.037 cal/sec cm °C	0.4 cal/sec cm °C
Heat sink (TCE)	7.2×10^{-06} °C	4.6×10^{-06} °C
Dielectric flexural strength	60,000 psi	20,000 psi
Heat sink attachment method	Copper-silver braze	Gold-tin solder
Lead frame attachment method	Copper-silver braze	Under development
Withstands assembly process	Yes	Yes
Withstands electronic environment	Yes	Yes
Hermetic seal method	Solder-glass	Solder
Dielectric thermal conductivity	0.050 cal/sec cm °C	0.004 cal/sec cm °C
Die attach techniques	Gold-tin or epoxy	Gold-tin or epoxy
Wire bonding techniques	Gold or aluminum wire	Gold or aluminum wire

HTCC (7.0 ppm/°C). However, the flexure strength of LTCC (2.1×10^3 kg/cm^2) is less than that of HTCC (3.8×10 kg/cm^2) due to the higher glass content in LTCC compared to HTCC. Typical mechanical properties of LTCC substrates are given in Table 14.4 and are also compared with HTCC substrates.

14.5 PHYSICAL PROPERTIES

14.5.1 Surface characteristics

The grain size of the substrate affects its mechanical properties, while the inclusions in the substrate affect its mechanical and electrical properties. The grain size and inclusions in the substrate can be determined using scanning electron microscopy (SEM) [25]. In the SEM, an accelerated beam of electrons is focused by electromagnetic lenses on a very small spot on the surface of the substrate. The beam is deflected by an electronic system to point the spot over an area of the specimen surface in a raster pattern of the type that is used in television systems. Radiation of various types is emitted from each spot on the surface as it is scanned, and the intensity of one or more ranges of radiations can be measured by a detector appropriate to each. The output of the detector is used to modulate a display unit or to operate a recording unit [26]. The whole system, with the exception of the display/recording unit, is enclosed in a chamber which is pumped out to a high vacuum.

Figure 14.23 is a scanning electron micrograph that shows the cross-sections of a fired laminate, made using a low temperature ceramic system. This micrograph illustrates the dense, pore-free microstructure which occurs during firing. The dark regions in Figure 14.23 are from ceramic fillers included in the tape formulation. For comparison, the surface of a standard 96% alumina substrate is also shown in Figure 14.24. The fully dense surface of the tape dielectric is apparent.

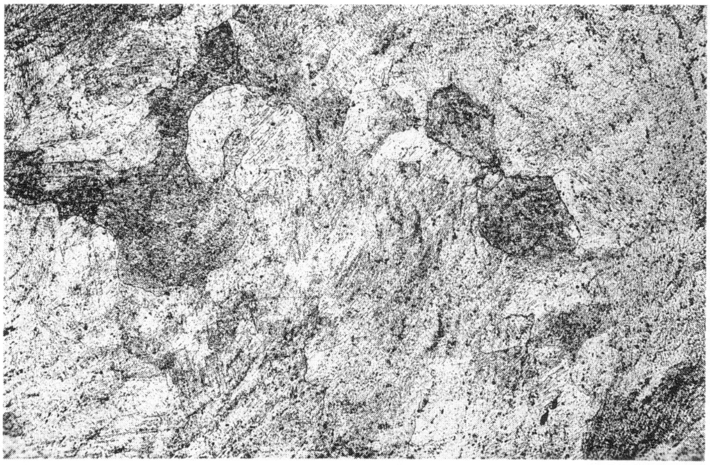

Figure 14.23 Fired microstructure of LTCC.

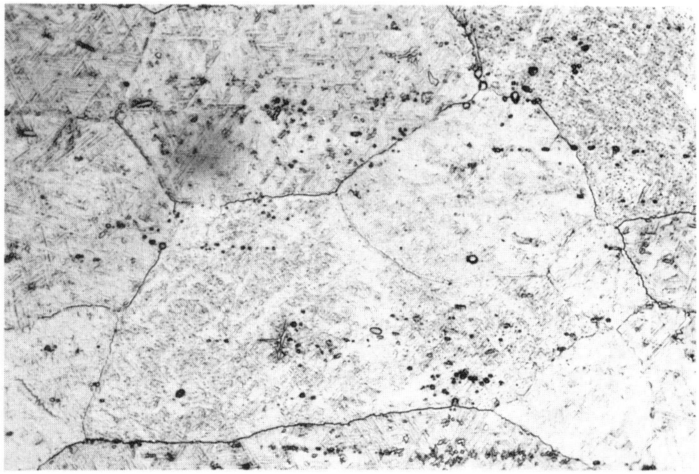

Figure 14.24 Fired microstructure of HTCC.

Some of the physical properties of the fired dielectric tape are given below:

- *Fired density*
  Theoretical 3.02 g/cm^3
  Actual > 2.89 g/cm^3
- *Camber*
  Fired ±75μm
  68 mm × 68 mm
- *Surface smoothness*
  Fired dielectric 0.8 μm
  50 mm × 50 mm (peak to peak)
- *Average shrinkage* in LTCC substrates during firing is about 12%, and lot-to-lot control is ±0.2%.

14.5.2 Surface impurities

The impurities on the surface of the substrate affect both the electrical and mechanical properties. Even a small impurity of two microns can cause an open in the electrical circuit of the substrate, since the conductor paths can be as small as two microns. The impurities on the surface of the substrate can be determined using an electron or auger microscope [25, 27]. In electron microscopy, a beam of electrons is generated and then focused by an electromagnetic condenser lens on the surface of the substrate, which must be thin enough to be reasonably transparent to electrons. A selected field of the specimen is then magnified by the electromagnetic objective and projector lenses, the image is projected on to either a fluorescent screen for direct viewing or a photographic film for recording. All of these units are located in a container which is pumped out to a high vacuum. The auger microscope is also based on the same principle, but the depth of field is higher than in the electron microscope. In the auger microscope a kinetic energy of 30–2000 eV is used for generation of electrons.

14.5.3 Bulk quantitative and qualitative analysis

The bulk quantitative and qualitative analyses of the substrate can be done by auger electron spectroscopy (AES), X-ray photoelectron spectroscopy (XPS), secondary ion mass spectroscopy (SIMS), and X-ray emission spectroscopy. The latter is more suitable for manufacturing control and for verifying that the product meets specifications. It provides rapid, multi-element determinations with sufficient accuracy to assure product quality and minimize production delays. In this analysis, the substrate, which is taken as the test specimen, is finished to a clean, uniform surface, and then irradiated with an X-ray beam of high energy. The secondary X-rays produced are dispersed by means of crystals, and the intensities are measured by suitable detectors at selected wavelengths. The outputs of the detectors are counted in voltage pulses. Radiation measurements are made based on the time required to reach a fixed number of counts, or on the total counts obtained for a fixed time. Concentrations of the elements are determined by relating the measured radiations of unknown specimens to analytical curves or charts prepared with suitable reference materials. Either a fixed-channel, polychromator system or a sequential, monochromator system may be used to provide simultaneous or sequential determinations of the elements. During the determination of the surface characteristics or impurities by the above methods, damage may be caused to the surface of the substrate. Liquid inspection and magnetic inspection do not cause damage to the substrate, but are limited to determination of large cracks on the surface.

14.6 FAILURE MECHANISMS AND MODELS

The failure mechanisms in the substrate are governed by the substrate materials, geometries, manufacturing and assembly processes, power cycles, and usage environments. The most critical aspects in the failure analysis of the substrate are associated with existing defects around the vias, which can, under severe stress conditions, initiate substrate fracture, and fatigue crack propagation. Following are the potential failure mechanisms of the substrate.

14.6.1 Substrate fracture

The materials in the package have different coefficients of thermal expansion, which induce mechanical stresses in the substrate under temperature changes. In die bonding, package sealing, and other processes, the thermal mismatch of different materials induces residual stresses. In field usage of the IC packages, power on and off of the device makes the junction temperature fluctuate. This temperature cycle mixed with the ambient temperature cycles, develop stresses in the substrate. These stresses may induce failure of the substrate. Especially microcracks or other defects, formed in different processing steps such as via drill and substrate cutting, may dramatically increase the local stresses and induce fracture. If a pre-existing initial crack, after the manufacture of the substrate, is equal to or greater than the critical crack size, the substrate would fracture at the first stress cycle.

The phenomenon of substrate fracture can be described by a fracture mechanics model. The allowable initial crack depth on the surface of the die and the allowable initial edge crack can be determined if the fracture toughness of the die is known. The maximum nominal thermal stress is calculated from a thermal stress model, for the specified type of crack. This theoretical allowable initial crack length also serves as a specification of substrate quality control.

14.6.2 Fatigue crack propagation in the substrate

A pre-existing microcrack may not initially be of critical size to cause catastrophic failure at the applied service stress, but it may grow to critical size gradually by stable fatigue propagation during power and temperature cycling.

The phenomenon of fatigue crack propagation can be described by Paris's law. The rate of fatigue crack propagation per cycle is determined by the cyclic change in stress intensity factor, a measure of the stress intensity around the crack tip. By integration of Paris's law, the number of cycles to failure can be estimated if the material constants in the equation of crack growth rate are known.

14.7 TESTS TO DETECT ADHESION FAILURES BETWEEN LAYERS OF LTCC SHEETS

After the ceramic materials are fired, problems of inadequate adhesion between layers are detected by the presence of blister-like defects [10]. The ceramic slurry is dried on a non-porous support and hence a subtle difference results in the texture of the surface, which is exposed to the atmosphere, compared to the surface which is against the support. The surface which is dried against the support has a smoother appearance, called the slick side. The other side is referred to as the dull side. In the fabrication of multilayer substrates, the blister-like defects occur only if two slick slides are placed adjacent to each other. The combinations of "dull-dull" and "slick-dull" do not result in adhesion failure. To detect early failures in the fabrication of multilayer ceramics, an electrical test method is used. This consists of measuring the constant current, low voltage resistance by the application of a high voltage pulse across the metal powders held together by an organic binder [28].

14.7.1 X-ray photo-electron spectroscopy (XPS or ESCA) analysis

Initially, adhesion failure was attributed to differences in surface chemistry. Thus ESCA was applied using a Perkin-Elmer/PSI Model 560 analytical system [10]. Specimens of unfired ceramic sheets were analyzed first. No consistent differences were found between the slick and dull sides. The data included peaks for carbon and oxygen but no peaks for aluminum and silicon. These results indicate that the alumina and silica powder particles are encapsulated by the organic binder. The relative concentrations of carbon and oxygen and their surface chemical states were found to be the same on both sides. Then, specimens of fired ceramic sheets were analyzed. Spectra from these samples included strong peaks from aluminum and silicon, as well as from carbon and oxygen. Once again, relative concentrations and chemical states were found to be the same on both faces. Differences were noted in the relative concentrations of aluminum and silicon from sample to sample, but there was no consistent pattern in the data to distinguish between slick and dull sides. Thus the data indicate that firing has removed most of the organic binder material. The blisters are broken in air immediately prior to insertion into the analytical chamber.

14.7.2 Microhardness analysis

Microhardness analysis can be performed using a Leco Model 900 to make Knop measurements [10]. Several fired specimens, from both slick and dull sides, when analyzed, indicate that the dull side is consistently harder than the slick slide. This is a result with the greater density of the dull side, as seen in the SEM micrographs.

14.8 APPLICATIONS

14.8.1 Application of LTCC substrates for microwave frequencies

The low dielectric constant of LTCC substrates provides benefits at microwave frequencies. The LTCC substrates also exhibit low loss tangents at microwave frequencies [29, 30].

14.8.1.1 Electrical benefits of LTCC

Low-fire multilayer layer ceramics (MLC) are electrically better than high fire MLC for three reasons [29, 31]:

- low conductor resistance,
- low substrate dielectric constant,
- the ability to incorporate thick film resistors.

The main reason that a low dielectric constant is desirable is that wider 50 Ω transmission lines are possible. Wider transmission lines mean fewer losses due to a reduced run resistance and skin effect. This in turn implies faster rise times. Low-fire MLC permit 75 Ω transmission lines that are greater in width and lower in resistance [32, 33]. A microstrip transmission line consists of a signal conductor separated from a ground plane by a dielectric material. The gold conductors and the low dielectric constant of low-fire MLC yield much better transmission lines, with a minimum of crosstalk and glitches. With LTCC it is possible to incorporate thick film resistors.

14.8.1.2 Mechanical benefits of LTCC

A heat sink material, made out of an alloy of tungsten and nickel, plated with gold over nickel, is attached to the substrate. This heat sink has high thermal conductivity, about 10 times better than Kovar. The TCE of this heat sink, $4.2 \times 10^{-6}/^{\circ}C$, is very close to that of Si, $3.4 \times 10^{-6}/^{\circ}C$, and GaAs, $6.5 \times 10^{-6}/^{\circ}C$.

14.8.1.3 Disadvantages of LTCC

LTCC is more brittle than HTCC, due to high glass content. Most uses of LTCC to date have incorporated a heat sink attached to the back of the substrate. Typically, microwave or higher frequency hybrids dissipate 8 to 12 W, which has dictated the use of heat sinks.

14.8.2 Other applications of LTCC substrates

Some of the other important applications of LTCC substrates are briefly discussed below:

- The LTCC substrate is a strong contender for ceramic printed circuit boards, as an alternative to organic boards, such as FR-4, polyimides, etc. [2].
- Improvements in lower interconnection delay for high performance circuits require lower dielectric constant ceramics. LTCC can fill this requirement. Some of the performance characteristics are also matched to silicon and gallium arsenide, which will greatly enhance the utilization of glass bonding and flip chip mounting of large area ICs.
- Systems with copper conductors are used as captively produced multilayer circuits in the Japanese market. Fujitsu's new supercomputer VP-2600 has a 61-layer LTCC substrate, with copper conductors functioning as a mother board PGas. Complicated multi-atmosphere processing of these copper co-fired systems may limit the board acceptance of this new technology for hybrid circuit manufacturers in the near term.
- From a materials and processing standpoint, there are potential developments on the horizon, such as thermal coefficient of expansion (TCE) matching to any component, dielectric tapes with increased mechanical strength, improved conductors for enhancement of power distribution, micro via generation with lasers, and prospects of buried passive components, including resistors and capacitors.
- LTCC offers a vast scope of packaging options. Three-dimensional structures and integral substrate packages offer the only viable solutions to problems of space. Tape technology represents a viable three-dimensional answer.

REFERENCES

[1] *Electronic Materials Handbook, Volume I Packaging*, ASM International, Materials Park, OH (1989).

[2] Patterson, F.K., "Ceramic tapes for multilayer fabrication", *ISHM Journal*, **17**(2), March/April 1990, 24–25.

[3] Shimada, Y., Yamashita, Y., and Takamizawa, H., "Low dielectric constant multilayer glass-ceramic substrate with Ag-Pd wiring for VLSI package", in *Multichip Modules*, IEEE Press, Piscataway, NJ, 1991, 464–471.

[4] Provo, M., "Hybrid revival with low-temperature co-fired ceramic modules", *Hybrid Circuit Technology*, November 1990, 41–43.

[5] Eustice, A.L., Horowitz, S.J., Stewart, J.J., Travis, A.R., and Sawhill, H.T., "Low

temperature co-fireable ceramics: a new approach for electronic packaging", in *Multichip Modules*, IEEE Press, Piscataway, NJ, 1991, 435–445.

[6] Nishigaki, S., Fukuta, J., Yano, S., Kawabe, H., Noda, K., and Fukaya, M., "A new low temperature fireable Ag multilayer ceramic substrate having post-fired Cu conductor (LFC-2)", in *Multichip Modules*, IEEE Press, Piscataway, NJ, 1991, pp. 446–449.

[7] Sawhill, H.T, Eustice, A.L., and Horowitz, S.J, "Low temperature co-fireable ceramics with co-fired resistors", in *Multichip Modules*, IEEE Press, Piscataway, NJ, 1991, pp. 472–478.

[8] Cadenhead, R.L., "A model to forecast the optimum selection of a substrate fabrication technology", *ISHM Journal*, **18**(5), September/October 1991, 26–27.

[9] Pecht, M., *Hand Book of Electronic Packaging*, Marcel Dekker, NY (1991).

[10] Lindfors, P.A., Kostec, A.D., and Bigaoutte, R.J., 'Study of adhesion failures between layers of co-fired ceramic sheets', *Proceedings of Electronic Packaging Materials and Processes*, American Society for Metals, 1985, 77–84.

[11] "Standard test methods for A-C loss characteristics and permittivity of solid electrical insulating materials", D 150-87, in *Annual Book of ASTM Standards*, **09.02**, American Society for Testing and Materials, Philadelphia, 1989, 27–45.

[12] "Standard terminology relating to electrical insulation", D 1711-90, in *Annual Book of ASTM Standards*, **10.02**, American Society for Testing and Materials, Philadelphia, 1990, 155–162.

[13] "Standard test method for relative permittivity (dielectric constant) and dissipation factor of polymer-based microwave circuit substrate", D 3380-90, in *Annual Book of ASTM Standards*, **10.02**, American Society for Testing and Materials, Philadelphia, 1990, 329–340.

[14] Rosenberg, H.M., *The Solid State*, Oxford Science Publications, 1988, pp. 223–225.

[15] "Standard test method for complex permittivity (dielectric constant) of solid electrical insulating materials at microwave frequencies and temperatures to 1650 °C", D 2520-86, in *Annual Book of ASTM Standards*, **10.02**, American Society for Testing and Materials, Philadelphia, 1990, 215–230.

[16] Hartshorn, L. and Ward, W.H., "The measurement of the permittivity and power factor of dielectrics at frequencies from 104 to 108 cycles per second", *Proceedings, PIEEA*, Institution of Electrical Engineers (London), **79**, 1936, 597–609.

[17] Gipprich, J.W., Leahy, A.L., and Martin, A.J., "Microwave dielectric constant of a low temperature co-fired ceramic", *IEEE Transactions on Components, Hybrids, and Manufacturing*, **14**(4), December 1991, 732–737.

[18] Troughton, P., "Measuring techniques in microstrip", *Electronics Letters*, **5**(2), January 23, 1969, 25–26.

[19] Tyler, J.R. and Gasper, D.J., "Evaluation and characterization of low-temperature co-fired ceramic structures for microwave applications", *ISHM '89 Proceedings*, Baltimore, MD, 390–399.

[20] "Standard test method for permittivity (dielectric constant) and dissipation factor of solid ceramic dielectrics at frequencies to 10 MHz and temperatures to 500 °C", D 2149-79, in *Annual Book of ASTM Standards*, **10.02**, American Society for Testing and Materials, Philadelphia, 1990, 178–181.

[21] Ginzton, E.L., *Microwave Measurements*, McGraw-Hill, New York, NY, 1959, Table 5.2.

[22] Lih-Tyng Hwang and Iwona Turlik, "A review of the skin effect as applied to thin film interconnections", *IEEE Transactions on Components, Hybrids, and Manufacturing Technology*, **15**(1), February 1992, 43–54.

[23] Jackson, J.D., *Classical Electrodynamics*, John Wiley & Sons, 1985, pp. 298.

[24] Ohanian, H.C., *Classical Electrodynamics*, Allyn and Bacon, 1988, pp. 396–400.

[25] Samuels, L.E., *Metals Engineering, a Technical Guide*, ASM International, Metals Park, Ohio, 183–203.

[26] Blodgett, A.J., Jr., "A multilayer ceramic multichip module", *IEEE Transactions on Components, Hybrids, and Manufacturing Technology*, 1980, 283–285.
[27] Gonser, U., *Microscopic Methods in Metals*, Springer-Verlag, 1991, pp. 219–227.
[28] Kiessling, J.B., Pelella, L.F., and Sands, S.A., "Electrical test method for unfired metal/ceramic substrates', *IBM Technical Disclosure Bulletin*, **24**(4), 9/81, 2171–2172.
[29] Berg, W.E., "Microwave hybrids using low-fire multilayer ceramics", *Hybrid Circuit Technology*, March 1991, 8–15.
[30] Cadenhead, R.L., "Assuring the integrity of multilayer substrates prior to assembly", *ISHM Technical Monograph Series*, 6984-001, 1984, 33–39.
[31] Burger, W.G., and Weigel, C.W., "Multilayer ceramics manufacturing", *IBM Journal of Research and Development*, **27**(1), January 1983, 11–19.
[32] Kata, K., Shimada, Y., and Takamizawa, H., "Low dielectric constant new materials for multilayer ceramic substrate", *Sixth IEEE/CHMT International Electronic Manufacturing Technology Symposium*, Japan, April 26–28, 1989, 24–27.
[33] McClandish, C.S., "The advantages of ceramic materials for military multilayer substrates", *Hybrid Circuit Technology*, July 1990, 14–17.

INDEX

Affix
stamp
here

Dept AC
John Wiley & Sons Ltd
Baffins Lane
Chichester
West Sussex PO19 1UD
UK